The Complete NHI Catalog
Generated on June 2, 2016
Sorted by Course Number

TABLE OF CONTENTS

INFORMATION

STRUCTURES

PAVEMENTS AND MATERIALS

GEOTECHNICAL

DESIGN AND TRAFFIC OPERATIONS

CONSTRUCTION AND MAINTENANCE

HYDRAULICS

ASSET MANAGEMENT

INTELLIGENT TRANSPORTATION SYSTEMS (ITS)

TRANSPORTATION PERFORMANCE MANAGEMENT

FREIGHT AND TRANSPORTATION LOGISTICS

REAL ESTATE

ENVIRONMENT

TRANSPORTATION PLANNING

FINANCIAL MANAGEMENT

BUSINESS, PUBLIC ADMIN, AND QUALITY

CIVIL RIGHTS

HIGHWAY SAFETY

SITE AND PERSONAL SAFETY

COMMUNICATIONS

ABOUT NHI

WHO WE ARE

The National Highway Institute (NHI) provides technical training to the highway transportation workforce to build skills and enhance job performance to improve the conditions and safety of our nations' roads, highways, and bridges.

As part of Federal Highway Administration's (FHWA) Office of Technical Services (OTS), NHI courses complement the targeted training and technical assistance of FHWA program offices, Resource Center, and Local and Tribal Technical Assistance Programs (LTAP/TTAP).

OUR TRAINING

NHI courses are instrumental in developing core competencies and new skills, as well as learning about leading technologies and current policies. Our instructors strive to ensure that participants leave training not only with additional knowledge, but also the ability to apply that knowledge directly to their work. NHI is an accredited training provider by the International Association of Continuing Education and Training (IACET), allowing participants to earn Continuing Education Units (CEUs) for completed coursework. NHI also is an approved provider of the American Institute of Certified Planners (AICP) certification maintenance (CM) credits.

NHI offers three types of training.

Instructor-led Training (ILT): These courses are held in-person and led by an instructor when an organization is available to host the session. Any organization may host a session by submitting a Host Request form on the NHI Web site.

Web-conference Training (WCT): These are live, online training sessions that take place at a set time. Web-conference Training sessions also require a host.

Web-based Training (WBT): These online courses are available 24/7 for six months after purchase by the registrant. Participants can control the pace at which they complete the course and may return to it as many times as they wish within the six-month access period.

LEARN MORE

For more information or to subscribe to our mailing list, please visit the NHI Web site at www.nhi.fhwa.dot.gov.

Customers with additional questions may also contact NHI Customer Service at NHICustomerService@dot.gov, or by phone during regular business hours, 7:30AM – 4:30PM Eastern Time, at (877) 558-6873.

NHI MAKES HOSTING EASY

HOSTING A COURSE

NHI partners with host organizations across the country to deliver training where it is needed most. NHI provides top-notch instructors and course materials, while hosting organizations provide the facilities and equipment.

WHO CAN HOST

Any United States-based organization can host Instructor-led Trainings (ILT), which are taught in classrooms, and/or Web-conference Trainings (WCT), which are taught online.

Our instructors may tailor individual sessions to meet the unique needs and array of experiences of the hosting organization, including covering local issues and topics of special interest. Instructors also may modify case studies and exercises based on their subject matter expertise to make them pertinent to the participant's experiences.

REQUESTING TO HOST

To host a course, domestic customers can go to the NHI Web site and complete the appropriate Host Request form (ILT or WCT). The process takes just a few minutes. First-time users will need to create a user profile and check the **INSTRUCTOR/ HOST BOX**.

If you run into any difficulty when you are logging in, filling out a Host Request form, or navigating the NHI Web site, please contact NHI Customer Service for help at (877) 558-6873 during normal business hours, 7:30am – 4:30pm Eastern time. Customers may also email NHI Customer Service at nhicustomerservice@dot.gov.

To assist the host in preparation for and coordination of the session, a hosting checklist is provided on the NHI Web site. This checklist includes important information about hosting your NHI training session, as well as valuable "best-practice" information based on NHI's 40 years of experience with our hosting partners.

CONFIRMING SESSION DATES/LOCATIONS/TIMES

After the Host Request form is received, an Instructor or a member of the NHI team will contact the host to discuss scheduling options. While preferred dates may be specified on the Host Request form, sessions are not official until the hosting organization receives formal confirmation from NHI. Once official, NHI will list the session publicly on its Web site.

Enrollment Options

The host's contact information is listed with the scheduled session. Interested participants from outside the host's organization may contact the host to enroll. Alternatively, the host may ask NHI to open public seats, which allow outside participants to enroll through NHI.

The NHI Scheduler will email all participant information to the host and instructor prior to the session start date.

HOSTING EXPENSES

To host a session, hosts are charged the per-participant price multiplied by the class-size minimum, or the host is charged per participant if the session class size exceeds the minimum. Pricing cannot be reduced if the minimum class size is not met. Therefore, if registration for a course is lower than anticipated, it is important for the host to contact NHI prior to the cancellation period (15 business days) to discuss a remedy. Please note that with sufficient notice, NHI may be able to offer marketing support for the session.

Three seats in every session are reserved for Federal Highway Administration (FHWA) employees until 15 days before the course begins. FHWA participants do not count toward the participant minimum, but should be considered in the course maximum. Hosts are not charged for FHWA personnel or participants who have paid via the NHI Web site. Hosts are not charged for any instructor expenses.

Course hosts may charge participants an additional fee to recover all or part of costs associated with hosting the course. However, we ask hosts to contact the NHI Scheduler at (703) 235-0534 with this information prior to the confirmation of the session.

Course fees, which include the cost of materials for each participant, are listed with every course description.

RECEIVING COURSE MATERIALS

NHI will ship course material to the host approximately three weeks prior to the session start date.

PROVIDING PAYMENT

Payment may be made to NHI by check, money order, or credit card. Checks and money orders must be made payable to the National Highway Institute. To make credit card payments, contact NHI Customer Service at NHICustomerService@dot.gov or 1-877-558-6873. You are not charged for any FHWA participants or for participants who paid via the NHI Web site.

CANCELLATION POLICY/REFUNDS

To avoid incurring the $1,500 cancellation fee, cancellation must be requested no later than 15 business days prior to the course start date. If a course must be cancelled, the host is required to contact NHI Customer Service at 1-877-558-6873 during normal business hours, 7:30AM – 4:30PM Eastern Time, or email NHICustomerService@dot.gov. If the course materials have been sent, the host must contact NHI Customer Service.

In the event of cancellation, it is the host's responsibility to contact all participants (including those registered for public seats). There must be verification that the registrants received the cancellation notice. Notice to out-of-state participants is especially important so that they may alter or cancel any travel arrangements.

In the case of an emergency or weather-related closing, the cancellation fee will not apply. NHI follows the host office's policy regarding weather and emergency closings.

RECEIVING COURSE CREDIT

Many of the courses offered at NHI can be used toward obtaining Continuing Education Units (CEUs), Certification Maintenance (CM) credits, and Professional Development Hours (PDHs). Please select the headers below for more information about receiving credits.

CONTINUING EDUCATION UNITS

NHI has been recognized as an Accredited Provider by the International Association for Continuing Education and Training (IACET). In obtaining this accreditation, NHI has demonstrated that it complies with the ANSI/IACET Standard which is recognized internationally as a standard of good practice. As a result of this Accredited Provider status, NHI is authorized to offer IACET CEUs for its programs that qualify under the ANSI/IACET Standard. IACET is an independent, non-profit association whose goal is to ensure quality continuing education for professionals. For an organization to become an IACET approved CEU Accredited Provider, it must demonstrate that it designs, develops, and delivers training in accordance with proven adult learning theory and recognizes instructional systems design practices. Each course description in the NHI catalog includes the number of CEUs offered upon successful completion of the course.

One CEU is offered for every ten contact hours of training led by a qualified instructor and qualified instruction. In order to be offered CEUs, a course participant must attend 100% of the course and must pass the course examination with a score of 70% or greater.

CEUs are offered to each course participant who fulfills the above stated requirement. NHI will maintain individual training records for seven years for the CEUs offered. Individuals and their employers are also encouraged to maintain their own training records including course name, class date(s), instructor name, class roster, and CEUs offered.

For proof of your CEU record, please contact NHI at NHICustomerService@dot.gov or 1-877-558-6873 and request your official transcript. Your official transcript displays a record of your NHI course history as well as the CEUs offered for each CEU-accredited course. Please allow at least one month after the completion of your course before requesting your official transcript.

CERTIFICATION MAINTENANCE CREDITS

NHI providers Certification Maintenance (CM) credits to assist professional planners become and maintain their membership as certified planners through the American Planning Association (APA).

American Institute of Certified Planners (AICP) is APA's professional institute. Certified Planners have demonstrated a commitment to high standards of professional practice and a mastery of theories and tools of planning.

NHI recognizes that the certification carries a high mark of distinction and requires planners to meet rigorous standards and maintain their expertise through continuing education. Planners must earn 32 CM continuing education credits every two years in order to stay up to date on the latest trends, technologies, and best practices. NHI courses will now help them achieve that requirement.

CM credits are measured in contact hours, so that 30 minutes of instructional time equals 30 minutes of CM credit (30 minutes contact = 0.5 CM credits; 1.0 contact hours = 1.0 CM credits). An event must be at least 30 minutes in duration to be eligible for CM credit.

Contact NHI Customer Service at NHICustomerService@dot.gov or 877-558-6873 to ask for an official transcript to be used by AICP to calculate CM credits. Please allow at least one month after the completion of your course before requesting your official transcript.

PROFESSIONAL DEVELOPMENT HOURS (PDHS)

NHI does not officially offer PDHs; however, it is possible to receive PDHs for your completed NHI training courses. To receive PDHs, please submit your course certificate (which indicates the contact hours assigned to the course) and/or your official transcript (which indicates the CEUs granted for a course) to the respective licensing agency. Upon consent, the licensing agency may convert your hours and/or CEUs into PDHs and proceed with the PDH awarding process.

PDHs are offered on a ratio of one contact hour to one PDH. When converting from CEU to PDH, please note that one CEU is equal to ten PDHs (or one PDH is equal to one-tenth of a CEU).

To request your official transcript with proof of CEU record and/or contact hours, please contact NHI at NHICustomerService@dot.gov or 1-877-558-6873. Your official transcript displays a record of your NHI course history as well as the CEUs offered for each CEU-accredited course. Please allow at least one month after the completion of your course before requesting your official transcript.

NHI CERTIFICATES OF ACCOMPLISHMENT

NHI's Certificate of Accomplishment program was designed to recognize individuals who have successfully enhanced their depth and breadth of knowledge and expertise in specific disciplines or topic areas. Students are eligible for the Certificate of Accomplishment when they have completed and passed a suite of related NHI course offerings.

Certificates of Accomplishment are available in the following disciplines, with more to come over the next year:

- Work Zone Safety
- Relocation Under The Uniform Act
- Incident Management
- Freight Management and Operations

For more information about the Certificates of Accomplishment, please visit the NHI Web site at www.nhi.fhwa.dot.gov/training/cert_programs.aspx.

FREE WEB-CONFERENCE TRAINING

NHI is excited to offer FREE Web-conference training. These trainings save both time and money, while covering the latest topics and techniques within the transportation industry. All transportation professionals in the public and private sectors are invited to participate in these trainings.

REAL SOLUTIONS SEMINAR SERIES

This series of free monthly Webinars features a guest speaker who presents problems or issues faced in the field and what steps were taken to solve them. In some sessions, additional panelists join the guest speaker to further discuss that seminar's topic.

Some past topics include:

- Best Practices for Integrating Climate Change Considerations in the Transportation Planning Process
- eLearning and Distance Learning within the Transportation Industry
- Smart Corridors and Complete Streets: A Look at Some Situations and Strategies
- Solving Old Traffic Noise Ills: Tennessee Type II Noise Abatement Program

Visit the *Real Solutions Seminar Series* section of the Web site to register for the next *Real Solutions* Web conference or to listen to past Web conferences.

LEARN MORE

For more information, please visit the NHI Web site at **www.nhi.fhwa.dot.gov**.

Want to be notified when a free Web conference is scheduled? Email nhimarketing@dot.gov.

Course Number

FHWA-NHI-130053

Course Title

Bridge Inspection Refresher Training

The major goals of this course are to refresh the skills of practicing bridge inspectors in fundamental visual inspection techniques; review the background knowledge necessary to understand how bridges function; communicate issues of national significance relative to the nations' bridge infrastructures; re-establish proper condition and appraisal rating practices; and review the professional obligations of bridge inspectors.

This course is based on the "Bridge Inspector's Reference Manual," 2002 (updated in 2006) with reference to the AASHTO Manual as defined by the National Bridge Inspection Standards regulation.

Core course topics include inspector qualifications and duties, bridge mechanics, record keeping and documentation, fatigue and fracture in steel bridges, traffic safety features, safety, National Bridge Inventory (NBI) component ratings, superstructure type identification, inspection techniques and case studies for decks, superstructures, bearings, substructures, channels and culverts, and a mock bridge inspection classroom exercise.

Optional topics include fiber reinforced polymer, inspection of truss gusset plates, inspection of adjacent box beams, bridge site signing, structure inventory and appraisal overview, common NBI miscodings, element level ratings and timber superstructures.

For this version of the course (3-day), the host agency will need to select four (4) desired optional topics. Course instructors will contact the host prior to the course to complete a pre-course questionnaire, determine optional topics to be taught, and discuss the course schedule.

Outcomes

Upon completion of the course, participants will be able to:

- Describe the current overall condition and condition trends for the nation's bridges
- Identify the recent National Bridge Inspection Standards (NBIS) revisions
- Accurately code National Bridge Inventory (NBI) items
- Identify and document inspection observations using standard methods
- Evaluate defects based on the 2008 AASHTO Manual for Bridge Evaluation
- Code NBI components using the Recording and Coding Guide for the Structure Inventory and Appraisal of the Nation's Bridges
- Determine if overall structure/structural member is fracture critical prone
- Accurately inspect and evaluate a bridge's four traffic safety features
- List the keys to ensuring a safe work environment
- Explain bridge responses and bridge mechanic principles

Target Audience

The target audience for this course includes Federal, State, and local agencies and private sector personnel employed in inspecting bridges or managing bridge inspection programs. The course is built to accommodate those that have completed comprehensive bridge inspection training (130055 or similar) or met the criteria for a bridge inspector under the State's procedures or requirements.

TRAINING LEVEL: Intermediate

FEE: 2016: $955 Per Person; 2017: $955 Per Person

LENGTH: 3 DAYS (CEU: 1.8 UNITS)

CLASS SIZE: MINIMUM: 20; MAXIMUM: 30

NHI Customer Service: (877) 558-6873 • nhicustomerservice@dot.gov

Course Number

FHWA-NHI-130053A

Course Title

Bridge Inspection Refresher Training

The major goals of this course are to refresh the skills of practicing bridge inspectors in fundamental visual inspection techniques; review the background knowledge necessary to understand how bridges function; communicate issues of national significance relative to the nations' bridge infrastructures; re-establish proper condition and appraisal rating practices; and review the professional obligations of bridge inspectors.

This course is based on the "Bridge Inspector's Reference Manual," 2002 (updated in 2006) with reference to the AASHTO Manual as defined by the National Bridge Inspection Standards regulation.

Core course topics include inspector qualifications and duties, bridge mechanics, record keeping and documentation, fatigue and fracture in steel bridges, traffic safety features, safety, National Bridge Inventory (NBI) component ratings, superstructure type identification, inspection techniques and case studies for decks, superstructures, bearings, substructures, channels and culverts, and two (2) mock bridge inspection classroom exercises.

Optional topics include fiber reinforced polymer, inspection of truss gusset plates, inspection of adjacent box beams, bridge site signing, structure inventory and appraisal overview, common NBI miscodings, element level ratings and timber superstructures.

For this version of the course (3.5-day), the host agency will need to select six to seven (6-7) desired optional topics. Course instructors will contact the host prior to the course to complete a pre-course questionnaire, determine optional topics to be taught, and discuss the course schedule.

Outcomes

Upon completion of the course, participants will be able to:

- Describe the current overall condition and condition trends for the nation's bridges
- Identify the recent National Bridge Inspection Standards (NBIS) revisions
- Accurately code National Bridge Inventory (NBI) items
- Identify and document inspection observations using standard methods
- Evaluate defects based on the 2008 AASHTO Manual for Bridge Evaluation
- Code NBI components using the Recording and Coding Guide for the Structure Inventory and Appraisal of the Nation's Bridges
- Determine if overall structure/structural member is fracture critical prone
- Accurately inspect and evaluate a bridge's four traffic safety features
- List the keys to ensuring a safe work environment
- Explain bridge responses and bridge mechanic principles

Target Audience

The target audience for this course includes Federal, State, and local agencies and private sector personnel employed in inspecting bridges or managing bridge inspection programs. The course is built to accommodate those that have completed comprehensive bridge inspection training (130055 or similar) or met the criteria for a bridge inspector under the State's procedures or requirements.

TRAINING LEVEL: Intermediate

FEE: 2016: $1025 Per Person; 2017: $1025 Per Person

LENGTH: 3 DAYS (CEU: 2.2 UNITS)

CLASS SIZE: MINIMUM: 20; MAXIMUM: 30

NHI Customer Service: (877) 558-6873 • nhicustomerservice@dot.gov

Course Number

FHWA-NHI-130054

Course Title

Engineering Concepts for Bridge Inspectors

This course was updated in 2010 and provides knowledge of the elementary concepts in bridge engineering for individuals involved with the inspection of in-service highway bridges. The course covers the purpose of highway bridge inspections and the roles of inspectors through the discussion of common bridge types and materials, material properties, and bridge components as well as details, loadings, stresses, strains, and deterioration of bridge materials and members. Participants will be asked to complete an exam at the end of the course, which they must earn a 70% or better on to successfully complete the course and receive a certificate of completion.

This course prepares participants for the 2-week, intensive Instructor-led course in bridge inspection, 130055 Safety Inspection of In-Service Bridges. Upon successful completion of 130054, participants will have met the prerequisite requirement for participation in the 130055 course.* If participants would like to enroll in the 130055 course, they will be required to demonstrate their certificate of completion for 130054 as proof that the prerequisite requirement has been fulfilled.

Participation in 130054 is not the only option to fulfill the prerequisite requirement for 130055.* Individuals have the option to 1) successfully complete the Web-based training and assessment (130101 Introduction to Safety of In-Service Bridges) or 2) for those with engineering backgrounds or prior knowledge and experience in the field of bridge inspection may "test-out" through a Web-based assessment (130101A Introduction to Safety Inspection of In-Service Bridges).

*Please note: Upon successful completion of this prerequisite course, you will be eligible to take the 130055 training course for up to 2 years.

Outcomes

Upon completion of the course, participants will be able to:

- Describe the basis for bridge inspection
- Describe the various roles of the bridge inspection team
- Identify common bridge types and major components, primary members, secondary members and features of highway bridges
- Name the common materials used in bridges
- Describe the basic properties, strengths, and weaknesses of each material
- Describe basic engineering concepts
- Describe standard highway bridge loadings
- Describe the types, signs, and causes of structural distress
- Identify other features associated with bridges
- Name protective measures required to mitigate hazards

Target Audience

This course is designed for Federal, State, and local technicians and inspectors who have limited experience with the inspection of in-service highway bridges. Engineers without bridge experience or those who need a refresher in basic bridge design concepts will also benefit from the course. Individuals completing this course could serve on a bridge inspection team, but would require additional experience and training to qualify as team leaders.

TRAINING LEVEL: Basic

FEE: 2016: $1175 Per Person; 2017: N/A

LENGTH: 5 DAYS (CEU: 3 UNITS)

CLASS SIZE: MINIMUM: 20; MAXIMUM: 30

NHI Customer Service: (877) 558-6873 • nhicustomerservice@dot.gov

COURSE NUMBER

FHWA-NHI-130055

COURSE TITLE

Safety Inspection of In-Service Bridges

NOTE: This course was updated in 2012 and 2015 and now contains mandatory prerequisite requirements for participants and host requirements in preparation for the field exercises. See details below.

This course is based on the 2015 FHWA "Bridge Inspector's Reference Manual (BIRM)" and provides training on the safety inspection of in-service highway bridges. The course includes two virtual bridge inspection exercises* facilitated using NHI's virtual bridge inspection (VBI) computer-based training (CBT) technology; instruction on critical findings, their identification and response; curriculum on the American Association of State Highway and Transportation Officials (AASHTO) element level inspection approach using the 2013 AASHTO Manual for Bridge Element Inspection 2015 Interim Revisions; and activities that maximize participant engagement throughout the course. This course does not go into depth on fracture critical, underwater, or complex bridge inspections. Other specialty courses, 130078 Fracture Critical Inspection Techniques for Steel Bridges and 130091 Underwater Bridge Inspection, cover these topics.

Participants will be asked to complete mid-term and end-of-course assessments each with a cumulative score of 70% or better to successfully complete the course and receive a certificate of completion. The sponsoring agency/State may monitor the examinations and retain the scores to qualify or certify bridge inspectors. Satisfactory completion of this course will fulfill the comprehensive bridge inspection training requirements of the National Bridge Inspection Standards. Note: Many States have additional requirements to become a bridge inspection team leader.

Participant Prerequisite Requirement: ALL participants must have met one of the three prerequisite requirements for participation in this course** and bring a course completion certificate bearing their name to the first day of the class. The passing score for all prerequisites is 70% or better. Individuals have the option to complete one of the following three prerequisite requirements: 1) 130054 Engineering Concepts for Bridge Inspectors, a 5-day Instructor-led course; 2) 130101 Introduction to Safety Inspection of In-Service Bridges, a 14-hour Web-based training and assessment; and/or 3) 130101a Prerequisite Assessment for Safety Inspection of In-Service Bridges, a Web-based assessment.

Host Requirements: Hosts must provide a training room large enough to accommodate at least 30 participants as well as the 15 NHI virtual bridge laptops (provided by NHI Instructors) that will be used for the virtual bridge exercises. Additionally, the host must ensure that ALL students have successfully met the prerequisite requirement** and have a valid course completion certificate for one of the three prerequisite options.

*Alternatively, the State can exercise the option to request to have a physical field trip in lieu of one or both virtual bridge exercises. If this option is exercised, the host/sponsoring agency is required to provide transportation for course participants to attend the field trip portion of this course at the host/sponsoring agency's own expense. The host must coordinate with the instructor to identify bridges for inspection during the field trip exercises, in advance of the course delivery.

**Please note: prerequisite must be completed within two years of the course start date. Additionally, it is recommended that prior to attending this course participants spend some time in the field, at bridge inspection sites, but not required.

OUTCOMES

Upon completion of the course, participants will be able to:

- Discuss the duties and responsibilities of a bridge inspector and define inspection concepts including personal and public safety issues associated with bridge inspections
- List the inspection equipment needs for various types of bridges and site conditions
- Describe, identify, evaluate, and document the various components and deficiencies that can exist on bridge components and elements
- List design characteristics and describe inspection methods and locations for common concrete, steel, and timber structures
- Identify and evaluate the various culvert and waterway deficiencies
- Discuss the need to inspect underwater portions of bridges
- Describe nondestructive evaluation methods for basic bridge materials
- Demonstrate how to field inspect and evaluate common concrete, steel, and timber bridges

TARGET AUDIENCE

Federal, State, and local highway agency employees; and consultants involved in inspecting bridges or in bridge inspection management and leadership positions. A background in bridge engineering is strongly recommended. All participants must successfully complete (score 70% or better) one of the following three prerequisite requirements within two years prior to attending this training: 1)130054 Engineering Concepts for Bridge Inspectors ; 2) 130101 Introduction to Safety Inspection of In-Service Bridges ; or 3) 130101a Prerequisite Assessment for Safety Inspection of In-Service Bridges .

TRAINING LEVEL: Intermediate

FEE: 2016: $2025 Per Person; 2017: $2025 Per Person

LENGTH: 10 DAYS (CEU: 6.7 UNITS)

CLASS SIZE: MINIMUM: 20; MAXIMUM: 30

NHI Customer Service: (877) 558-6873 • nhicustomerservice@dot.gov

Course Number

FHWA-NHI-130078

Course Title

Fracture Critical Inspection Techniques for Steel Bridges

The course curriculum for this training reflects current practices, while addressing new and emerging technologies available to bridge inspectors. In addition, the course features exemplary training; hands-on workshops for popular types of nondestructive evaluation (NDE) equipment; and a case study of an inspection plan for a fracture critical bridge.

The first day of the training focuses on the concept of fracture critical members (FCMs), FCM identification, failure mechanics, fatigue in metal, and an overview of NDE methods. Day two includes demonstration sessions and hands-on applications of NDE techniques for dye penetrant, magnetic particle testing, Eddy current testing, and ultrasonic testing. Days three and four emphasize inspection procedures and reporting for common FCMs, including problematic details, I-girders, floor beams, trusses, box girders, pin and hanger assemblies, arch ties, eyebars, and cross girders/pier caps. The course will conclude with a case study detailing the preparation of an inspection plan of a fracture critical bridge. Additionally, the course instructors will tailor discussions of topics based on State needs and requirements.

"This training will help inspectors evaluate bridges more thoroughly and will provide them with additional knowledge in how structures work and what can take place when they don't work," states Bill Drosehn, district bridge inspection engineer for the Massachusetts DOT.

Note: Hosts are required to provide safety goggles for all course participants as well as a well-ventilated space for conducting the dye penetrant demonstration.

Outcomes

Upon completion of the course, participants will be able to:

- Identify fracture critical members (FCMs)
- Identify problematic details
- Identify areas most susceptible to fatigue and fracture
- Record defects
- Evaluate defects
- Evaluate nondestructive evaluation (NDE) methods
- Evaluate retrofit details

Target Audience

Those who will benefit most from this training are public and private sector bridge inspectors, supervisors, project engineers, and others responsible for field inspection of fracture critical steel bridge members. Prior to taking this course, participants should have completed NHI course 130055, Safety Inspection of In-Service Bridges, or possess equivalent field experience relative to bridges. Participants also should have a thorough understanding of bridge mechanics and bridge safety inspection procedures as required by the National Bridge Inspection Standards.

Training Level: Intermediate

Fee: 2016: $1025 Per Person; 2017: $1025 Per Person

Length: 3.5 DAYS (CEU: 2.5 UNITS)

Class Size: MINIMUM: 20; MAXIMUM: 30

NHI Customer Service: (877) 558-6873 • nhicustomerservice@dot.gov

Course Number

FHWA-NHI-130081

STRUCTURES

Course Title

LRFD for Highway Bridge Superstructures - (NEW 4-Day ILT)

This updated course describes Load and Resistance Factor Design (LRFD) for steel and concrete highway bridge superstructures. It provides a combination of instructor-led discussions and workshop exercises. The course also includes LRFD theory applied to design examples and illustrates step-by-step LRFD design procedures. The curriculum follows the AASHTO LRFD Bridge Design Specifications, 7th Edition, 2014 (AASHTO LRFD), including the approved 2015 Interims. The training includes the extensive use of student exercises and example problems to demonstrate overall design, detailing, and construction principles addressed in the reference materials. It affords hands-on experience in LRFD design and detailing of highway bridge superstructures. The curriculum materials are comprised of a comprehensive reference manual (FHWA Publication No. FHWA-NHI-15-047), lecture and workshop exercises intended to promote or enhance a working knowledge of AASHTO LRFD, and a participant workbook for lecture notes and exercises.

The curriculum includes the following major topics:

*Generals superstructure design considerations

*Preliminary design concepts for steel I-girder superstructures

*Steel I-girder design

*Preliminary design concepts for prestressed concrete superstructures

*Prestressed concrete I-girder design

*Spliced prestressed concrete girder bridges

Outcomes

Upon completion of the course, participants will be able to:

- Describe the bridge superstructure design and construction process in accordance with the current AASHTO LRFD specifications.
- Apply the appropriate current AASHTO LRFD specification articles dealing with selection of bridge type, size, and location.
- Apply the appropriate current AASHTO LRFD specification articles dealing with bridge economics.
- Apply the appropriate current AASHTO LRFD specification articles dealing with bridge materials.
- Describe the appropriate current AASHTO LRFD specification articles dealing with evolution of bridge design codes.
- Apply the appropriate current AASHTO LRFD specification articles dealing with bridge loads and load combinations.
- Apply the appropriate current AASHTO LRFD specification articles dealing with structural analysis.
- Apply the appropriate current AASHTO LRFD specification articles dealing with concrete bridge superstructure design.
- Apply the appropriate current AASHTO LRFD specification articles dealing with steel bridge superstructure design.
- Demonstrate the use of the current AASHTO LRFD specification requirements for superstructure design through the completion of step-by-step procedures, participant exercises, and design examples.

Target Audience

This course has been developed for the needs of practicing public and private sector structural engineers with one to ten years of experience. The primary audience is Agency and consultant structural designers. Pre-training Competencies: Individuals attending this course should have a minimum BSCE degree and should complete the Web-based Training Course NHI-130081P prior to the first day of class. They should also have a working knowledge of the current AASHTO LRFD and should have relevant design experience using this specification on at least one bridge superstructure.

TRAINING LEVEL: Intermediate

FEE: 2016: $1250 Per Person; 2017: $1250 Per Person

LENGTH: 4 DAYS (CEU: 2.5 UNITS)

CLASS SIZE: MINIMUM: 20; MAXIMUM: 30

NHI Customer Service: (877) 558-6873 • nhicustomerservice@dot.gov

COURSE NUMBER

FHWA-NHI-130081P

STRUCTURES

COURSE TITLE

General Superstructure Design Considerations (Web-based)

This course, 130081P, serves as a WBT prerequisite to the following ILTs:

130081

130081A, and

130081B

This course provides training on the fundamentals for LRFD highway superstructure design. This includes a basic understanding of LRFD development and implementation, general design and location features related to superstructure design, and the primary loads and load combinations used for superstructure design. This course is a prerequisite to the Instructor-Led Training (ILT) Courses 130081 LRFD for Highway Bridge Superstructures - Steel and Concrete (4-Day ILT), 130081A LRFD for Highway Bridge Superstructures - Steel (2-Day ILT), and 130081B LRFD for Highway Bridge Superstructures - Concrete (2-Day ILT), and it covers only general sections of the LRFD Specifications.

OUTCOMES

Upon completion of the course, participants will be able to:

- Describe the fundamentals of LRFD, the historical background of LRFD, and the basic components of LRFD for superstructure design
- Describe location features, basic design objectives, principles of bridge aesthetics, and constructability issues for superstructure design
- Describe the primary loads, load combinations, and load factors used for steel and concrete superstructure design

TARGET AUDIENCE

The target audience for this course is practicing public and private sector structural and bridge engineers with 0 to 20 years of experience. This includes agency and consultant structural designers, as well as project managers. Individuals taking this course should have a minimum Bachelor of Science in Civil Engineering (BSCE) or equivalent degree.
This course is intended for engineers that require experience with AASHTO bridge design provisions and updates. Additionally, participants wishing to take 130081, 130081A, or 130081B should have taken this WBT in advance of the first day of the ILT.

TRAINING LEVEL: Basic

FEE: 2016: $0 Per Person; 2017: $0 Per Person

LENGTH: 3 HOURS (CEU: .3 UNITS)

CLASS SIZE: MINIMUM: 20; MAXIMUM: 40

NHI Customer Service: (877) 558-6873 • nhicustomerservice@dot.gov

Course Number

FHWA-NHI-130087

Course Title

Inspection and Maintenance of Ancillary Highway Structures

This course provides training in the inspection and maintenance of ancillary structures, such as structural supports for highway signs, luminaries, and traffic signals. Its goal is to provide agencies with information to aid in establishing and conducting an inspection program in accordance with the FHWA "Guidelines for the Installation, Inspection, Maintenance, and Repair of Structural Supports for Highway Signs, Luminaries, and Traffic Signals."

Outcomes

Upon completion of the course, participants will be able to:

- List and identify common visible weld defects
- Identify appropriate nondestructive testing techniques
- Identify factors that lead to corrosion and explain mitigation methods used in ancillary structures
- Define the severity of observed defects in accordance with the FHWA guidelines
- Identify defects in base/anchor rod installations
- List key issues in construction inspection of ancillary structures
- Identify repair techniques and discuss their use

Target Audience

Structural engineers, material engineers, traffic engineers, field inspectors, construction supervisors, maintenance personnel, and other technical personnel involved in the installation, inspection, maintenance, and repair of ancillary highway structures. This course is not a design course; however, the information should be helpful to those working in design and specification of ancillary structures.

Training Level: Basic

Fee: 2016: $750 Per Person; 2017: $750 Per Person

Length: 2 DAYS (CEU: 1.1 UNITS)

Class Size: MINIMUM: 20; MAXIMUM: 30

NHI Customer Service: (877) 558-6873 • nhicustomerservice@dot.gov

Course Number

FHWA-NHI-130088

Course Title

Bridge Construction Inspection

The Bridge Construction Inspection Course (BCIC) is one of the core curriculum initiatives cited by AASHTO, FHWA, and the five regional organizations. These core curriculum initiatives are being pursued in order to maximize regional, public, and industry resources in the development of core training and qualification-based certification programs, improve the quality of bridge construction, and promote uniformity in training content and qualification requirements.

Overall, the BCIC improves quality, ensures uniformity, and establishes minimum competencies for bridge construction inspection. The underlying themes of the course can be broken down into key segments. The BCIC will provide the construction inspector with:

1. The requisite knowledge of construction that will make him/her an effective inspector

2. An overall awareness of the problems and consequences that can arise during construction and how these factors will impact the safety and service life of the structure

3. A knowledge of the inspections that should be performed to confirm conformance to the contract documents, or document contract nonconformance

Outcomes

Upon completion of the course, participants will be able to:

- Explain the role of the construction inspector as part of the overall project team
- Interpret drawings and specifications
- Anticipate possible construction and materials problems
- Maintain bridge controls for location and elevation
- Describe construction sequence for various bridge systems (e.g. foundations, substructures, superstructures, and miscellaneous systems), bridge types and materials
- Conduct regular systematic inspections of materials and standards of construction, through the use of job aids, such as checklists
- Explain and perform basic inspection and testing of materials
- Perform accurate surveys and checking of dimensions
- Make and maintain sufficient records

Target Audience

Construction supervisors, transportation department field inspectors, field engineers, resident engineers, structural engineers, materials engineers, and other technical personnel involved in the construction inspection of bridges. The course is developed for participants without an in-depth engineering background. However, more knowledgeable persons can attend and will add to the overall effectiveness of the training through their active participation.

Training Level: Basic

Fee: 2016: $1100 Per Person; 2017: $1100 Per Person

Length: 4.5 DAYS (CEU: 2.7 UNITS)

Class Size: MINIMUM: 20; MAXIMUM: 30

NHI Customer Service: (877) 558-6873 • nhicustomerservice@dot.gov

Course Number

FHWA-NHI-130091

Course Title

Underwater Bridge Inspection

The latest changes to the National Bridge Inspection Standards (NBIS), which became effective January 13, 2005, require FHWA-approved bridge inspection training for all divers conducting underwater inspections. One method of meeting this requirement is the completion of an FHWA-approved underwater diver bridge inspection training course. Satisfactory completion of this 4-day course will fulfill the NBIS requirement.

This course provides an overview of diving operations that will be useful to agency personnel responsible for managing underwater bridge inspections.

Course topics include: methods of underwater inspection, underwater material deterioration mechanisms and inspection techniques, scour inspection techniques, underwater element-level rating, and underwater bridge inspection training. A final examination based on course content will be administered to participants.

Outcomes

Upon completion of the course, participants will be able to:

- Specifically, upon completion of the course, participants should be able to:
- Explain the need and benefits of inspecting the underwater portions of bridge structures.
- Describe typical underwater defects and deterioration, and identify conditions contributing to rates of deterioration.
- Identify the types of inspection equipment available, and the advantages and limitations of each.
- Identify procedures for planning and performing thorough and safe underwater bridge inspections.
- Assign component and element level condition ratings for underwater components in accordance with NBIS and agency requirements.

Target Audience

The course is intended for trained divers who require a knowledge base of underwater bridge inspection and evaluation techniques in order to meet the educational requirements of the NBIS for underwater bridge inspection training. The course would also be of interest to non-diver bridge inspectors, and FHWA, state, and local agency structural engineers.

Training Level: Basic

Fee: 2016: $1050 Per Person; 2017: $1050 Per Person

Length: 4 DAYS (CEU: 2.4 UNITS)

Class Size: MINIMUM: 20; MAXIMUM: 30

NHI Customer Service: (877) 558-6873 • nhicustomerservice@dot.gov

Course Number

FHWA-NHI-130091B

Course Title

Underwater Bridge Repair, Rehabilitation, and Countermeasures

Underwater Bridge Repair, Rehabilitation, and Countermeasures is a two-day course that will provide training to design engineers, construction inspectors, resident engineers and inspection divers in techniques for selecting and executing repairs to below water bridge elements. The primary goal of this course is to enable design engineers to select, design, and specify appropriate and durable repairs to below water bridge elements. A secondary goal of this course is to train staff in effective construction inspection of below water repairs. This course may be presented as a follow-up to NHI Course No. 130091A, Underwater Bridge Inspections.

Outcomes

Upon completion of the course, participants will be able to:

- Determine whether below water repairs can be completed "in the wet", or require a cofferdam (or similar).
- Describe typical environmental constraints to performing repairs below water.
- Describe three methods of achieving a dry construction site within a body of water.
- List three attributes of good concrete repair mix designs.
- Describe the differences between flexible and rigid concrete forming systems.
- Describe underwater concrete placement techniques.
- Write installation procedures for pile jackets.
- Describe three methods for repair of pier scour.
- Describe the benefits of cathodic protection for bridge substructures.
- Describe four stages of underwater repair activities for underwater construction inspection.

Target Audience

The course is intended for design engineers, construction inspectors, resident engineers and inspection divers who may be engaged in the design, specifications or inspection of repairs to bridge elements located in and below water. The course may be of interest to contract administrators responsible for bridge repair or rehabilitation projects. It is expected that participants will have a working knowledge of bridge terminology, construction materials, and traditional repair techniques. Participants may also have backgrounds in bridge maintenance, repair, or construction. The audience will include persons with a range of education and technical backgrounds.

Training Level: Basic

Fee: 2016: $750 Per Person; 2017: N/A

Length: 2 DAYS (CEU: 1.4 UNITS)

Class Size: MINIMUM: 20; MAXIMUM: 30

NHI Customer Service: (877) 558-6873 • nhicustomerservice@dot.gov

Course Number

FHWA-NHI-130092

Course Title

Fundamentals of LRFR and Applications of LRFR for Bridge Superstructures

This course provides novice and experienced bridge engineers with the fundamental knowledge necessary to apply the most recent AASHTO LRFR Specifications to bridge ratings. This course introduces participants to applications of LRFR specifications that can be used to enhance bridge safety and to identify and discuss the steps to ensure successful transition to this new state-of-the art methodology.

Load Rating of Concrete and Steel Superstructure Bridges will provide participants with in-depth training in evaluating reinforced and prestressed concrete bridges and steel bridges using LRFR methodology. This course will illustrate the use of the current AASHTO evaluation specifications and state-of-the art evaluation methods with step-by-step examples.

Outcomes

Upon completion of the course, participants will be able to:

- Describe the purpose of performing a load rating
- Identify the benefits of the LRFR methodology
- Demonstrate the LRFR process and the general load rating equations
- Explain legal loads and their use in load rating
- Determine distribution factors for load rating
- State the LRFR limit states
- Select evaluation factors for rating
- Describe the process for load posting and importance of load posting
- Describe the procedure for checking overload permits
- Demonstrate the application of LRFR requirements by completing load rating exercises
- Identify material deteriorations that affect load capacity of bridge components
- Calculate the flexural resistances of a prestressed concrete girder for load rating
- Calculate the shear resistance of a prestressed concrete girder for load rating
- Apply the load rating procedures for concrete slab bridges
- Calculate the flexural and shear resistance of a steel I-girder bridge for load rating
- Evaluate fatigue for load rating a steel girder bridge
- Apply LRFR requirements by completing load rating exercises

Target Audience

Bridge engineers with 0-20 years of experience.

Training Level: Basic

Fee: 2016: $1250 Per Person; 2017: $1250 Per Person

Length: 4 DAYS (CEU: 2.4 UNITS)

Class Size: MINIMUM: 20; MAXIMUM: 40

NHI Customer Service: (877) 558-6873 • nhicustomerservice@dot.gov

Course Number

FHWA-NHI-130092A

Course Title

Load and Resistance Factor Rating for Highway Bridges

This course provides novice and experienced bridge engineers with the fundamental knowledge necessary to apply the most recent AASHTO Load and Resistance Factor Rating (LRFR) Specifications to bridge load rating.

Outcomes

Upon completion of the course, participants will be able to:

- Describe the purpose of performing a load rating
- Identify the benefits of the LRFR methodology
- Demonstrate the LRFR process and the general load rating equations
- Explain legal loads and their use in load rating
- Determine distribution factors for load rating
- State the LRFR limit states
- Select evaluation factors for rating
- Describe the process for load posting and importance of load posting
- Describe the procedure for checking overload permits
- Demonstrate the application of LRFR requirements by completing load rating exercises

Target Audience

Bridge engineers with 0-20 years of experience.

Training Level: Basic

Fee: 2016: $875 Per Person; 2017: $875 Per Person

Length: 2 DAYS (CEU: 1.2 UNITS)

Class Size: MINIMUM: 20; MAXIMUM: 40

NHI Customer Service: (877) 558-6873 • nhicustomerservice@dot.gov

Course Number

FHWA-NHI-130092B

Course Title

Fundamentals of LRFR and Applications of LRFR for Bridge Superstructures

This course provides novice and experienced bridge engineers with the fundamental knowledge necessary to apply the most recent AASHTO LRFR Specifications to bridge ratings. This course introduces participants to applications of LRFR specifications that can be used to enhance bridge safety and to identify and discuss the steps to ensure successful transition to this new state-of-the art methodology.

This 2-day course (130092B) is the second half of the 4-day 130092 course.

Outcomes

Upon completion of the course, participants will be able to:

- Demonstrate the application of LRFR requirements by completing load rating exercises
- Identify material deteriorations that affect load capacity of bridge components
- Calculate the flexural resistances of a prestressed concrete girder for load rating
- Calculate the shear resistance of a prestressed concrete girder for load rating
- Apply the load rating procedures for concrete slab bridges
- Calculate the flexural and shear resistance of a steel I-girder bridge for load rating
- Evaluate fatigue for load rating a steel girder bridge
- Apply LRFR requirements by completing load rating exercises

Target Audience

Bridge engineers with 0-20 years of experience.

Training Level: Basic

Fee: 2016: $875 Per Person; 2017: N/A

Length: 2 DAYS (CEU: 1.1 UNITS)

Class Size: MINIMUM: 20; MAXIMUM: 40

NHI Customer Service: (877) 558-6873 • nhicustomerservice@dot.gov

Course Number

FHWA-NHI-130093

Course Title

LRFD Seismic Analysis and Design of Bridges

This course is a comprehensive and practical training course that addresses the requirements and recommendations of the seismic provisions in both the AASHTO LRFD Bridge Design Specifications and the AASHTO Guide Specifications for LRFD Seismic Bridge Design. The course reviews the fundamental principles of seismic design including engineering seismology, seismic and geotechnical hazards, and methods for modeling and analyzing bridges subject to earthquake ground motions. The course also discusses seismic capacity design methods of piers, foundations, superstructures and connections. Additionally, the course presents the principles and pros and cons of common seismic isolation techniques, typical isolation hardware, and construction and testing requirements consistent with the recently updated AASHTO Guide Specifications for Seismic Isolation Design. Lastly, the final lesson of the course addresses screening, evaluation, and selection of retrofit strategies and measures following closely to the philosophy and process described in the FHWA Seismic Retrofitting Manual for Highway Structures.

Outcomes

Upon completion of the course, participants will be able to:

- Identify geotechnical hazards and their impact on structural design
- Discuss what Earthquake Resisting Elements (ERE) are and explain why some are preferred and why some are not
- List three Describe the essential parts of the capacity design process
- Describe strategies for protecting bridge superstructures and methods for accommodating lateral displacements
- List the steps of foundation seismic design
- Describe the seismic analysis and design process in accordance with the AASHTO LRFD Bridge Design Specifications (LS) and AASHTO Seismic Guide Specifications (GS).
- Develop design response spectrum
- Describe common processes embedded in both the LS and GS and explain the key differences between the Force-Based (LS) and
- Displacement-Based (GS) Methods.
- Describe the key difference between the LS and GS seismic design methods
- List basic purposes, component and testing requirements for a seismic isolation system
- Describe common retrofitting measures for bridge superstructures, columns and foundations

Target Audience

This course is intended to engage a target audience of bridge engineers with zero and up to 20 years of experience, through instructor-led presentations, discussions, Q&A, group activities, walkthrough examples, hands-on student exercises, and demonstrations.

Training Level: Intermediate

Fee: 2016: $1325 Per Person; 2017: $1325 Per Person

Length: 5 DAYS (CEU: 3 UNITS)

Class Size: MINIMUM: 20; MAXIMUM: 30

NHI Customer Service: (877) 558-6873 • nhicustomerservice@dot.gov

Course Number

FHWA-NHI-130093A

Course Title

Displacement-Based Seismic Design of Bridges

This 3-day NHI training course 130093A entitled "Displacement-Based Seismic Analysis and Design of Bridges" is a shortened version of the 5-day NHI 130093 Course "LRFD Seismic Analysis and Design of Bridges" focusing specifically on the displacement-based design philosophies. It is a comprehensive and practical training course that addresses the requirements and recommendations of the seismic provisions in the AASHTO Guide Specifications for LRFD Seismic Bridge Design.

The 130093A course reviews the fundamental principles of seismic design including engineering seismology, structural dynamics (SDOF and MDOF), seismic and geotechnical hazards, and methods for modeling and analyzing bridges subject to earthquake ground motions. The 130093A course then discusses the principles and applications of capacity design to piers, foundations, superstructures and connections, and a brief introduction to the principles and some application of seismic isolation.

The course is accompanied by a prerequisite Web-based Training (WBT) 130093W Course "Introduction to Earthquake Engineering". The participants are highly recommended to complete the WBT course prior to the Instructor Led course. The WBT prerequisite course consists of 5 lessons including Introduction to Earthquake Seismology (L1); Damages to Bridges due to Strong Motion (L2); Single Degree of Freedom (SDOF) Systems and Response Spectra (L3); AASHTO Design Ground Motion Characterization (L4); and Introduction to Geotechnical Hazards (L5).

Outcomes

Upon completion of the course, participants will be able to:

- Identify types of bridge damage to avoid
- Use acceleration and displacement response spectra to estimate peak forces and displacements
- List three elements of Capacity Design
- Describe the most common method for determining dynamic seismic response (i.e. multi-mode response spectrum)
- Calculate, by hand, inelastic displacements of simple pier systems
- Compare and contrast various bridge modeling techniques from stick models to finite element models
- Describe the relationship between detailing of transverse steel and ductility demand on a column
- Develop the design overstrength forces for a column
- Explain how liquefaction affects the seismic design process
- Describe strategies for protecting superstructures from damage
- Compute required support lengths in accordance with AASHTO design specifications
- Describe common processes embedded in both the LS and GS
- List the four seismic design categories in the GS and the key requirements for each category
- Describe the basic purpose of seismic isolation

Target Audience

This course is intended to engage a target audience of bridge engineers with zero and up to 20 years of experience, through instructor-led presentations, discussions, Q&A, group activities, walkthrough examples, and hands-on student exercises and design example practices.

TRAINING LEVEL: Intermediate

FEE: 2016: $1020 Per Person; 2017: $1020 Per Person

LENGTH: 3 DAYS (CEU: 1.8 UNITS)

CLASS SIZE: MINIMUM: 20; MAXIMUM: 30

NHI Customer Service: (877) 558-6873 • nhicustomerservice@dot.gov

Course Number

FHWA-NHI-130093W

Course Title

Introduction to Earthquake Engineering

130093W Introduction to Earthquake Engineering is a Web-based Training (WBT) prerequisite to the 3-day 130093A Displacement-Based LRFD Seismic Analysis and Design of Bridges Instructor-led Training (ILT). The participants will generally be notified to take the WBT about 1 month before the 130093A ILT session and must complete it before the start of Day 1 of the ILT. This WBT consists of 5 lessons including: Introduction to Earthquake Seismology (Lesson 1); Damages to Bridges due to Strong Motion (Lesson 2); Single Degree-of-Freedom (SDOF) Systems and Response Spectra (Lesson 3); AASHTO Design Ground Motion Characterization (Lesson 4); and Introduction to Geotechnical Hazards (Lesson 5).

Outcomes

Upon completion of the course, participants will be able to:

- Describe basic concepts of plate tectonics and seismology
- Explain fundamental concepts of modern seismic design
- Identify parameters used to characterize earthquake ground motions
- Recognize the steps employed in a probabilistic seismic hazard analysis
- Characterize design ground motions in accordance with AASHTO
- List the different types of geotechnical hazards

Target Audience

The target audience for this course includes bridge and geotechnical engineers with 0 to 20 years of experience that are preparing to attend the 130093A Instructor-led Training.

Training Level: Basic

Fee: 2016: $0 Per Person; 2017: $0 Per Person

Length: 4 HOURS (CEU: .4 UNITS)

Class Size: MINIMUM: 0; MAXIMUM: 0

NHI Customer Service: (877) 558-6873 • nhicustomerservice@dot.gov

Course Number

FHWA-NHI-130095

Course Title

LRFD and Analysis of Curved Steel Highway Bridges

This five-day course expands the suite of FHWA services to assist State and local governments in a successful implementation of Load and Resistance Factor Design (LRFD). This course applies the principles of LRFD to the analysis and design of skewed and horizontally curved steel bridges. For structural applications, the curriculum follows the AASHTO LRFD Bridge Design Specifications, 5th Edition, 2010 (AASHTO LRFD Specifications). The training course focuses primarily on the analysis and design of skewed and horizontally curved steel I-girder bridges. However, the accompanying Reference Manual also includes design examples for horizontally curved steel box-girder bridges.

This course provides a combination of instructor-led discussions and workshop exercises. It includes LRFD theory applied to design examples, and it illustrates step-by-step LRFD design procedures for skewed and curved steel bridges. The course includes participant exercises in which students apply the LRFD principles to specific applications, guided walk-throughs in which the instructor guides the participants through design examples, case studies in which real-life examples are used to illustrate the principles being learned, as well as models to help participants observe firsthand the behavior of skewed and curved bridges.

The curriculum materials are comprised of a comprehensive Reference Manual, lecture and workshop exercises intended to promote and enhance a working knowledge of the AASHTO LRFD Specifications as they apply to skewed and curved steel bridges, and a Participant Workbook containing slides, design examples, exercises, narrative descriptions and room for participant notes.

The curriculum material contains the following major topics:

1. General introduction (course introduction and overview)
2. Fundamentals (system behavior, torsion and live load force effects)
3. Structural analysis (general analysis considerations, bearing constraints, approximate methods, 2D refined methods, 3D refined methods and recommended level of analysis)
4. Design (preliminary design decisions, girder design verifications and design detail items)
5. Fabrication and construction

Outcomes

Upon completion of the course, participants will be able to:

- Describe the bridge superstructure analysis, design, fabrication and construction process for skewed or horizontally curved steel I-girder superstructures and for horizontally curved steel box-girder superstructures in accordance with the AASHTO LRFD Specifications
- Illustrate the application of the AASHTO LRFD Specifications to the analysis and design process for skewed and curved steel-bridge superstructures, taking into account erection and construction considerations
- Demonstrate understanding of analysis and design specification requirements for skewed and curved steel girder bridges through the completion of participant exercises and guided walk-throughs and the review of design examples

Target Audience

This course has been developed for the needs of practicing public and private sector structural and bridge engineers with 0 to approximately 20 years of experience. The primary audience is Host Agency and consultant structural designers. Pre-training Competencies: Individuals attending this course should have a minimum BSCE degree and have a working knowledge of the current AASHTO LRFD Specifications or the AASHTO Standard Specifications for Highway Bridges. They should also have relevant design experience using either of these specifications on at least one bridge superstructure.

TRAINING LEVEL: Basic

FEE: 2016: $1325 Per Person; 2017: $1325 Per Person

LENGTH: 5 DAYS (CEU: 3 UNITS)

CLASS SIZE: MINIMUM: 20; MAXIMUM: 30

NHI Customer Service: (877) 558-6873 • nhicustomerservice@dot.gov

Course Number

FHWA-NHI-130095A

Course Title

Fundamental and Structural Analysis for Curved and Skewed Steel Bridges

This 2½-day course presents the first half of the five-day course (Course No. FHWA-NHI-130095). It expands the suite of FHWA services to assist State and local governments in a successful implementation of Load and Resistance Factor Design (LRFD). This course applies the principles of LRFD to the analysis of skewed and horizontally curved steel bridges. For structural applications, the curriculum follows the AASHTO LRFD Bridge Design Specifications, 5th Edition, 2010 (AASHTO LRFD Specifications). The training course focuses primarily on the analysis of skewed and horizontally curved steel I-girder bridges. However, the accompanying Reference Manual also includes design examples for horizontally curved steel box-girder bridges.

This course provides a combination of instructor-led discussions and workshop exercises. It includes LRFD theory applied to analysis examples, and it illustrates step-by-step LRFD analysis procedures for skewed and curved steel bridges. The course includes participant exercises in which students apply the LRFD principles to specific applications, guided walk-throughs in which the instructor guides the participants through analysis examples, case studies in which real-life examples are used to illustrate the principles being learned, as well as models to help participants observe firsthand the behavior of skewed and curved bridges.

The curriculum materials are comprised of a comprehensive Reference Manual, lecture and workshop exercises intended to promote and enhance a working knowledge of the AASHTO LRFD Specifications as they apply to skewed and curved steel bridges, and a Participant Workbook containing slides, analysis examples, exercises, narrative descriptions and room for participant notes.

The curriculum material contains the following major topics:

1. General introduction (course introduction and overview)
2. Fundamentals (system behavior, torsion and live load force effects)
3. Structural analysis (general analysis considerations, bearing constraints, approximate methods, 2D refined methods, 3D refined methods and recommended level of analysis)

Outcomes

Upon completion of the course, participants will be able to:

- Describe the bridge superstructure analysis process for skewed or horizontally curved steel I-girder superstructures and for horizontally curved steel box-girder superstructures in accordance with the AASHTO LRFD Specifications
- Illustrate the application of the AASHTO LRFD Specifications to the analysis process for skewed and curved steel-bridge superstructures
- Demonstrate understanding of analysis specification requirements for skewed and curved steel girder bridges through the completion of participant exercises and guided walk-throughs and the review of analysis examples

Target Audience

This course has been developed for the needs of practicing public and private sector structural and bridge engineers with 0 to approximately 20 years of experience. The primary audience is Host Agency and consultant structural designers. Pre-training Competencies: Individuals attending this course should have a minimum BSCE degree and have a working knowledge of the current AASHTO LRFD Specifications or the AASHTO Standard Specifications for Highway Bridges. They should also have relevant design experience using either of these specifications on at least one bridge superstructure.

TRAINING LEVEL: Basic

FEE: 2016: $935 Per Person; 2017: $935 Per Person

LENGTH: 2.5 DAYS (CEU: 1.6 UNITS)

CLASS SIZE: MINIMUM: 20; MAXIMUM: 30

NHI Customer Service: (877) 558-6873 • nhicustomerservice@dot.gov

Course Number

FHWA-NHI-130095B

Course Title

Design and Fabrication of Curved and Skewed Steel Bridges

This 2.5-day course presents the second half of the five-day course (Course No. FHWA-NHI-130095). It expands the suite of FHWA services to assist State and local governments in a successful implementation of Load and Resistance Factor Design (LRFD). This course applies the principles of LRFD to the design of skewed and horizontally curved steel bridges. For structural applications, the curriculum follows the AASHTO LRFD Bridge Design Specifications, 5th Edition, 2010 (AASHTO LRFD Specifications). The training course focuses primarily on the design of skewed and horizontally curved steel I-girder bridges. However, the accompanying Reference Manual also includes design examples for horizontally curved steel box-girder bridges.

This course provides a combination of instructor-led discussions and workshop exercises. It includes LRFD theory applied to design examples, and it illustrates step-by-step LRFD design procedures for skewed and curved steel bridges. The course includes participant exercises in which students apply the LRFD principles to specific applications, guided walk-throughs in which the instructor guides the participants through design examples, case studies in which real-life examples are used to illustrate the principles being learned, as well as models to help participants observe firsthand the behavior of skewed and curved bridges.

The curriculum materials are comprised of a comprehensive Reference Manual, lecture and workshop exercises intended to promote and enhance a working knowledge of the AASHTO LRFD Specifications as they apply to skewed and curved steel bridges, and a Participant Workbook containing slides, design examples, exercises, narrative descriptions and room for participant notes.

The curriculum material contains the following major topics:

1. Design (preliminary design decisions, girder design verifications and design detail items)
2. Fabrication and construction

Outcomes

Upon completion of the course, participants will be able to:

- Describe the bridge superstructure design, fabrication and construction process for skewed or horizontally curved steel I-girder superstructures and for horizontally curved steel box-girder superstructures in accordance with the AASHTO LRFD Specifications
- Illustrate the application of the AASHTO LRFD Specifications to the design process for skewed and curved steel-bridge superstructures, taking into account erection and construction considerations
- Demonstrate understanding of design specification requirements for skewed and curved steel girder bridges through the completion of participant exercises and guided walk-throughs and the review of design examples
- Successfully complete applicable Learning Outcome Assessments with a combined score of 70 percent or higher

Target Audience

This course has been developed for the needs of practicing public and private sector structural and bridge engineers with 0 to approximately 20 years of experience. The primary audience is Host Agency and consultant structural designers. Pre-training Competencies: Individuals attending this course should have a minimum BSCE degree and have a working knowledge of the current AASHTO LRFD Specifications or the AASHTO Standard Specifications for Highway Bridges. They should also have relevant design experience using either of these specifications on at least one bridge superstructure.

TRAINING LEVEL: Basic

FEE: 2016: $935 Per Person; 2017: $935 Per Person

LENGTH: 2.5 DAYS (CEU: 1.5 UNITS)

CLASS SIZE: MINIMUM: 20; MAXIMUM: 30

NHI Customer Service: (877) 558-6873 • nhicustomerservice@dot.gov

Course Number

FHWA-NHI-130096

Course Title

Cable-Stayed Bridge Seminar

The National Highway Institute's (NHI) one-day Cable-Stayed Signature Bridge Seminar is intended to provide participants with an introduction to planning, design, and construction of long-span, cable-stayed bridges. The seminar provides an overview of the features of cable-stayed bridges; their construction and maintenance considerations; and analyses needed to design these highly redundant structures including special aerodynamic studies.

This seminar will engage participants through Instructor-led presentations, discussions, Q&A, group activities, and walkthrough examples. Participants will review a case study to help them understand how the curriculum can be applied to making basic design decisions. Major topics covered include: bridge configurations, construction methodology, component details, analysis, aerodynamics, design methodology, construction engineering, and maintenance and inspection. As part of the seminar, participants will receive a copy of FHWA Design Guidelines for the Arch and Cable-Supported Signature Bridges.

As a result of the seminar, participants will become familiar with the features of, construction and maintenance considerations; and analyses needed to design cable-stayed bridges.

Outcomes

Upon completion of the course, participants will be able to:

- Describe the benefits of the cable-stayed bridge as a structure type over other alternatives
- Identify possible span and cable arrangements
- Compare steel, concrete or composite superstructure types
- Select possible pylon shape
- Define the general approaches for erecting steel and concrete cable-stayed bridges
- Define the roles and responsibilities of the owner, contractor and construction engineer
- Identify the needs for aerodynamics studies, testing and evaluation, and discuss practical solutions to mitigate wind effects

Target Audience

The primary target audience includes bridge engineers with 10 to 30 years of expericence.

Training Level: Basic

Fee: 2016: $645 Per Person; 2017: $645 Per Person

Length: 1 DAYS (CEU: .6 UNITS)

Class Size: MINIMUM: 20; MAXIMUM: 30

NHI Customer Service: (877) 558-6873 • nhicustomerservice@dot.gov

Course Number

FHWA-NHI-130099A

Course Title

Bridge Inspection Nondestructive Evaluation Seminar (BINS)

The FHWA Office of Infrastructure R&D, in cooperation with the FHWA Office of Bridge Technology and the FHWA Resource Center, has identified a need for training in select nondestructive evaluation (NDE) methods that can be used to assess existing conditions on highway bridge structures during routine inspections. These NDE methods can also be used to supplement visual inspections of highway bridge structures.

The Bridge Inspector Nondestructive Evaluation Seminar (BINS) is a two-day course which provides bridge inspectors and managers the ability to learn about the latest in commercially available nondestructive tools and systems for use on bridges. The seminar is presented through a series of slides, instructional videos, and video demonstrations showing basic operation of the equipment. The training has been fully developed in conjunction with the FHWA's NDE Validation Center and is delivered by qualified instructors experienced in using NDE equipment on bridges.

This seminar is designed to provide bridge inspection staff the opportunity to view efficient and effective inspection tools and techniques with the ultimate goal of achieving safer bridges through more reliable bridge inspections. The following NDE methods are discussed: Eddy Current, Ultrasonic Testing, Infrared Thermography, Impact Echo, Ultrasonic Surface Waves, Ground Penetrating Radar, Acoustic Emission, Magnetic Particle, Radiographic, Pulse Velocity, Pulse Echo, Pachometers, Physical Sounding Methods, and Electrical Methods. Additionally, other commonly used equipment will be briefly introduced with basic information provided about attributes in an easy to use reference table and select extra information in the appendix.

Outcomes

Upon completion of the course, participants will be able to:

- Summarize the National Bridge Inspection Program (NBIP) expectations as they relate to NDE
- Compare the various stress wave NDE methods as used in steel bridge inspection
- Demonstrate understanding of stress wave and electromagnetic methods by choosing applicable NDE methods for specific defects
- Summarize how NDE was used to assist decision makers in the repair of the Sherman Minton Bridge
- Restate the theories, applications, advantages and limitations of various NDE testing methods
- Compare the theories and applications of various acoustic stress wave testing methods for concrete and timber inspections
- Demonstrate an understanding of electromagnetic and electric NDE methods in bridge inspection programs
- Summarize feasible methods used to evaluate the deck on the Arlington Memorial Bridge (AMB)

Target Audience

The primary target audience for the Bridge Inspection Non-Destructive Evaluation Seminar (BINS) course is federal, state, and local highway bridge inspectors, bridge management staff, and consultants. Individuals involved in material testing, as well as transportation structure design and construction, will find the information useful to ensure quality. Prior to taking this course, participants should have a broad basic knowledge of physics and engineering principles, a knowledge of the basic bridge inspection fundamentals, a background in bridge engineering or completion of NHI course FHWA-NHI-130054 Engineering Concepts for Bridge Inspectors (strongly recommended), and experience with bridge inspection.

TRAINING LEVEL: Basic

FEE: 2016: $750 Per Person; 2017: $750 Per Person

LENGTH: 2 DAYS (CEU: 1.3 UNITS)

CLASS SIZE: MINIMUM: 20; MAXIMUM: 30

NHI Customer Service: (877) 558-6873 • nhicustomerservice@dot.gov

Course Number

FHWA-NHI-130101

Course Title

Introduction to Safety Inspection of In-Service Bridges - WEB-BASED

This training is a prerequisite of another NHI training and is offered at no cost.

Introduction to Safety Inspection of In-Service Bridges is designed to prepare participants with the necessary fundamentals required for a more intensive course in bridge inspection. This WBT introduces the elementary concepts of bridge inspection, bridge functions, and bridge inspection terminology. Participants who complete this WBT will be prepared for more intensive courses in bridge inspection, which focus on documentation, rating, assessment, and field inspection.

Introduction to Safety Inspection of In-Service Bridges covers bridge components and elements, bridge mechanics, design features, bridge materials, decks, superstructures, bearings, substructures, channels, inspection preparations, inspection reporting activities, and work area safety.

This course prepares participants for the 2-week, intensive Instructor-led course in bridge inspection, 130055 Safety Inspection of In-Service Bridges.

Upon successful completion of 130101, participants will have met the prerequisite requirement for participation in the 130055 course (for sessions beginning March 5, 2012 or later).* If participants would like to enroll in the 130055 course, they will be required to demonstrate their certificate of completion for 130101 as proof that the prerequisite requirement has been fulfilled.

Participation in 130101 is not the only option to fulfill the prerequisite requirement for 130055.* Individuals have the option to 1) successfully complete NHI-130054 Engineering Concepts for Bridge Inspectors (Instructor-led course) or 2) for those with engineering backgrounds or prior knowledge and experience in the field of bridge inspection may "test-out" through a Web-based assessment (130101A Introduction to Safety Inspection of In-Service Bridges).

*Please note: Upon successful completion of this prerequisite course, you will be eligible to take the 130055 training course for up to 2 years.

Outcomes

Upon completion of the course, participants will be able to:

- Describe the basis for bridge inspection
- Identify the three major bridge components and various culvert types
- Identify the various elements that comprise bridge components
- Describe standard highway bridge loadings
- Describe the basic concepts of elasticity of materials, response of materials to an applied force, response of structural members to a variety of loadings, the relationship between stresses and strains, and load rating
- Describe span arrangements, deck-superstructure interaction, and redundancy
- Describe the basic properties, strengths and weaknesses of steel, concrete, and timber
- Describe the types, signs and causes of structural distress in steel, concrete, and timber
- Describe the general purpose of decks, superstructures, and bearings
- Describe the general purpose and function of substructure units
- Describe waterway features and the effect of scour
- Describe the requirements for preparing for an inspection
- Describe the basic bridge inspection reporting requirements
- Name protective measurements to mitigate the hazards involved when working in the field performing bridge inspection

Target Audience

This training has been developed for Federal, State, and local highway agency employees and consultants involved in inspecting bridges or in charge of a bridge inspection unit. A background in bridge engineering is strongly recommended.

TRAINING LEVEL: Basic

FEE: 2016: $0 Per Person; 2017: N/A

LENGTH: 14 HOURS (CEU: 1.4 UNITS)

CLASS SIZE: MINIMUM: 0; MAXIMUM: 0

NHI Customer Service: (877) 558-6873 • nhicustomerservice@dot.gov

COURSE NUMBER

FHWA-NHI-130101A

COURSE TITLE

Prerequisite Assessment for Safety Inspection of In-Service Bridges - WEB-BASED

This training is a prerequisite of another NHI training and is offered at no cost.

Prerequisite Assessment for Safety Inspection of In-Service Bridges (FHWA-NHI-130101A) is a required prerequisite necessary for those interested in taking the course Safety Inspection of In-Service Bridges (FWHA-NHI-130055). The assessment is divided into three sections; participants are given three opportunities to pass each section with a score of 70% or better. Passing all three assessment sections signifies successful completion.

The assessment covers a range of topics that includes the bridge inspection program, bridge components and elements, bridge mechanics, design features, bridge materials, decks, superstructures, bearings, substructures, channels, inspection preparations, inspection reporting activities, and work area safety. To access this online assessment, enroll in NHI 130101A "Prerequisite Assessment for Safety Inspection of In-Service Bridges" via the NHI Web site.

Upon successful completion of 130101A, participants will have met the prerequisite requirement for participation in the 130055 Safety Inspection of In-Service Bridges course (for sessions beginning March 5, 2012 or later).* If participants would like to enroll in the 130055 course, they will be required to demonstrate their certificate of completion for 130101A as proof that the prerequisite requirement has been fulfilled.

Participation in 130101A is not the only option to fulfill the prerequisite requirement for 130055.* Individuals have the option to 1) successfully complete NHI-130054 Engineering Concepts for Bridge Inspectors (Instructor-led course) or 2) successfully complete the Web-based training and assessment (130101 Introduction to Safety of In-Service Bridges)

*Please note: Upon successful completion of this prerequisite course, you will be eligible to take the 130055 training course for up to 2 years.

OUTCOMES

Upon completion of the course, participants will be able to:

- There are no course outcomes associated with this prerequisite assessment.

TARGET AUDIENCE

This assessment has been developed for Federal, State, and local highway agency employees and consultants involved in inspecting bridges or in charge of a bridge inspection unit. A background in bridge engineering is strongly recommended.

TRAINING LEVEL: Basic

FEE: 2016: $0 Per Person; 2017: N/A

LENGTH: 1 HOURS (CEU: 0 UNITS)

CLASS SIZE: MINIMUM: 0; MAXIMUM: 0

NHI Customer Service: (877) 558-6873 • nhicustomerservice@dot.gov

Course Number

FHWA-NHI-130102

Course Title

Engineering for Structural Stability in Bridge Construction (2.5 Day)

The objective of this course is to train participants on the behavior of steel and concrete girder bridges during construction and teach them to identify vulnerabilities and engineering methods to investigate the structure's strength and stability at each critical stage. This is done within the practical context of engineering, development, verification, and/or review of erection plans.

Starting with basic structural stability principles, course participants are introduced to stability analysis methods and how they should be applied to properly engineer a bridge erection plan. The role of both permanent and temporary bracing in achieving structural stability is covered, and methods for bracing design presented. Behavior and design considerations for construction phases are provided through presentation of case studies, demonstrations, design examples, and guided walk-throughs. The impacts of construction practices, means, and methods are explored and demonstrated.

During bridge erection, the member support conditions, loads, stresses, strength, and stability are affected by the erection practices such as lifting, installation of bracing, bearing conditions, temporary supports, and placing sequence. Deck placing equipment, overhang brackets and staging can also have significant effects on girder stability. Thus, this course presents information on construction practices as it relates to these considerations.

Engineering criteria for use in evaluating bridges during erection are presented. Loading criteria and load factors for analysis are provided along with discussion of their applicability. Equations for checking member conditions during erection are included. Participants learn how loads during construction differ from final design conditions and appropriate methods to compute and apply those loads. The required contents of erection engineering plans, procedures, and submittals are presented in the course. Check lists are included to assist both the erection engineer and submittal reviewer.

The extended Course 130102A (3.5 days) provides an additional 8 hours of hands-on practicum where participants are given opportunity to apply advanced stability analysis on real-world examples, using software executed on laptop computers. This provides a valuable "capstone" experience to solidify their understanding, relate curriculum to practice, apply the concepts presented, and engage in self-discovery.

Outcomes

Upon completion of the course, participants will be able to:

- Explain the fundamentals of stability theory and how they affect bridge strength and performance during construction
- Describe the differences between local, girder, and system (global) stability limit states
- Recognize the potential for stability-related failures that have occurred in past bridges and how to effectively avoid similar results
- Select loads, load combinations, and factors that are appropriate for the construction plan verification
- Explain common techniques for evaluating the stability of bridge member and components
- Choose an appropriate advanced stability analysis for a critical construction stage where stability is in question
- Describe the role of bracing and shoring and how to use for providing stability
- Assess procedures and details for a construction plan that will be safe and economical

Target Audience

This course has been developed for the needs of practicing public and private sector structural engineers with zero to approximately twenty years of experience. The primary audience is Host Agency and consultant bridge structural engineers and project managers, particularly those who prepare and/or review erection plans and procedures. The course will also be of benefit to bridge contractors and erectors as well as those Agency staff overseeing bridge erection.

TRAINING LEVEL: Basic

FEE: 2016: $935 Per Person; 2017: $935 Per Person

LENGTH: 2.5 DAYS (CEU: 1.7 UNITS)

CLASS SIZE: MINIMUM: 20; MAXIMUM: 30

NHI Customer Service: (877) 558-6873 • nhicustomerservice@dot.gov

COURSE NUMBER

FHWA-NHI-130102A

COURSE TITLE

Engineering for Structural Stability in Bridge Construction (3.5 day)

The objective of this course is to train participants on the behavior of steel and concrete girder bridges during construction and teach them to identify vulnerabilities and engineering methods to investigate the structure's strength and stability at each critical stage. This is done within the practical context of engineering, development, verification, and/or review of erection plans.

Starting with basic structural stability principles, course participants are introduced to stability analysis methods and how they should be applied to properly engineer a bridge erection plan. The role of both permanent and temporary bracing in achieving structural stability is covered, and methods for bracing design presented. Behavior and design considerations for construction phases are provided through presentation of case studies, demonstrations, design examples, and guided walk-throughs. The impacts of construction practices, means, and methods are explored and demonstrated.

During bridge erection, the member support conditions, loads, stresses, strength, and stability are affected by the erection practices such as lifting, installation of bracing, bearing conditions, temporary supports, and placing sequence. Deck placing equipment, overhang brackets and staging can also have significant effects on girder stability. Thus, this course presents information on construction practices as it relates to these considerations.

Engineering criteria for use in evaluating bridges during erection are presented. Loading criteria and load factors for analysis are provided along with discussion of their applicability. Equations for checking member conditions during erection are included. Participants learn how loads during construction differ from final design conditions and appropriate methods to compute and apply those loads. The required contents of erection engineering plans, procedures, and submittals are presented in the course. Check lists are included to assist both the erection engineer and submittal reviewer.

This extended Course 130102A (3.5 days) provides an additional 8 hours of hands-on practicum where participants are given opportunity to apply advanced stability analysis on real-world examples, using software executed on laptop computers. This provides a valuable "capstone" experience to solidify their understanding, relate curriculum to practice, apply the concepts presented, and engage in self-discovery.

OUTCOMES

Upon completion of the course, participants will be able to:

- Explain the fundamentals of stability theory and how they affect bridge strength and performance during construction
- Describe the differences between local, girder, and system (global) stability limit states
- Employ lessons learned from past stability-related failures to avoid similar results
- Explain common techniques for evaluating the stability of bridge members and components
- Choose an appropriate advanced stability analysis for a critical construction stage where stability is in question
- Describe the role of bracing and shoring and how to use them to provide stability
- Select loads, load combinations, and factors that are appropriate for the construction plan verification
- Assess procedures and details for a construction plan that will be safe and economical
- Employ stability evaluation techniques to conduct an erection analysis for steel girder and concrete splice girder bridges (3 ½ day course)

TARGET AUDIENCE

This course has been developed for the needs of practicing public and private sector structural engineers with zero to approximately twenty years of experience. The primary audience is Host Agency and consultant bridge structural engineers and project managers, particularly those who prepare and/or review erection plans and procedures. The course will also be of benefit to bridge contractors and erectors as well as those Agency staff overseeing bridge erection.

TRAINING LEVEL: Basic

FEE: 2016: $1100 Per Person; 2017: $1100 Per Person

LENGTH: 3.5 DAYS (CEU: 2.4 UNITS)

CLASS SIZE: MINIMUM: 0; MAXIMUM: 0

NHI Customer Service: (877) 558-6873 • nhicustomerservice@dot.gov

COURSE NUMBER

FHWA-NHI-130103

COURSE TITLE

Post-Tensioning Tendon Installation and Grouting - WBT

Post-Tensioning Tendon Installation and Grouting Web-based Training (WBT) delivers content on post-tensioning principles, system components, and installation procedures - including quality control procedures - which will assist supervisors, inspectors, and construction inspectors in the performance of their job. This WBT provides guidance to individuals involved in the design, installation, grouting, and inspection of post-tensioning tendons for prestressed concrete bridges and is intended to be an online complement to the Post-Tensioning Tendon Installation and Grouting Manual. Participants who complete this WBT will have a general understanding of post-tensioning components, construction, as well as testing and acceptance procedures. This WBT will better prepare individuals for more intensive certification courses in post-tensioning installation and grouting (PTI Level 1 & 2 PT Field Specialist and ASBI Grouting Certification Training).

OUTCOMES

Upon completion of the course, participants will be able to:

- Describe the use of post-tensioning to prestress concrete bridges
- Describe the composition and essential features of prestressing steel and anchorages
- Describe the composition and essential features of ducts and grout
- Describe the testing and acceptance procedures for post-tensioning system materials and components
- Describe post-tensioning tendon component installation, including the role of post-tensioning shop drawings in the construction process
- Describe the operations required to stress post-tensioning tendons
- Describe the importance and proper methods for calibrating jacks and their role in on-site testing for friction and modulus of elasticity
- Describe the elements of grouting operations
- Identify the methods, materials, and details that provide satisfactory corrosion protection

TARGET AUDIENCE

This training is targeted at owners and private company personnel that may be involved in the design, inspection, and construction of bridges that contain PT tendons. This course is intended for those with beginner to intermediate knowledge and/or skills in the area post-tensioning tendon installation and grouting principles and practices.

TRAINING LEVEL: Basic

FEE: 2016: $50 Per Person; 2017: N/A

LENGTH: 6 HOURS (CEU: .6 UNITS)

CLASS SIZE: MINIMUM: 1; MAXIMUM: 1

NHI Customer Service: (877) 558-6873 • nhicustomerservice@dot.gov

COURSE NUMBER

FHWA-NHI-130105A

COURSE TITLE

Introduction to FRP Materials and Applications for Concrete Structures, WEB-BASED

Introduction to FRP Materials and Applications for Concrete Structures is designed to assist State Department of Transportation (DOT) construction and maintenance operation staff develop knowledge of the types of FRP Composite material, form, and properties used in the repair and retrofit of concrete structures, as well as versatility in applications of FRP in the repair of concrete structures.

Topics covered in this course include:

Background of FRP material development in bridge applications

Different types of FRP Composite material (Fiber and Resin)

Common concrete superstructure and substructure defects that are candidates for FRP repair and retrofit

Versatility in the application of FRP in the repair and retrofit of common concrete structure defects

Benefits of FRP repairs and retrofits for concrete structures over traditional methods

The success of repairs of concrete structures using FRP Composites is dependent on choosing FRP material suitable for the application. It is essential to develop knowledge of FRP material, properties, and suitable application.

OUTCOMES

Upon completion of the course, participants will be able to:

- Describe the application of FRP materials for concrete structures.
- Describe the different methods of repairing and retrofitting concrete structures using FRP materials.

TARGET AUDIENCE

This training is appropriate for persons with minimal or no experience in bonded repair and retrofit of concrete structures using FRP Composites, as well as those experienced with using FRP Composite. The course focuses on construction areas, however, bridge designers as well as field personnel will benefit from the content.

TRAINING LEVEL: Basic

FEE: 2016: $50 Per Person; 2017: N/A

LENGTH: 3 HOURS (CEU: .3 UNITS)

CLASS SIZE: MINIMUM: 0; MAXIMUM: 0

NHI Customer Service: (877) 558-6873 • nhicustomerservice@dot.gov

Course Number

FHWA-NHI-130105B

Course Title

Construction Procedures and Specifications for Bonded Repair and Retrofit of Concrete Structures

Construction Procedures and Specifications for Bonded Repair and Retrofit of Concrete Structures using FRP Composites is designed to assist State Department of Transportation (DOT) construction and maintenance operation staff develop knowledge of project requirements of FRP repairs, substrate surface preparation methods, and procedures and steps for installation of FRP systems.

Topics covered in this course include:

Specifications, including scope, definitions, tolerances, and site considerations

Submittal requirements, including working drawings and quality control/quality assurance plans

Storage, handling, and disposal requirements, including shelf life, safety hazards, personnel and work place protection, and clean up

Various aspects of substrate repairs and surface preparation of concrete structures

Use of externally-bonded and near-surface mounted FRP systems for repairs

Procedures and steps for installation of externally bonded FRP systems

Procedures and steps for installation of near-surface mounted FRP systems

Environmental considerations for FRP installation

Identification of defects and appropriate solutions of FRP applications

The success of repairs and retrofit of concrete structures using FRP Composite is dependent on State Department of Transportation (DOT) construction personnel taking an active role in ensuring construction procedures and specifications are adhered to. Hence, knowledge of proper construction procedures and specifications for FRP projects is necessary to control quality of work.

Outcomes

Upon completion of the course, participants will be able to:

- Identify the general project requirements for FRP repair and retrofit of concrete structures.
- Explain the general procedures for FRP repair and retrofit of concrete structures.
- Describe the general installation procedures of FRP systems for repair and retrofit of concrete structures.

Target Audience

This training is appropriate for persons with minimal or no experience in bonded repair and retrofit of concrete structures using FRP Composites, as well as those experienced with using FRP Composite. The course focuses on construction areas, however, bridge designers as well as field personnel will benefit from the content.

Training Level: Basic

Fee: 2016: $50 Per Person; 2017: N/A

Length: 5 HOURS (CEU: .5 UNITS)

Class Size: MINIMUM: 0; MAXIMUM: 0

NHI Customer Service: (877) 558-6873 • nhicustomerservice@dot.gov

COURSE NUMBER

FHWA-NHI-130105C

COURSE TITLE

Quality Control of Repair and Retrofit of Concrete Structures Using FRP Composites

Quality Assurance and Construction Process Control of Bonded Repair and Retrofit of Concrete Structures Using FRP Composites is designed to assist State Department of Transportation (DOT) construction and maintenance operation staff develop knowledge of the requirements of quality assurance and quality control during construction, and equip them with the necessary means to control the application of the repair system and the adequacy of the construction process.

This course covers the following topics:

Responsibilities and qualifications of personnel implementing Quality Control and Quality Assurance (QC/QA) program for FRP application

Requirements of Quality control and Quality Assurance (QC/QA) for FRP applications

Application of Quality Control and Quality Assurance (QC/QA) for FRP application

Inspection methods and acceptance criteria for FRP application

Threshold values of concrete surface preparations and construction tolerances

Key elements of Process Control Manual and checklists for inspection of FRP systems

Examples of defective work, repair for defects, and acceptance criteria for repairs

The success of repairs of concrete structures using FRP Composite is dependent on quality control of materials and workmanship, secured by quality assurances processes.

OUTCOMES

Upon completion of the course, participants will be able to:

- Explain the quality assurance methods of FRP repair and retrofit of concrete structures.
- Explain the inspection methods for FRP repair and retrofit of concrete structures.
- Describe the procedures for repairs of defective FRP work.

TARGET AUDIENCE

This training is appropriate for persons with minimal or no experience in bonded repair and retrofit of concrete structures using FRP Composites, as well as those experienced with using FRP Composite. The course focuses on construction areas, however, bridge designers as well as field personnel will benefit from the content.

TRAINING LEVEL: Basic

FEE: 2016: $50 Per Person; 2017: N/A

LENGTH: 5 HOURS (CEU: .5 UNITS)

CLASS SIZE: MINIMUM: 0; MAXIMUM: 0

NHI Customer Service: (877) 558-6873 • nhicustomerservice@dot.gov

Course Number

FHWA-NHI-130106A

Course Title

Bridge Preservation Fundamentals

Bridge Preservation Fundamentals (130106A) provides the participant key bridge preservation strategies that can help assist in the planning and implementation of their own bridge preservation program. It is a six lesson course that starts off with introducing definitions, terminology, and categories of bridge action. It also shares details on the benefits of timely bridge preservation and the consequences of deferred maintenance. This course discusses at length user best practices and activities related to deck preservation, superstructure preservation, and substructure preservation. This course also includes a lesson with detail on cost-effective culvert preservation practices.

This course is the first course in the three-course Bridge Preservation Web-based Training (WBT) series which includes Establishing a Bridge Preservation Program (130106B) and Communication Strategies for Bridge Preservation (130106C). This course series covers areas such as concepts of bridge preservation; how to establish and maintain a good bridge preservation program; best practices; common treatments and strategies; and resource management strategies (in-house vs. contract). The goal of the Bridge Preservation WBT Series is to provide training to bridge owners and those that are responsible for managing and maintaining the bridge inventory on the principles of planning and implementing successful bridge management and preservation programs.

Outcomes

Upon completion of the course, participants will be able to:

- Define activities and classifications related to bridge preservation, and associated work categories of rehabilitation, preventive maintenance, and systematic preventive maintenance
- Identify the benefits of timely bridge preservation activities, consequences of deferred maintenance, and strategies to transition bridge programs from reactive to proactive
- Determine cost-effective deck preservation practices and activities
- Determine cost-effective superstructure preservation practices and activities
- Determine cost-effective substructure preservation practices and activities
- Determine cost-effective culvert preservation practices and activities

Target Audience

The target audience for the Bridge Preservation Fundamentals WBT course is individuals involved in the development, implementation, and delivery of a bridge preservation program. This course is intended for those with general knowledge and/or skills in the area of bridge maintenance and management principles and practices.

Training Level: Basic

Fee: 2016: $50 Per Person; 2017: N/A

Length: 5 HOURS (CEU: .5 UNITS)

Class Size: MINIMUM: 0; MAXIMUM: 0

NHI Customer Service: (877) 558-6873 • nhicustomerservice@dot.gov

Course Number

FHWA-NHI-130106B

Course Title

Establishing a Bridge Preservation Program

Establishing a Bridge Preservation Program (130106B) focuses on efforts in developing a bridge preservation program. This course includes a lesson on the establishment of goals, objective and performance measures in a bridge preservation program. The course goes in-depth into the needs assessment and data management process, including the creation of a bridge preservation strategy, and it outlines the prioritization process. A lesson on budgeting and resource allocation describes the linkage between data to budgeting and resource allocation activities. The course also includes a lesson on work plan development and implementation with specific details on implementing network, corridor, and site specific strategies. The course concludes with a lesson on program monitoring.

This course is the second course in the three-course Bridge Preservation Web-based Training (WBT) series which includes Bridge Preservation Fundamentals (130106A) and Communication Strategies for Bridge Preservation (130106C). This course series covers areas such as concepts of bridge preservation; how to establish and maintain a good bridge preservation program; best practices; common treatments and strategies; and resource management strategies (in-house vs. contract). The goal of the Bridge Preservation WBT Series is to provide training to bridge owners and those that are responsible for managing and maintaining the bridge inventory on the principles of planning and implementing successful bridge management and preservation programs.

Outcomes

Upon completion of the course, participants will be able to:

- Summarize the process of forming goals, objectives and performance measures for a bridge preservation program
- Determine the condition and needs assessment activities involved in a bridge preservation program
- Determine the budgeting and resource allocation activities involved in a bridge preservation program
- Determine the work plan development and implementation strategies involved in a bridge preservation program
- Determine program monitoring activities that are part of an effective bridge preservation program

Target Audience

The target audience for the Establishing a Bridge Preservation Program WBT course is key individuals involved in managing the development, implementation, and delivery of a bridge preservation program within a transportation agency. This course is intended for those with working knowledge and/or skills in the area of highway bridge infrastructure program management principles.

Training Level: Basic

Fee: 2016: $50 Per Person; 2017: N/A

Length: 4 HOURS (CEU: .4 UNITS)

Class Size: MINIMUM: 0; MAXIMUM: 0

NHI Customer Service: (877) 558-6873 • nhicustomerservice@dot.gov

STRUCTURES

WBT

Course Number

FHWA-NHI-130106C

Course Title

Communication Strategies for Bridge Preservation

Communication Strategies for Bridge Preservation (130106C) is a course that shares details on communication of bridge preservation program values, benefits and needs to stakeholders and the general public. This course starts with a lesson on identifying customers and stakeholders, specifically the identification of potential audience members and dividing these members into segments and the identifying what motivates them to action and assessing these segments. A lesson shares details on developing the message with a breakdown of the process into identifying critical activities in message design, the steps involved in designing a message, and strategies on avoiding common message design mistakes. In the lesson on communicating the message, critical activities in message delivery are identified, specifically the four steps in delivering a message and identifying common mistakes in message delivery. In the final lesson on performing market research, it shares typical methods transportation agencies use to track stakeholder opinions, details the phases in market research, and identifies effective marketing research techniques.

This course is the third course in the three-course Bridge Preservation Web-based Training (WBT) series which includes Bridge Preservation Fundamentals (130106A) and Establishing a Bridge Preservation Program (130106B). This course series covers areas such as concepts of bridge preservation; how to establish and maintain a good bridge preservation program; best practices; common treatments and strategies; and resource management strategies (in-house vs. contract). The goal of the Bridge Preservation WBT Series is to provide training to bridge owners and those that are responsible for managing and maintaining the bridge inventory on the principles of planning and implementing successful bridge management and preservation programs.

Outcomes

Upon completion of the course, participants will be able to:

- Determine the strategies required to identify an agency champion and the target customers and stakeholders for a bridge preservation program
- Recognize strategies for developing bridge preservation messages that capture values, benefits and needs, intended for highway infrastructure stakeholders and the general public
- Determine strategies for communicating bridge preservation messages that capture values, benefits and needs, intended for highway infrastructure stakeholders and the general public
- Summarize key activities involved in performing market research, as it applies to a bridge preservation program

Target Audience

The target audience for the Communication Strategies for Bridge Preservation WBT course is individuals involved in communications with highway infrastructure stakeholders and the general public.

Training Level: Basic

Fee: 2016: $50 Per Person; 2017: N/A

Length: 3 HOURS (CEU: .3 UNITS)

Class Size: MINIMUM: 0; MAXIMUM: 0

NHI Customer Service: (877) 558-6873 • nhicustomerservice@dot.gov

Course Number

FHWA-NHI-130107A

Course Title

Fundamentals of Bridge Maintenance WBT

Fundamentals of Bridge Maintenance (NHI-130107A) teaches the participant the fundamental aspects of an effective bridge maintenance program. Module 1 - Introduction to Bridge Maintenance explains the importance of a balanced bridge maintenance program and the organizational structure, roles, and responsibilities of a bridge maintenance unit. Module 2 - Bridge Maintenance Management provides basic information about bridge inspections, reviews the general concept of Maintenance Management Systems (MMS) and Bridge Management Systems (BMS), reviews the various steps and activities involved in the proper planning and implementation of bridge maintenance program activities, discusses commonly used contracting bridge maintenance methods, and describes the principles of quality assurance and quality control measures used in bridge maintenance. Module 3 - Bridge Anatomy introduces bridge components, associated elements, and their intended functions, and also reviews common bridge types. Module 4 - Bridge Mechanics explains the bridge mechanics as it relates to different bridge components, introduces concepts such as redundancy and fracture critical details, and reviews basic hydraulic, scour and channel erosion concepts. Module 5 - Concrete Basics addresses the basic material properties of concrete; describes proper concrete mixing and testing processes; summarizes proper concrete placement, finishing and curing processes; and reviews proper methods for locating and removing unsound concrete. Module 6 - Maintenance of Bridge Ancillary Items examines general maintenance considerations and practices related to ancillary items often attached to bridges, such as utilities, and sign and lighting structures. This web-based training serves as a prerequisite to the 4-day instructor-led training NHI-130108 Bridge Maintenance.

Outcomes

Upon completion of the course, participants will be able to:

- Describe common organizational structures of transportation agencies, the role of the bridge maintenance unit and where it fits within such organizations, and the various cost-effective maintenance and preservation activities that these units perform
- Review various bridge maintenance program management activities and tools used to facilitate the accomplishment of these activities
- Classify bridge components, associated elements, and their intended function for commonly used materials
- Review the fundamentals of bridge mechanics and behaviors
- Review the fundamental steps involved in using concrete as a repair material
- Describe general maintenance practices associated with bridge mounted sign and lighting structures

Target Audience

The target audience for course 130107A, Fundamentals of Bridge Maintenance Web-Based Training is primarily members of Federal, State, and Local Departments of Transportation, as well as those contractors that perform work on behalf of these agencies. This training is primarily geared for individuals involved in onsite bridge maintenance activities and those that supervise the activities. This training is appropriate for those with basic knowledge of bridge maintenance and repair activities.

Training Level: Basic

Fee: 2016: $0 Per Person; 2017: N/A

Length: 7 HOURS (CEU: .7 UNITS)

Class Size: MINIMUM: 500; MAXIMUM: 500

NHI Customer Service: (877) 558-6873 • nhicustomerservice@dot.gov

Course Number

FHWA-NHI-130108

Course Title

Bridge Maintenance (ILT)

Replacing the original Bridge Maintenance course (FHWA-NHI-134029), this entirely new Instructor-led Training (ILT) course will provide participants with knowledge regarding common deficiencies that occur in bridges, common defects in bridge elements, preventive maintenance techniques, and protective systems intended to prevent deterioration and deficiencies in bridges. With this knowledge, this course will enable participants to investigate proper bridge maintenance procedures using bridge maintenance resources and apply these practices on-the-job.

Outcomes

Upon completion of the course, participants will be able to:

- Identify key steps involved in the development and implementation efforts of a cost-effective preservation strategy for a group of bridges.
- Identify maintenance and/or repair needs and select the best remedial strategy.
- Discuss properties and preservation options involving common bridge materials such as concrete, steel and timber.
- Describe the step-by-step tasks required to accomplish proven preservation procedures on the various bridge elements.
- Identify critical members and avoid procedures that might result in damage such as field welding repairs on fracture critical tension members.
- Recognize problems that warrant specialized expertise, for example, soliciting the involvement of a qualified structural engineer when repairing structural damage.
- Apply effective management techniques (such as planning, scheduling, monitoring and reporting) during daily bridge maintenance operations.

Target Audience

This course is primarily for members of State and Local Departments of Transportation, as well as those contractors that perform work on behalf of these agencies. This training is primarily geared for individuals involved in on-site bridge maintenance and preservation activities and those that supervise and manage these activities. This training is appropriate for those with intermediate to advanced experience in bridge maintenance and repair activities. This training is also suitable for those with intermediate/advanced knowledge of general maintenance and repair activities that have successfully completed the prerequisite, FHWA-NHI-130107A Fundamentals of Bridge Maintenance WBT course. Those that are not involved in on-site bridge maintenance activities, such as designers and construction personnel, may also benefit from this training.

Training Level: Intermediate

Fee: 2016: $1050 Per Person; 2017: N/A

Length: 4 DAYS (CEU: 2.3 UNITS)

Class Size: MINIMUM: 0; MAXIMUM: 0

NHI Customer Service: (877) 558-6873 • nhicustomerservice@dot.gov

COURSE NUMBER

FHWA-NHI-130109A

COURSE TITLE

Bridge Management Fundamentals

When the average citizen commutes to work or runs errands, they are relying on us, public transportation agencies, to keep their bridges safe and available for use. It is their expectation that we keep their bridges serviceable and at the lowest life-cycle cost possible. Bridge management systems will help your agency to efficiently balance the various bridge needs against available resources. The Bridge Management Fundamentals course describes a bridge management system and walks through the process of selecting and implementing the right bridge management software for your agency. Throughout the course, you will learn direct from agencies with mature and successful bridge management systems about how they get the most utility from their system.

OUTCOMES

Upon completion of the course, participants will be able to:

- Explain the need for a BMS
- Describe a typical BMS organizational structure
- Describe the seven components of a BMS
- Describe tools that are used as part of the bridge management process
- Describe an implementation plan for a comprehensive BMS
- Describe effective practices when using BMSs
- Identify successful applications of BMS components by agencies
- Describe the bridge management process as it relates to an agency business model
- Describe how to address risk

TARGET AUDIENCE

The target audience includes Federal, State, and local bridge program managers; bridge management engineers; bridge management practitioners; transportation planners; and project planning and programming personnel. Additionally, transportation performance management team members, transportation asset management team members, bridge preservation and maintenance engineers, the financial management team, bridge inspectors, and bridge designers may benefit from this training.All participants should have knowledge of basic bridge terminology.

TRAINING LEVEL: Basic

FEE: 2016: $50 Per Person; 2017: N/A

LENGTH: 4 HOURS (CEU: .4 UNITS)

CLASS SIZE: MINIMUM: 0; MAXIMUM: 0

NHI Customer Service: (877) 558-6873 • nhicustomerservice@dot.gov

COURSE NUMBER

FHWA-NHI-130109B

COURSE TITLE

Performance-Based Management of Highway Bridges

The traditional approach to bridge management has focused on identifying the worst performing structures in the inventory and addressing their deficiencies before anything else. But as inventories expand and age and as budgets shrink, most agencies discover that even as they address the worst bridges in the inventory, other bridges that could have been saved through preservation activities slip into deficiency. Today, the public expects transportation agencies to adopt a performance-based management approach that will achieve the highest level of performance possible and make the most effective use of available funds. The Performance-based Management of Bridges course uses video-based testimonies from transportation professionals to illustrate the ways in which their agencies have used performance-based management to estimate the cost-effectiveness of decisions and assess risk.

OUTCOMES

Upon completion of the course, participants will be able to:

- Describe how a bridge management system supports a performance-based bridge program.
- Identify framework for a performance-based management business model;
- Describe the development of performance measures;
- Describe methods for determining cost-effectiveness of actions;
- Describe considerations when assessing risk; and
- Describe strategies for communicating and reporting highway bridge performance-based management actions and results to other agency stakeholders and the public

TARGET AUDIENCE

The target audience includes Federal, State, and local bridge program managers; bridge management engineers; bridge management practitioners; transportation planners; and project planning and programming personnel. Additionally, transportation performance management team members, transportation asset management team members, bridge preservation and maintenance engineers, the financial management team, bridge inspectors, and bridge designers may benefit from this training.All participants should have knowledge of basic bridge terminology.

TRAINING LEVEL: Basic

FEE: 2016: $50 Per Person; 2017: N/A

LENGTH: 4 HOURS (CEU: .4 UNITS)

CLASS SIZE: MINIMUM: 0; MAXIMUM: 0

NHI Customer Service: (877) 558-6873 • nhicustomerservice@dot.gov

Course Number

FHWA-NHI-130110

Course Title

Tunnel Safety Inspection

This 5-day, Instructor-led Training (ILT) is highly interactive and builds upon participants' prior knowledge of tunnel and/or bridge inspection. This course covers the entire breadth of knowledge necessary to manage or execute a successful tunnel inspection based on the National Tunnel Inspection Standards (NTIS), Tunnel Operations, Maintenance, Inspection and Evaluation (TOMIE) Manual and Specifications for the National Tunnel Inventory (SNTI). However, it does not replace the need for specialized experts to assist in inspections. There are nine instructional modules. Once participants display achievement of the learning outcomes of one module, the class will progress to the next module. During the course, the instructor will lead participants through a series of case studies giving them an opportunity to practice and apply their knowledge in real-life tunnel inspection situations. The capstone case study will be a virtual tunnel inspection that takes place in a computer-simulated, 3D environment. Using this tool, participants will be able to perform a tunnel inspection and demonstrate their achievement of all learning outcomes.

*Participant Prerequisite Requirement: ALL participants should successfully complete one of the following three prerequisite requirements:

-130054 Engineering Concepts for Bridge Inspectors; or

-130101 Introduction to Safety Inspection of In-Service Bridges; or

-130101A Prerequisite Assessment for Safety Inspection of In-Service Bridges.

Prior to taking this course, it is strongly recommended that participants complete 130055 Safety Inspection of In-Service Bridges, or possess equivalent field experience.

It is not required, but strongly recommended that participants possess some design or safety inspection experience of in-service tunnels or bridges.

Host Requirements: Hosts must provide a training room large enough to accommodate at least 30 participants as well as the 15 NHI virtual tunnel laptops (provided by NHI Instructors) that will be used for the virtual tunnel exercises. Additionally, the host must ensure that ALL students have successfully met the prerequisite requirement* and have a valid course completion certificate for one of the three prerequisite options.

Outcomes

Upon completion of the course, participants will be able to:

- Articulate the importance and purpose of tunnel inspection
- Apply the fundamentals of tunnel inspection
- Demonstrate the inspection and evaluation of tunnel structural, civil, mechanical, electrical, signage and lighting, and fire/life safety/security elements
- Use tunnel inspection references

Target Audience

The target audience for the Tunnel Safety Inspection ILT course is primarily members of Federal, State, local (Authority or Commission) and Tribal highway agency employees, who are involved with tunnel design, inspection and maintenance, as well as consultants involved in inspecting tunnels or in tunnel inspection management and leadership positions.

Training Level: Basic

Fee: 2016: $1450 Per Person; 2017: $1450 Per Person

Length: 5 DAYS (CEU: 3.2 UNITS)

Class Size: MINIMUM: 20; MAXIMUM: 30

NHI Customer Service: (877) 558-6873 • nhicustomerservice@dot.gov

Course Number

FHWA-NHI-130111

Course Title

Nondestructive Evaluation Fundamentals for Bridge Inspection (Web-based)

130111 Nondestructive Evaluation (NDE) Fundamentals for Bridge Inspection is an introductory course that exposes bridge inspectors to NDE technologies. This course defines and describes the progression of nondestructive evaluation bridge inspection, overview explanations of NDE techniques, and descriptions of the NDE approaches in terms of their applicability to the primary bridge materials of concrete, steel, and timber. The goal of 130111 Nondestructive Evaluation Fundamentals for Bridge Inspection is to provide learners with the necessary background to identify the primary NDE technologies to supplement bridge inspection, and the materials for which they are best suited. A secondary goal of this course is to provide a foundation for more in-depth study of the NDE topics covered in the WBT Course Series, Practical Applications of Nondestructive Evaluation for Bridge Inspection, which includes 130112A NDE for Concrete Bridge Elements, 130112B NDE for Steel Bridge Elements, and 130112C NDE for Timber and Other Bridge Elements.

Outcomes

Upon completion of the course, participants will be able to:

- Describe the application of NDE technology to corrosion and related flaws.
- Describe the application of NDE technology to construction flaws including honeycombing, voids, and inadequate rebar cover
- Explain NDE investigation techniques of concrete bridge elements

Target Audience

The target audience for course 130112A includes public and private sector bridge inspectors, supervisors, project engineers, and others responsible for field inspection of in-service bridges. This will include personnel who may be engineers or technicians in positions such as bridge inspection program manager, bridge inspection project manager, bridge inspection team leader, bridge inspection team member, and FHWA Structural/Bridge Engineers.

Training Level: Basic

Fee: 2016: $50 Per Person; 2017: N/A

Length: 6 HOURS (CEU: .6 UNITS)

Class Size: MINIMUM: 0; MAXIMUM: 0

NHI Customer Service: (877) 558-6873 • nhicustomerservice@dot.gov

COURSE NUMBER

FHWA-NHI-130112A

COURSE TITLE

NDE for Concrete Bridge Elements (Web-based)

130112A Nondestructive Evaluation (NDE) for Concrete Bridge Elements explains the "why" behind the approaches with theoretical explanations of the techniques, comparative costs of each approach, and their applicability to concrete as a primary bridge material. This course is the first of three courses in the WBT Course Series, Practical Applications of Nondestructive Evaluation for Bridge Inspection, which also includes 130112B NDE for Steel Bridge Elements and 130112C NDE for Timber and Other Bridge Elements. This Course Series (130112A, 130112B, 130112C) is a follow up to introductory course 130111 providing a more in-depth study of NDE topics.

OUTCOMES

Upon completion of the course, participants will be able to:

- Describe the application of NDE technology to corrosion and related flaws
- Describe the application of NDE technology to construction flaws including honeycombing, voids, and inadequate rebar cover
- Explain NDE investigation techniques of concrete bridge elements

TARGET AUDIENCE

The target audience for course 130112A includes public and private sector bridge inspectors, supervisors, project engineers, and others responsible for field inspection of in-service bridges. This will include personnel who may be engineers or technicians in positions such as bridge inspection program manager, bridge inspection project manager, bridge inspection team leader, bridge inspection team member, and FHWA Structural/Bridge Engineers.

TRAINING LEVEL: Basic

FEE: 2016: $50 Per Person; 2017: N/A

LENGTH: 5 HOURS (CEU: .5 UNITS)

CLASS SIZE: MINIMUM: 0; MAXIMUM: 0

NHI Customer Service: (877) 558-6873 • nhicustomerservice@dot.gov

COURSE NUMBER

FHWA-NHI-130112B

COURSE TITLE

NDE for Steel Bridge Elements (Web-based)

130112B Nondestructive Evaluation (NDE) for Steel Bridge Elements explains the "why" behind the approaches with theoretical explanations of the techniques, comparative costs of each approach, and their applicability to steel as a primary bridge material. This course is the second of three courses in the WBT Course Series, Practical Applications of Nondestructive Evaluation for Bridge Inspection, which also includes 130112B NDE for Steel Bridge Elements and 130112C NDE for Timber and Other Bridge Elements. This Course Series (130112A, 130112B, 130112C) is a follow up to introductory course 130111 providing a more in-depth study of NDE topics.

OUTCOMES

Upon completion of the course, participants will be able to:

- Describe the application of NDE technology to evaluate the remaining section of steel
- Describe the application of NDE technology to detect cracks in steel
- Explain NDE investigation techniques of steel bridge elements

TARGET AUDIENCE

The target audience for course 130112A includes public and private sector bridge inspectors, supervisors, project engineers, and others responsible for field inspection of in-service bridges. This will include personnel who may be engineers or technicians in positions such as bridge inspection program manager, bridge inspection project manager, bridge inspection team leader, bridge inspection team member, and FHWA Structural/Bridge Engineers.

TRAINING LEVEL: Basic

FEE: 2016: $50 Per Person; 2017: N/A

LENGTH: 5 HOURS (CEU: .5 UNITS)

CLASS SIZE: MINIMUM: 0; MAXIMUM: 0

NHI Customer Service: (877) 558-6873 • nhicustomerservice@dot.gov

Course Number

FHWA-NHI-130112C

Course Title

NDE for Timber and Other Material Bridge Elements (Web-based)

130112C Nondestructive Evaluation (NDE) for Timber and other Material Bridge Elements explains the "why" behind the approaches with theoretical explanations of the techniques, comparative costs of each approach, and their applicability to timber and other bridge materials. This course is the third of three WBTs in the WBT Course Series, Practical Applications of Nondestructive Evaluation for Bridge Inspection, which also includes 130112A NDE for Concrete Bridge Elements and 130112B NDE for Steel Bridge Elements. This Course Series (130112A, 130112B, 130112C) is a follow up to introductory course 130111 providing a more in-depth study of NDE topics.

Outcomes

Upon completion of the course, participants will be able to:

- Describe the application of NDE technology to decay and other voids of timber bridge elements
- Describe the application of NDE technology to delamination and cracks of FRP bridge elements

Target Audience

The target audience for course 130112A includes public and private sector bridge inspectors, supervisors, project engineers, and others responsible for field inspection of in-service bridges. This will include personnel who may be engineers or technicians in positions such as bridge inspection program manager, bridge inspection project manager, bridge inspection team leader, bridge inspection team member, and FHWA Structural/Bridge Engineers.

Training Level: Basic

Fee: 2016: $50 Per Person; 2017: N/A

Length: 4 HOURS (CEU: .4 UNITS)

Class Size: MINIMUM: 0; MAXIMUM: 0

NHI Customer Service: (877) 558-6873 • nhicustomerservice@dot.gov

STRUCTURES

Course Number

FHWA-NHI-132012

Course Title

Soils and Foundations Workshop

This course is geared toward practicing design and construction engineers who routinely deal with soil and foundation problems but have little theoretical background in soil mechanics or foundation engineering. The course takes a project-oriented approach whereby the soils input to a bridge project is followed from conception to completion. In each phase of the project, the soil concepts will be developed into specific foundation designs and recommendations. The classroom presentation includes a variety of exercises to verify achievement of learning objectives. Each participant will take away a comprehensive reference manual on soils and foundations and a participant workbook containing a copy of all slides presented and completed exercises.

NOTE TO PARTICIPANT: All participants should bring calculators that perform trigonometric calculations, a note pad, and a pencil.

NOTE TO HOST: In addition to the typical host requirements of NHI courses, for this course the host is asked to arrange for the state's geotechnical engineering group to conduct a short presentation (usually on the second day of the course) summarizing the administrative and technical procedures followed by the host state.

Outcomes

Upon completion of the course, participants will be able to:

- Identifying the minimum level of geotechnical input in various project phases of a highway project
- Recalling the equipment and procedures used to implement a subsurface investigation of soil and rock conditions
- Demonstrating basic skills in visual description of soils native to the host state
- Recalling geotechnical facilities and personnel in the host state
- Recalling the basic soil test procedures and how the results of the various soil tests are applied results to highway projects
- Listing procedures used for both settlement and stability analysis, and recalling design solutions to stability and settlement problems for approach roadway embankments
- Listing procedures used for determining bearing capacity and settlement of shallow foundations such as spread footings
- Identifying the basic skills needed in the design and construction management of driven pile and drilled shaft foundations
- Recalling the driven pile and drilled shaft foundation construction equipment and construction inspection procedures
- Description static load testing and recalling the basic skills needed to interpret static load test results
- Recalling the basic skills needed in the design and construction of earth retaining structures
- Discussing the format and minimum content of an adequate foundation report

Target Audience

Personnel from the following units at the transportation agency could benefit from this workshop: geotechnical, bridge design, roadway design, materials, construction, and maintenance. The personnel who will benefit the most are the first-line supervisors involved in the design of highway structures and embankments. The greatest impact will be achieved by convincing structural, design, and construction engineers to use procedures from this course as a guide for routine geotechnical work. All attendees should be encouraged to attend the entire course, not just sections that are in their specialty. One of the major benefits of this course is to give engineers an appreciation of activities outside their specialties that influence, or are influenced by, the work of the geotechnical engineer.

TRAINING LEVEL: Basic

FEE: 2016: $1075 Per Person; 2017: N/A

LENGTH: 4 DAYS (CEU: 2.4 UNITS)

CLASS SIZE: MINIMUM: 20; MAXIMUM: 30

NHI Customer Service: (877) 558-6873 • nhicustomerservice@dot.gov

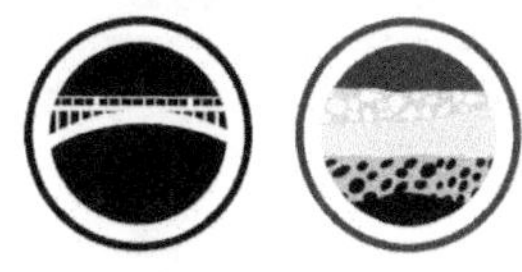

Course Number

FHWA-NHI-132014

Course Title

Drilled Shafts

Drilled shafts are an alternate type of deep foundation that may be more cost effective and perform better than other types of deep foundations in bridge piers at river crossings and in retrofit operations, high-mast lighting, earth retaining structures, single-column piers, and similar applications. This course provides participants with specific technical guidance on all aspects of designing, installing, and monitoring the construction of drilled shafts. The lessons address the following topics: applications, advantages, and disadvantages of drilled shafts for transportation structure foundations; general requirements for subsurface investigations; construction methods; construction case histories; construction specifications; principles of designing drilled shafts for axial and lateral loading; expansive soils, downdrag, and similar effects; load testing; inspection; integrity testing; repair and retrofit of defective shafts; and cost estimation. The participants will receive a comprehensive reference manual on drilled shaft construction and design used by engineers who perform detailed designs of drilled shafts, write construction specifications, and evaluate the performance of contractors through a comprehensive inspection program.

Outcomes

Upon completion of the course, participants will be able to:

- Describe the various drilling rigs and tools that are available to construct drilled shafts under varied subsurface soil and rock conditions
- Recognize the basic features of drilling aids, such as casings and drilling slurries, and the reasons for certain fundamental requirements for these aids
- Design drilled shafts for axial loading in simple soil and rock profiles
- Demonstrate a general understanding of the elements of designing drilled shafts for lateral loads
- Demonstrate an understanding of the need for load tests and available methods for performing the tests
- Formulate the basic elements of construction specifications for drilled shafts
- Demonstrate an understanding of integrity testing, repair, and retrofit of defective shafts
- Estimate costs for drilled shafts

Target Audience

The target audience for this course includes geotechnical engineers, bridge designers, and resident engineers. The course embraces both construction and design, and it is important that all participants attend all lessons, not just those in their immediate areas of interest. A key issue is how the details of construction affect the way in which a drilled shaft should be designed and how the intent of the design affects inspection. Participants are expected to have a degree in engineering for which they have passed an undergraduate course in soil mechanics and/or have successfully completed NHI course FHWA-NHI-132012 Soils and Foundations Workshop.

Training Level: Intermediate

Fee: 2016: $925 Per Person; 2017: N/A

Length: 3 DAYS (CEU: 1.8 UNITS)

Class Size: MINIMUM: 20; MAXIMUM: 30

NHI Customer Service: (877) 558-6873 • nhicustomerservice@dot.gov

 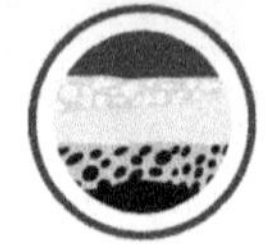 

Course Number

FHWA-NHI-132036

Course Title

Earth Retaining Structures

The goal of this course is to provide agencies with state-of-the-practice design tools and construction techniques to expand implementation of safe and cost-effective earth retention technologies. This course addresses the selection, design, construction, and performance of earth retaining structures used for support of fills and excavations or cut slopes. Instructors cover factors that affect wall selection, including contracting approaches with an emphasis on required bidding documents for each approach. Class discussions will include design procedures and case histories, demonstrating the selection, design, and performance of various earth retaining structures. Detailed information on subsurface investigation, soil and rock property design parameter selection, lateral earth pressures for wall system design, and load and resistance factor design (LRFD) for retaining walls are provided.

Outcomes

Upon completion of the course, participants will be able to:

- Describe potential applications for Earth Retaining Structures (ERS)
- Select a technically appropriate and cost-effective ERS
- Select appropriate material properties, soil design parameters, and earth pressure diagrams
- Perform design analysis and prepare conceptual designs
- Review contractor submitted documents
- Discuss contracting methods
- Describe construction and inspection activities for ERS

Target Audience

The primary audience for this course is agency and consultant bridge/structures, geotechnical, and roadway design engineers; engineering geologists; and consultant review specialists. In addition, management, specification, and contracting specialists and construction engineers involved in design and contracting aspects of retaining structures are encouraged to attend. Attendees should have a basic knowledge of soil mechanics and structural engineering, including some understanding of LRFD concepts.

Training Level: Intermediate

Fee: 2016: $925 Per Person; 2017: N/A

Length: 3 DAYS (CEU: 1.8 UNITS)

Class Size: MINIMUM: 20; MAXIMUM: 30

NHI Customer Service: (877) 558-6873 • nhicustomerservice@dot.gov

 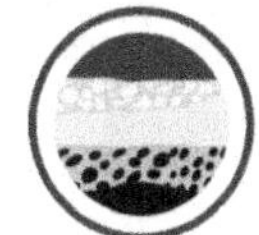 

COURSE NUMBER

FHWA-NHI-132040

COURSE TITLE

Geotechnical Aspects of Pavements

This course covers the latest methods and procedures to address the geotechnical issues in pavement design, construction, and performance for new construction, reconstruction, and rehabilitation pavement projects. The course content includes geotechnical exploration and characterization of in-place and constructed subgrades; design and construction of subgrades and unbound layers for paved and unpaved roads, with emphasis on the American Association of State Highway Transportation Officials (AASHTO) 1993 empirical design procedure and on the new Mechanistic-Empirical Pavement Design Guide (MEPDG); drainage of bases, subbases, and subgrades and its impact on providing safe, cost-effective, and durable pavements; problematic soils, soil improvement, stabilization, and other detailed geotechnical issues in pavement design and construction; and construction methods, specifications, and QC/QA (quality control/quality assurance) inspection for pavement projects.

The goal of the course is for each participant to recognize the importance of the geotechnical aspects relevant to the design, construction, and performance of a pavement system. Participants will develop an appreciation for the importance of adequate subsurface exploration and laboratory characterization of subgrade soils as well as the requisite pavement design parameters for subgrades, unbound base and subbase layers, including drainage features. The course is designed to elicit maximum input from participants, particularly regarding an understanding of the impact of geotechnical features on the long-term performance of pavement systems.

NOTE TO PARTICIPANT: Please bring a calculator that can perform trigonometric, log, and other engineering calculations, a note pad, and a pencil.

NOTE TO HOST: For this course, the host is asked to identify a state speaker to conduct a host state presentation. The presentation is usually on the first day of the class and lasts approximately 25 minutes with an additional 15 minutes of discussion. The objective of the presentation is to communicate the state's current practices and experience to the course participants. The state representative should have experience in geotechnical pavement activities. A detailed list of issues to be addressed in the host presentation will be provided. Also for this course, the host is asked to secure at least 6 laptop computers to be used during team exercises. The host can request that at least 6 participants bring their laptops to the course. The machines must have Microsoft Excel (Office 97 or later) and the optional Solver add-in tool installed. Lastly, the host state is asked to complete a "Questionnaire on Geotechnical Practices in Pavement Design" and provide policies and special provisions for (1.) obtaining subsurface information and laboratory testing in relation to pavement design, (2.) pavement design along with any agency design guides, (3.) field construction monitoring for subgrade approval and pavement component approval as well as contractors QC requirements for pavement component construction.

OUTCOMES

Upon completion of the course, participants will be able to:

- Explain the geotechnical parameters of interest in pavement design and their effects on the performance of different types of pavements
- Explain the influence of climate, moisture, and drainage on pavement performance
- Identify and explain the impact of unsuitable subgrades on pavement performance
- Determine the geotechnical inputs needed for design of pavements, both for the AASHTO 93 empirical design procedure and the new MEPDG
- Evaluate and select appropriate remediation measures for pavement subgrades
- Explain the geotechnical aspects of construction specifications and inspection requirements
- Identify subgrade problems during construction and develop recommended solutions

TARGET AUDIENCE

Many groups within an agency are involved with different aspects of definition, design, use, and construction verification of pavement geomaterials. These groups include pavement design engineers, geotechnical engineers, materials engineers, specification writers, and construction engineers who are or will be involved in the design, evaluation, and construction (or reconstruction or rehabilitation) of pavements. This course was developed as a forum for these various personnel to work together to enhance current procedures for building and maintaining more cost-efficient pavement

structures.

TRAINING LEVEL: Basic

FEE: 2016: $925 Per Person; 2017: N/A

LENGTH: 3 DAYS (CEU: 1.8 UNITS)

CLASS SIZE: MINIMUM: 20; MAXIMUM: 30

NHI Customer Service: (877) 558-6873 • nhicustomerservice@dot.gov

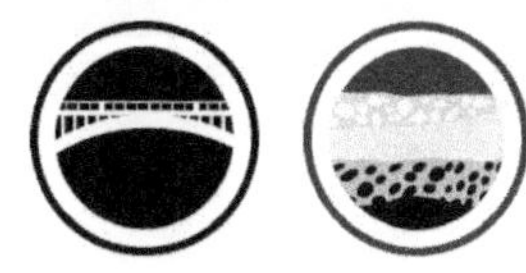

Course Number

FHWA-NHI-132042

Course Title

Design of Mechanically Stabilized Earth Walls and Reinforced Soil Slopes

Mechanically stabilized earth walls (MSEWs) are commonly used on roadway projects and are typically cost effective and aesthetically pleasing. The basic concept behind MSEWs is to combine soil, reinforcing materials made of steel or polymers, and appropriate facing to produce a composite system with engineering properties that are ideal for most roadway applications. Reinforced soil slopes (RSS) utilize the same types of reinforcement for the construction of steep embankments. Both MSEWs and RSS structures can provide substantial savings in construction time and costs when compared with other types of earth retaining systems.

The goal of the course is to educate agencies about state-of-the-practice design tools. This includes comprehensive instruction on the design of MSEWs using load resistance factor design (LRFD). The course also presents construction practices to promote implementation of mechanically stabilized earth technology in cost effective earth retention structures. This course would most benefit persons who are involved in the design and construction of earth retention structures for surface transportation projects.

NOTE TO PARTICIPANT: Please bring a calculator that performs trigonometric calculations, a note pad, and a pencil.

NOTE TO HOST: In addition to the typical host requirements of NHI courses, for this course the host state technical contact is asked to bring 30 copies of the standard MSE wall and the RSS specifications (or special provisions), a complete set of applicable state DOT state construction specifications, standard plates, standard details, inspection guidelines, etc. pertaining to earth retaining structures. Copies should be forwarded to the instructors a month before the course. The host agency is also asked to provide approximately 20-25 pounds of dry sand. About 1/2 bag of "play" sand from a hardware store will suffice.

Outcomes

Upon completion of the course, participants will be able to:

- Recognize potential applications for MSEWs and RSS structures in transportation facilities
- Prepare conceptual and basic (i.e., for simple geometry) designs, and be able to check contractor-submitted designs for walls and slopes
- Examine and select appropriate material properties and parameters used in design
- Calculate the cost of conceptual MSEWs and RSS structures and determine if construction is a cost-effective option
- Select appropriate specification/contracting method(s) and prepare detailed specifications for materials and methods of construction
- Define and communicate major components of construction inspection of MSEWs and RSS structures to confirm compliance with design

Target Audience

The primary audience for this course is agency and consultant bridge/structures, geotechnical, and roadway design engineers; engineering geologists; and consultant review specialists. In addition, management, specification and contracting specialists, and construction engineers interested in design and contracting aspects of MSEWs and RSS structures are encouraged to attend. Attendees should have a basic knowledge of soil mechanics and structural engineering. (Note that NHI offers a 1-day course, FHWA-NHI-132043 Construction of MSEW and RSS.)

Training Level: Intermediate

Fee: 2016: $925 Per Person; 2017: N/A

Length: 3 DAYS (CEU: 1.8 UNITS)

Class Size: MINIMUM: 20; MAXIMUM: 30

NHI Customer Service: (877) 558-6873 • nhicustomerservice@dot.gov

Course Number

FHWA-NHI-132078

Course Title

Micropile Design and Construction

The primary goal of this course is to provide the target audience with guidance on when and where it is appropriate to use micropiles, and educate engineers about the state of the practice in the design and construction of micropiles. The course covers stepwise procedures for the design of micropiles for structural support and for slope stability applications. Construction, inspection and integrity-testing aspects and issues are discussed as well. Classroom presentations include exercises that will lead participants through the technical and cost feasibility aspects of structural support and slope stability design with micropiles. Each participant will receive a workbook and reference manual containing detailed micropile design examples for various applications.

FHWA-NHI-132012 Soils and Foundations course is a recommended prerequisite.

Outcomes

Upon completion of the course, participants will be able to:

- Briefly describe the history and current status of the micropile industry
- Identify potential micropile applications
- Explain construction constraints, techniques, and performance
- Assess feasibility of micropiles for a given application
- Prepare conceptual and basic designs, and evaluate contractor-submitted designs
- Select appropriate specification/contracting method(s) and prepare contract documents
- Describe construction monitoring and inspection requirements

Target Audience

This course is directed toward practicing geotechnical, foundation, construction and bridge/structural engineers who have knowledge and experience in the design and construction of driven piles and drilled shaft foundations. Engineers involved with the design and construction of structure foundations will all benefit from this training, which builds upon the basic concepts presented in NHI courses FHWA-NHI-132012, FHWA-NHI-132014, and FHWA-NHI-132021.

Training Level: Intermediate

Fee: 2016: $775 Per Person; 2017: N/A

Length: 2 DAYS (CEU: 1.2 UNITS)

Class Size: MINIMUM: 20; MAXIMUM: 30

NHI Customer Service: (877) 558-6873 • nhicustomerservice@dot.gov

Course Number

FHWA-NHI-134062

Course Title

Bridge Evaluation for Rehabilitation Design Considerations 4.5 Day

The ultimate goal of this effort is the development of a nationally accepted program that will serve to improve quality, ensure uniformity, and establish a minimum standard for bridge rehabilitation. The course will present innovative and state-of-the-art bridge rehabilitation technologies and procedures for a broad array of structural elements including bridge decks, girders, piers, and abutments.

Core curriculum for the course is 4.5 days and covers the outcomes listed below.

Outcomes

Upon completion of the course, participants will be able to:

- Describe conditions that suggest the need for rehabilitation
- Identify the need for, and capacity of, destructive and/or non destructive testing (NDT) for assessment of existing conditions
- Prescribe analysis and load testing to determine the effect of existing conditions on the structure
- Distinguish root causes of distress and deterioration
- Formulate appropriate rehabilitation strategies
- Select procedures and materials for rehabilitation
- Develop effective rehabilitation construction documents
- Prepare and implement quality assurance for construction
- Monitor and resolve construction and material problems

Target Audience

The target audience includes design engineers, field engineers, resident engineers, structural engineers, materials engineers, and other technical personnel involved in the construction and rehabilitation design of bridges. Participants with an engineering background are expected to constitute the target audience. People knowledgeable in new bridge design, but not necessarily bridge rehabilitation, should attend.

Training Level: Intermediate

Fee: 2016: $1100 Per Person; 2017: $1100 Per Person

Length: 4.5 DAYS (CEU: 2.7 UNITS)

Class Size: MINIMUM: 20; MAXIMUM: 30

NHI Customer Service: (877) 558-6873 • nhicustomerservice@dot.gov

STRUCTURES

Course Number

FHWA-NHI-134062A

Course Title

Bridge Evaluation for Rehabilitation Design Considerations 5-Day

The ultimate goal of this effort is the development of a nationally accepted program that will serve to improve quality, ensure uniformity, and establish a minimum standard for bridge rehabilitation. The course will present innovative and state-of-the-art bridge rehabilitation technologies and procedures for a broad array of structural elements including bridge decks, girders, piers, and abutments.

The 5-day version of this course includes two additional modules on the rehabilitation of timber and masonry structures.

Outcomes

Upon completion of the course, participants will be able to:

- Describe conditions that suggest the need for rehabilitation
- Identify the need for, and capacity of, destructive and/or non destructive testing (NDT) for assessment of existing conditions
- Prescribe analysis and load testing to determine the effect of existing conditions on the structure
- Distinguish root causes of distress and deterioration
- Formulate appropriate rehabilitation strategies
- Select procedures and materials for rehabilitation
- Develop effective rehabilitation construction documents
- Prepare and implement quality assurance for construction
- Monitor and resolve construction and material problems

Target Audience

The target audience includes design engineers, field engineers, resident engineers, structural engineers, materials engineers, and other technical personnel involved in the construction and rehabilitation design of bridges. Participants with an engineering background are expected to constitute the target audience. People knowledgeable in new bridge design, but not necessarily bridge rehabilitation should attend.

Training Level: Intermediate

Fee: 2016: $1175 Per Person; 2017: $1175 Per Person

Length: 5 DAYS (CEU: 3 UNITS)

Class Size: MINIMUM: 20; MAXIMUM: 30

NHI Customer Service: (877) 558-6873 • nhicustomerservice@dot.gov

Course Number

FHWA-NHI-134067

Course Title

Construction Inspection of Bridge Rehabilitation Projects

This 4-day course has been designed to improve quality, ensure uniformity, and establish a minimum standard for bridge rehabilitation.

The keys to successfully ensuring quality on rehab jobs are: knowing what should happen on a given job; identifying problems when they do happen; and correctly using available resources to solve the problem. This course presents innovative and best practice inspection techniques for each structural element of a bridge.

This course will introduce participants to distress and deterioration they may encounter when working with concrete or steel that requires repair. It is essential to identify the issues that harm these materials because it is often poor construction techniques that lead to reduced structural condition or shortened service life. The focus then turns to construction and inspection practices pertaining to concrete decks, steel superstructures, concrete superstructures and substructures, joints, and bearings.

The course is activity-rich, using discussions of best practices, small and large group activities for identifying critical inspection moments, and a wide array of case studies from real projects to emphasize the importance of applying these techniques in the field.

Outcomes

Upon completion of the course, participants will be able to:

- Relate observable deterioration of bridge structural elements to distress mechanisms
- Associate potential construction and materials problems
- Explain the role of the construction inspector as part of the overall project team
- Interpret drawings and specifications
- Describe rehabilitation sequences for various bridge systems, bridge types, and materials
- Explain basic inspection and testing of materials
- Make and maintain sufficient records

Target Audience

This course will be appropriate for inspectors with 1-5 years of experience who are seeking a better foundation in bridge rehabilitation techniques. They will likely have a basic grasp of construction and inspection methods, bridge terminology, and causes of distress and deterioration, although this information will be reviewed at the beginning of the course.The course will be appropriate for experienced bridge inspectors who are seeking to learn about innovative methods in bridge rehabilitation and obtain a refresher on familiar inspection methods. Construction supervisors, transportation department field inspectors, construction inspectors, field engineers, resident engineers, structural engineers, materials engineers, and other technical personnel involved in the inspection of bridge rehabilitation projects will benefit from this course. The course is designed for participants without an in-depth engineering background. However, those with engineering backgrounds are welcome to attend and can provide valuable perspective in the context of group activities and discussions.

Training Level: Basic

Fee: 2016: $1050 Per Person; 2017: $1050 Per Person

Length: 4 DAYS (CEU: 2.4 UNITS)

Class Size: MINIMUM: 20; MAXIMUM: 30

NHI Customer Service: (877) 558-6873 • nhicustomerservice@dot.gov

Course Number

FHWA-NHI-131050

Course Title

Asphalt Pavement In-Place Recycling Techniques

Transportation agencies focusing on the use of sustainable, cost effective, and environmentally conscious construction practices often consider in-place recycling techniques as a viable alternative to the more traditional rehabilitation techniques used on asphalt-surfaced pavements. NHI training 131050 Asphalt Pavement In-place Recycling Techniques is designed to help participants acquire necessary skills for selecting the appropriate in-place recycling technique for a given set of conditions, choosing the appropriate materials for the project, developing suitable specifications, and constructing those projects effectively.

The Asphalt Pavement In-place Recycling Techniques course includes two brief Web-based training (WBT) modules, and two days of instructor-led, classroom-based training (ILT). Through independent study, classroom interaction, and workshop activities, participants explore the current technologies available in the area of asphalt pavement in-place recycling. Two WBT lessons introduce pavement evaluation techniques and the three potential recycling techniques, along with the types of equipment commonly used for each. The classroom session focuses on project and technique selection and justification, materials considerations and mix design, construction specifications, and project control considerations during construction.

Outcomes

Upon completion of the course, participants will be able to:

- Describe the economic, environmental, and engineered performance benefits associated with using in-place asphalt recycling
- Identify the key factors that contribute to the selection of appropriate in-place asphalt recycling techniques under different traffic levels, pavement conditions, and environments
- Identify the key requirements in developing effective in-place asphalt recycling construction specifications, including method specification and end-result or performance specifications
- Demonstrate the ability to select the appropriate new materials and additives needed for each of three HMA pavement in-place recycling techniques
- List steps that can be taken to address a variety of issues that may impact the constructability of a project

Target Audience

This course is intended for State and local transportation agency engineers, such as pavement managers and maintenance engineers, and other agency personnel who are responsible for selecting, designing, or constructing the agency's asphalt pavement maintenance, resurfacing, rehabilitation, and reconstruction alternatives. The course particularly benefits those individuals responsible for selecting and designing asphalt in-place recycling projects, for writing effective specifications, or for inspecting asphalt in-place recycling projects during their construction. Contractors, consulting engineers, and industry representatives involved in asphalt pavement in-place recycling also will benefit from this course.

Training Level: Intermediate

Fee: 2016: $225 Per Person; 2017: $225 Per Person

Length: 2 DAYS (CEU: 1.3 UNITS)

Class Size: MINIMUM: 20; MAXIMUM: 30

NHI Customer Service: (877) 558-6873 • nhicustomerservice@dot.gov

COURSE NUMBER

FHWA-NHI-131050A

COURSE TITLE

Asphalt Pavement In-Place Recycling Techniques--WEB-BASED

This training is a prerequisite of another NHI training and is offered at no cost.

Transportation agencies focusing on the use of sustainable, cost-effective, and environmentally conscious construction practices often consider in-place recycling techniques as a viable alternative to the more traditional rehabilitation techniques used on asphalt-surfaced pavements. NHI training 131050 Asphalt Pavement In-place Recycling Techniques is designed to help participants acquire necessary skills for selecting the appropriate in-place recycling technique for a given set of conditions, choosing the appropriate materials for the project, developing suitable specifications, and constructing those projects effectively.

The Asphalt Pavement In-place Recycling Techniques course includes two brief Web-based training (WBT) modules, and two days of instructor-led, classroom-based training (ILT). Through independent study, classroom interaction, and workshop activities, participants explore the current technologies available in the area of asphalt pavement in-place recycling. Two WBT lessons introduce pavement evaluation techniques and the three potential recycling techniques, along with the types of equipment commonly used for each. The classroom session focuses on project and technique selection and justification, materials considerations and mix design, construction specifications, and project control considerations during construction.

OUTCOMES

Upon completion of the course, participants will be able to:

- Describe the economic, environmental, and engineered performance benefits associated with using in-place asphalt recycling
- Identify the key factors that contribute to the selection of appropriate in-place asphalt recycling techniques under different traffic levels, pavement conditions, and environments
- Identify the key requirements in developing effective in-place asphalt recycling construction specifications, including method specification and end-result or performance specifications
- Demonstrate the ability to select the appropriate new materials and additives needed for each of three HMA pavement in-place recycling techniques
- List steps that can be taken to address a variety of issues that may impact the constructability of a project

TARGET AUDIENCE

This course is intended for State and local transportation agency engineers, such as pavement managers and maintenance engineers, and other agency personnel who are responsible for selecting, designing, or constructing the agency's asphalt pavement maintenance, resurfacing, rehabilitation, and reconstruction alternatives. The course particularly benefits those individuals responsible for selecting and designing asphalt in-place recycling projects, for writing effective specifications, or for inspecting asphalt in-place recycling projects during their construction. Contractors, consulting engineers, and industry representatives involved in asphalt pavement in-place recycling also will benefit from this course.

TRAINING LEVEL: Basic

FEE: 2016: $0 Per Person; 2017: $0 Per Person

LENGTH: 2 HOURS (CEU: 0 UNITS)

CLASS SIZE: MINIMUM: 20; MAXIMUM: 30

NHI Customer Service: (877) 558-6873 • nhicustomerservice@dot.gov

Course Number

FHWA-NHI-131100

Course Title

Pavement Smoothness: Use of Inertial Profiler Measurements for Construction Quality Control

Studies have shown that roughness is one of the biggest priorities of highway users. Additional studies have shown that pavements that are built smooth stay smoother longer and provide a longer pavement life. Most State highway agencies (SHAs) have some type of smoothness specification that is used to evaluate the smoothness of newly constructed or rehabilitated pavements during acceptance testing. Many agencies also have incentives or disincentives for new construction and rehabilitation, which are based on pavement smoothness.

Increasingly these agencies are turning to inertial profilers as the most reliable instrument for construction acceptance testing and verifying pavement smoothness. The intent of this course is to train inertial profiler operators in the basics of performing construction acceptance testing and to train those reviewing the data to comprehend how those data were obtained and what they represent in order to build smoother riding roadways.

The course has been developed to be delivered in a single day of instructor-led training. In order to keep the instructor-led portion of the training to a single day, the training includes two hours of independent study that should be completed prior to attending the instructor-led session.

Outcomes

Upon completion of the course, participants will be able to:

- Perform checks of the inertial profiler components to identify that the equipment is in proper working order.
- Determine the impact of current surface and environmental conditions on data collection.
- Collect profile data using appropriate operating techniques.
- Calculate a smoothness index using appropriate data processing techniques and computational procedures for use in construction quality control and specification compliance.
- Identify what features in a collected profile are manifested in a smoothness or roughness index.

Target Audience

The course was designed for an audience directly involved in the use of inertial profilers and the application of the data obtained from inertial profilers. This includes State and contractor road profiler operators who perform data collection, initial processing, and reporting of smoothness data. Paving superintendents, project engineers, pavement engineers, and inspectors who are performing data analysis, quality control, and acceptance will also benefit from this course. Ideally, each session of the course will include a mixture of State and contractor personnel, including those who collect data, those performing data processing, and those making decisions based upon data. ASSUMED TRAINING COMPETENCIESThe participants should have a basic understanding of how to operate a computer including turning it on and off, running programs, and saving data.

Training Level: Intermediate

Fee: 2016: $150 Per Person; 2017: $150 Per Person

Length: 1 DAYS (CEU: .6 UNITS)

Class Size: MINIMUM: 20; MAXIMUM: 30

NHI Customer Service: (877) 558-6873 • nhicustomerservice@dot.gov

Course Number

FHWA-NHI-131110

Course Title

Pavement Preservation Treatment Construction - WEB-BASED

FHWA, in partnership with Caltrans, the National Center for Pavement Preservation, and the Transportation Curriculum Coordination Council (TCCC) created the Pavement Preservation Treatment Construction Guide (PPTCG) as a resource for agency and industry pavement preservation practitioners. This course is designed to provide participants with an introduction to the PPTCG, so that they can better use it to familiarize themselves with general information on pavement preservation concepts and techniques. The guide covers basic pavement preservation concepts, as well as information on specific treatments to extend the life of asphalt pavements. The module topics are:

1. Introduction to Pavement Preservation (NHI-131110A)
2. Materials (NHI-131110B)
3. Crack Sealing, Crack Filling and Joint Sealing of Flexible and Rigid Pavements (NHI-131110C)
4. Patching and Edge Repairs (NHI-131110D)
5. Chip Seals (NHI-131110E)
6. Fog Seals (NHI-131110F)
7. Slurry Seals (NHI-131110G)
8. Micro-surfacing Projects (NHI-131110H)
9. Thin Functional and Maintenance Overlay Projects (NHI-131110I)
10. Ultra Thin, Hot-Mixed, Bonded Overlay Projects (NHI-131110J)
11. Selecting a Pavement Presentation Treatment (NHI-131110K)

Each of the modules is also offered as individual trainings and can be accessed by registering for the course number listed with each module.

Outcomes

Upon completion of the course, participants will be able to:

- Identify the components and value of a Pavement Preventive Maintenance (PPM) program
- Identify pavement conditions and other attributes that suggest whether preventive maintenance is appropriate
- Identify various pavement preservation strategies, techniques and materials
- State the performance characteristics of various pavement preservation strategies, techniques and materials
- Select the appropriate strategy(ies), technique(s) and material to extend the service life and retard the development of pavement distress

Target Audience

The primary audience for the Pavement Preservation Treatment Construction WBT course is Federal, State, and local highway construction and maintenance teams, specifically the highway workers and inspectors involved in the placement of pavement preservation treatments. Although not in the primary audience, design engineers will also benefit from the online guide and the associated training. The training course is primarily targeted at individuals unfamiliar with pavement preservation policy and technical information.

TRAINING LEVEL: Intermediate

FEE: 2016: $50 Per Person; 2017: N/A

LENGTH: 6.5 HOURS (CEU: 0 UNITS)

CLASS SIZE: MINIMUM: 1; MAXIMUM: 1

NHI Customer Service: (877) 558-6873 • nhicustomerservice@dot.gov

Course Number

FHWA-NHI-131110A

Course Title

Pavement Preservation Treatment Series: Introduction to Pavement Preservation - WEB-BASED

This training is part of the "Pavement Preservation Treatment" series and is designed to provide participants with an introduction to the Pavement Preservation Treatment Construction Guide (PPTCG) and the basics of pavement preservation. Topics include: pavement structure, distresses, and differentiating pavement preservation from preventive maintenance.

As stated, this training draws on the PPTCG, which was created by FHWA, in partnership with Caltrans, the National Center for Pavement Preservation, and the Transportation Curriculum Coordination Council (TCCC) as a resource for agency and industry pavement preservation practitioners. It provides information on basic pavement preservation concepts and the different treatments available and how they should be applied, so agencies can make informed decisions when determining which treatments best fit their pavement preservation needs. The training is primarily targeted at individuals unfamiliar with pavement preservation policy and technical information.

To take the entire series of trainings for the PPTCG, access the NHI website and register for NHI-131110.

Outcomes

Upon completion of the course, participants will be able to:

- Identify common surface distresses in pavements.
- Distinguish between distresses caused by surface failure and those caused by subsurface layer failure.
- Recognize the difference between pavement preservation and pavement maintenance.

Target Audience

The primary audience for the Pavement Preservation Treatment Construction WBT course is Federal, State, and local highway construction and maintenance teams, specifically the highway workers and inspectors involved in the placement of pavement preservation treatments. Although not in the primary audience, design engineers will also benefit from the online guide and the associated training. The training course is primarily targeted at individuals unfamiliar with pavement preservation policy and technical information.

Training Level: Basic

Fee: 2016: $25 Per Person; 2017: N/A

Length: .5 HOURS (CEU: 0 UNITS)

Class Size: MINIMUM: 1; MAXIMUM: 1

NHI Customer Service: (877) 558-6873 • nhicustomerservice@dot.gov

Course Number

FHWA-NHI-131110B

Course Title

Pavement Preservation Treatment Series: Materials - WEB-BASED

This training is part of the "Pavement Preservation Treatment" series and is designed to provide participants with information on the materials used for preventive maintenance treatments. Topics include: materials comprising maintenance treatments, emulsions, and aggregates. This course is primarily intended for inspectors and technicians.

This training draws on the Pavement Preservation Treatment Construction Guide (PPTCG), which was created by FHWA, in partnership with Caltrans, the National Center for Pavement Preservation, and the Transportation Curriculum Coordination Council (TCCC) as a resource for agency and industry pavement preservation practitioners. It provides information on basic pavement preservation concepts and the different treatments available and how they should be applied, so agencies can make informed decisions when determining which treatments best fit their pavement preservation needs. The training is primarily targeted at individuals unfamiliar with pavement preservation policy and technical information.

To take the entire series of trainings for the PPTCG, access the NHI website and register for NHI-131110.

Outcomes

Upon completion of the course, participants will be able to:

- List the materials used in preventive maintenance treatments for flexible and rigid pavements.
- Recognize the differences between asphalt cement and emulsions and their use in pavement preservation treatments.
- List the six physical properties of aggregates that affect the performance of preservation treatments.

Target Audience

The primary audience for the Pavement Preservation Treatment Construction WBT course is Federal, State, and local highway construction and maintenance teams, specifically the highway workers and inspectors involved in the placement of pavement preservation treatments. Although not in the primary audience, design engineers will also benefit from the online guide and the associated training. The training course is primarily targeted at individuals unfamiliar with pavement preservation policy and technical information.

Training Level: Intermediate

Fee: 2016: $25 Per Person; 2017: $25 Per Person

Length: 1 HOURS (CEU: 0 UNITS)

Class Size: MINIMUM: 1; MAXIMUM: 1

NHI Customer Service: (877) 558-6873 • nhicustomerservice@dot.gov

Course Number

FHWA-NHI-131110C

Course Title

Pavement Preservation Treatment Series: Crack Sealing & Filling, and Joint Sealing - WEB-BASED

This training is part of the "Pavement Preservation Treatment" series and is designed to provide participants with information on crack sealing, crack filling, and joint sealing of flexible and rigid pavements. Topics include: working and non-working cracks, fatigue and longitudinal cracks, correct temperatures for crack sealant, crack repair sequence, hot sealant, and crack sealing or filling criteria. This course is primarily intended for inspectors and technicians.

This training draws on the Pavement Preservation Treatment Construction Guide (PPTCG), which was created by FHWA, in partnership with Caltrans, the National Center for Pavement Preservation, and the Transportation Curriculum Coordination Council (TCCC) as a resource for agency and industry pavement preservation practitioners. It provides information on basic pavement preservation concepts and the different treatments available and how they should be applied, so agencies can make informed decisions when determining which treatments best fit their pavement preservation needs. The training is primarily targeted at individuals unfamiliar with pavement preservation policy and technical information.

To take the entire series of trainings for the PPTCG, access the NHI website and register for NHI-131110.

Outcomes

Upon completion of the course, participants will be able to:

- Describe the difference between a working crack and a nonworking crack.
- List the types of distresses that crack sealing, crack filling, and joint sealing treatments will repair.
- Describe how proper storage and handling of sealants and fillers affect their constructability and performance.
- Describe the procedure of repairing surface cracks and rigid joints.
- Identify common problems associated with crack sealing, crack filling, and joint sealing treatments and recognize their solutions.
- List the capabilities and limitations of crack sealing, crack filling, and joint sealing treatments.

Target Audience

The primary audience for the Pavement Preservation Treatment Construction WBT course is Federal, State, and local highway construction and maintenance teams, specifically the highway workers and inspectors involved in the placement of pavement preservation treatments. Although not in the primary audience, design engineers will also benefit from the online guide and the associated training. The training course is primarily targeted at individuals unfamiliar with pavement preservation policy and technical information.

Training Level: Intermediate

Fee: 2016: $25 Per Person; 2017: $25 Per Person

Length: 1 HOURS (CEU: 0 UNITS)

Class Size: MINIMUM: 1; MAXIMUM: 1

NHI Customer Service: (877) 558-6873 • nhicustomerservice@dot.gov

Course Number

FHWA-NHI-131110D

Course Title

Pavement Preservation Treatment Series: Localized Pavement Repair - WEB-BASED

This training is part of the "Pavement Preservation Treatment" series and is designed to provide participants with information on localized pavement repair. Topics include: pothole formation and edge failure, seal or fill decisions, construction of, and problems with, pothole patching, dig outs, edge repairs, and skin patching, and capabilities and limitations of localized repairs. This course is primarily intended for inspectors and technicians.

This training draws on the Pavement Preservation Treatment Construction Guide (PPTCG), which was created by FHWA, in partnership with Caltrans, the National Center for Pavement Preservation, and the Transportation Curriculum Coordination Council (TCCC) as a resource for agency and industry pavement preservation practitioners. It provides information on basic pavement preservation concepts and the different treatments available and how they should be applied, so agencies can make informed decisions when determining which treatments best fit their pavement preservation needs. The training is primarily targeted at individuals unfamiliar with pavement preservation policy and technical information.

To take the entire series of trainings for the PPTCG, access the NHI website and register for NHI-131110.

Outcomes

Upon completion of the course, participants will be able to:

- Describe the mechanisms of pothole formation and edge failure.
- Select the type of localized pavement repair best suited to a given condition.
- Describe the process of pothole patching, dig outs, edge repairs, and skin patching.
- Identify common problems associated with pothole patching, dig outs, edge repairs, and skin patching and recognize their solutions.
- List the key capabilities and limitations of localized pavement repairs.

Target Audience

The primary audience for the Pavement Preservation Treatment Construction WBT course is Federal, State, and local highway construction and maintenance teams, specifically the highway workers and inspectors involved in the placement of pavement preservation treatments. Although not in the primary audience, design engineers will also benefit from the online guide and the associated training. The training course is primarily targeted at individuals unfamiliar with pavement preservation policy and technical information.

Training Level: Intermediate

Fee: 2016: $25 Per Person; 2017: $25 Per Person

Length: 1 HOURS (CEU: 0 UNITS)

Class Size: MINIMUM: 1; MAXIMUM: 1

NHI Customer Service: (877) 558-6873 • nhicustomerservice@dot.gov

COURSE NUMBER

FHWA-NHI-131110E

COURSE TITLE

Pavement Preservation Treatment Series: Chip Seals - WEB-BASED

This training is part of the "Pavement Preservation Treatment" series and is designed to provide participants with information on chip seals. Topics include: project selection, pavement and weather condition requirements, storage, traffic control, construction sequence, aggregate spreading distance, brooming, chip spreading process, distributor preparation, and troubleshooting.

This training draws on the Pavement Preservation Treatment Construction Guide (PPTCG), which was created by FHWA, in partnership with Caltrans, the National Center for Pavement Preservation, and the Transportation Curriculum Coordination Council (TCCC) as a resource for agency and industry pavement preservation practitioners. It provides information on basic pavement preservation concepts and the different treatments available and how they should be applied, so agencies can make informed decisions when determining which treatments best fit their pavement preservation needs. The training is primarily targeted at individuals unfamiliar with pavement preservation policy and technical information.

To take the entire series of trainings for the PPTCG, access the NHI website and register for NHI-131110.

OUTCOMES

Upon completion of the course, participants will be able to:

- Recognize pavement conditions best suited to the chip seal treatment.
- Identify how proper storage and handling of chip seal materials affect their constructability and performance.
- Describe the construction of chip seals.
- Identify common problems associated with chip seals and recognize their solutions.
- Recognize key capabilities and limitations of chip seals.

TARGET AUDIENCE

The primary audience for the Pavement Preservation Treatment Construction WBT course is Federal, State, and local highway construction and maintenance teams, specifically the highway workers and inspectors involved in the placement of pavement preservation treatments. Although not in the primary audience, design engineers will also benefit from the online guide and the associated training. The training course is primarily targeted at individuals unfamiliar with pavement preservation policy and technical information.

TRAINING LEVEL: Intermediate

FEE: 2016: $25 Per Person; 2017: $25 Per Person

LENGTH: 1 HOURS (CEU: 0 UNITS)

CLASS SIZE: MINIMUM: 1; MAXIMUM: 1

NHI Customer Service: (877) 558-6873 • nhicustomerservice@dot.gov

COURSE NUMBER

FHWA-NHI-131110F

COURSE TITLE

Pavement Preservation Treatment Series: Fog Seals - WEB-BASED

This training is part of the "Pavement Preservation Treatment" series and is designed to provide participants with information on fog seals. Topics include: uses of fog seals, suitable pavement surfaces, storage and handling of materials, application process, and problems and causation. This course is primarily intended for inspectors and technicians.

This training draws on the Pavement Preservation Treatment Construction Guide (PPTCG), which was created by FHWA, in partnership with Caltrans, the National Center for Pavement Preservation, and the Transportation Curriculum Coordination Council (TCCC) as a resource for agency and industry pavement preservation practitioners. It provides information on basic pavement preservation concepts and the different treatments available and how they should be applied, so agencies can make informed decisions when determining which treatments best fit their pavement preservation needs. The training is primarily targeted at individuals unfamiliar with pavement preservation policy and technical information.

To take the entire series of trainings for the PPTCG, access the NHI website and register for NHI-131110.

OUTCOMES

Upon completion of the course, participants will be able to:

- Recognize pavement conditions most suitable for a fog seal.
- Describe how proper storage and handling of fog seal materials affect their constructability and performance.
- Describe the construction of a fog seal.
- Identify common problems associated with fog seals and recognize their solutions.
- List the key capabilities and limitations of fog seal treatments.

TARGET AUDIENCE

The primary audience for the Pavement Preservation Treatment Construction WBT course is Federal, State, and local highway construction and maintenance teams, specifically the highway workers and inspectors involved in the placement of pavement preservation treatments. Although not in the primary audience, design engineers will also benefit from the online guide and the associated training. The training course is primarily targeted at individuals unfamiliar with pavement preservation policy and technical information.

TRAINING LEVEL: Intermediate

FEE: 2016: $25 Per Person; 2017: $25 Per Person

LENGTH: 1 HOURS (CEU: 0 UNITS)

CLASS SIZE: MINIMUM: 1; MAXIMUM: 1

NHI Customer Service: (877) 558-6873 • nhicustomerservice@dot.gov

Course Number

FHWA-NHI-131110G

Course Title

Pavement Preservation Treatment Series: Slurry Seals - WEB-BASED

This training is part of the "Pavement Preservation Treatment" series and is designed to provide participants with information on slurry seals. Topics include: reasons to use slurry seals, gradations of slurry seal aggregate, preparation and application process, and problems and solutions. This course is primarily intended for inspectors and technicians.

This training draws on the Pavement Preservation Treatment Construction Guide (PPTCG), which was created by FHWA, in partnership with Caltrans, the National Center for Pavement Preservation, and the Transportation Curriculum Coordination Council (TCCC) as a resource for agency and industry pavement preservation practitioners. It provides information on basic pavement preservation concepts and the different treatments available and how they should be applied, so agencies can make informed decisions when determining which treatments best fit their pavement preservation needs. The training is primarily targeted at individuals unfamiliar with pavement preservation policy and technical information.

To take the entire series of trainings for the PPTCG, access the NHI website and register for NHI-131110.

Outcomes

Upon completion of the course, participants will be able to:

- Identify the type of slurry seal appropriate to various traffic conditions.
- Describe the construction of slurry seals.
- Identify common problems associated with slurry seals and recognize their solutions.
- List the key capabilities and limitations of slurry seals.

Target Audience

The primary audience for the Pavement Preservation Treatment Construction WBT course is Federal, State, and local highway construction and maintenance teams, specifically the highway workers and inspectors involved in the placement of pavement preservation treatments. Although not in the primary audience, design engineers will also benefit from the online guide and the associated training. The training course is primarily targeted at individuals unfamiliar with pavement preservation policy and technical information.

Training Level: Intermediate

Fee: 2016: $25 Per Person; 2017: $25 Per Person

Length: 1 HOURS (CEU: 0 UNITS)

Class Size: MINIMUM: 1; MAXIMUM: 1

NHI Customer Service: (877) 558-6873 • nhicustomerservice@dot.gov

Course Number

FHWA-NHI-131110H

Course Title

Pavement Preservation Treatment Series: Micro-Surfacing - WEB-BASED

This training is part of the "Pavement Preservation Treatment" series and is designed to provide participants with information on micro-surfacing. Topics include: pavement and traffic condition considerations, construction, and troubleshooting.

This training draws on the Pavement Preservation Treatment Construction Guide (PPTCG), which was created by FHWA, in partnership with Caltrans, the National Center for Pavement Preservation, and the Transportation Curriculum Coordination Council (TCCC) as a resource for agency and industry pavement preservation practitioners. It provides information on basic pavement preservation concepts and the different treatments available and how they should be applied, so agencies can make informed decisions when determining which treatments best fit their pavement preservation needs. The training is primarily targeted at individuals unfamiliar with pavement preservation policy and technical information.

To take the entire series of trainings for the PPTCG, access the NHI website and register for NHI-131110.

Outcomes

Upon completion of the course, participants will be able to:

- Identify pavement conditions most suitable for a micro-surfacing treatment.
- Describe the construction of micro-surfacing.
- Identify common problems associated with micro-surfacing and recognize their solutions.
- List the key capabilities and limitations of micro-surfacing relative to various traffic conditions.

Target Audience

The primary audience for the Pavement Preservation Treatment Construction WBT course is Federal, State, and local highway construction and maintenance teams, specifically the highway workers and inspectors involved in the placement of pavement preservation treatments. Although not in the primary audience, design engineers will also benefit from the online guide and the associated training. The training course is primarily targeted at individuals unfamiliar with pavement preservation policy and technical information.

Training Level: Intermediate

Fee: 2016: $25 Per Person; 2017: $25 Per Person

Length: 1 HOURS (CEU: 0 UNITS)

Class Size: MINIMUM: 1; MAXIMUM: 1

NHI Customer Service: (877) 558-6873 • nhicustomerservice@dot.gov

COURSE NUMBER

FHWA-NHI-131110I

COURSE TITLE

Pavement Preservation Treatment Series: Thin Functional HMA Overlay - WEB-BASED

This training is part of the "Pavement Preservation Treatment" series and is designed to provide participants with information on thin functional hot-mix asphalt overlays. Topics include: proper usage, suitable pavement conditions, construction, and troubleshooting. This course is primarily intended for inspectors and technicians.

This training draws on the Pavement Preservation Treatment Construction Guide (PPTCG), which was created by FHWA, in partnership with Caltrans, the National Center for Pavement Preservation, and the Transportation Curriculum Coordination Council (TCCC) as a resource for agency and industry pavement preservation practitioners. It provides information on basic pavement preservation concepts and the different treatments available and how they should be applied, so agencies can make informed decisions when determining which treatments best fit their pavement preservation needs. The training is primarily targeted at individuals unfamiliar with pavement preservation policy and technical information.

To take the entire series of trainings for the PPTCG, access the NHI website and register for NHI-131110.

OUTCOMES

Upon completion of the course, participants will be able to:

- Identify pavement conditions best suited for a thin hot mix asphalt overlay.
- Describe the construction process for a thin hot mix asphalt overlay.
- Identify common problems associated with a thin hot mix asphalt overlay and recognize their solutions.
- List the key capabilities and benefits of a thin hot mix asphalt overlay relative to various traffic conditions.

TARGET AUDIENCE

The primary audience for the Pavement Preservation Treatment Construction WBT course is Federal, State, and local highway construction and maintenance teams, specifically the highway workers and inspectors involved in the placement of pavement preservation treatments. Although not in the primary audience, design engineers will also benefit from the online guide and the associated training. The training course is primarily targeted at individuals unfamiliar with pavement preservation policy and technical information.

TRAINING LEVEL: Intermediate

FEE: 2016: $25 Per Person; 2017: $25 Per Person

LENGTH: 1 HOURS (CEU: 0 UNITS)

CLASS SIZE: MINIMUM: 1; MAXIMUM: 1

NHI Customer Service: (877) 558-6873 • nhicustomerservice@dot.gov

COURSE NUMBER

FHWA-NHI-131110J

COURSE TITLE

Pavement Preservation Treatment Series: Ultra Thin HMA Bonded Wearing Course - WEB-BASED

This training is part of the "Pavement Preservation Treatment" series and is designed to provide participants with information on ultra thin, hot-mixed asphalt bonded wearing course. Topics include: usage, distresses and application considerations, construction, and troubleshooting. This course is primarily intended for inspectors and technicians.

This training draws on the Pavement Preservation Treatment Construction Guide (PPTCG), which was created by FHWA, in partnership with Caltrans, the National Center for Pavement Preservation, and the Transportation Curriculum Coordination Council (TCCC) as a resource for agency and industry pavement preservation practitioners. It provides information on basic pavement preservation concepts and the different treatments available and how they should be applied, so agencies can make informed decisions when determining which treatments best fit their pavement preservation needs. The training is primarily targeted at individuals unfamiliar with pavement preservation policy and technical information.

To take the entire series of trainings for the PPTCG, access the NHI website and register for NHI-131110.

OUTCOMES

Upon completion of the course, participants will be able to:

- Identify pavement conditions best suited to ultra thin, hot-mixed asphalt bonded wearing course.
- Describe the construction of ultra thin, hot-mixed, asphalt bonded wearing course.
- Identify common problems associated with ultra thin, hot-mixed, asphalt bonded wearing course and recognize their solutions.
- List key capabilities and benefits of ultra thin, hot-mixed, asphalt bonded wearing course relative to various traffic conditions.

TARGET AUDIENCE

The primary audience for the Pavement Preservation Treatment Construction WBT course is Federal, State, and local highway construction and maintenance teams, specifically the highway workers and inspectors involved in the placement of pavement preservation treatments. Although not in the primary audience, design engineers will also benefit from the online guide and the associated training. The training course is primarily targeted at individuals unfamiliar with pavement preservation policy and technical information.

TRAINING LEVEL: Intermediate

FEE: 2016: $25 Per Person; 2017: $25 Per Person

LENGTH: 1 HOURS (CEU: 0 UNITS)

CLASS SIZE: MINIMUM: 1; MAXIMUM: 1

NHI Customer Service: (877) 558-6873 • nhicustomerservice@dot.gov

Course Number

FHWA-NHI-131110K

Course Title

Pavement Preservation Treatment Series: Selecting the Right Treatment - WEB-BASED

This training is part of the "Pavement Preservation Treatment" series and is designed to provide participants with information on preservation treatment selection. This course is primarily intended for inspectors and technicians.

The training draws on the Pavement Preservation Treatment Construction Guide (PPTCG), which was created by FHWA, in partnership with Caltrans, the National Center for Pavement Preservation, and the Transportation Curriculum Coordination Council (TCCC) as a resource for agency and industry pavement preservation practitioners. It provides information on basic pavement preservation concepts and the different treatments available and how they should be applied, so agencies can make informed decisions when determining which treatments best fit their pavement preservation needs. The training is primarily targeted at individuals unfamiliar with pavement preservation policy and technical information.

To take the entire series of trainings for the PPTCG, access the NHI website and register for NHI-131110.

Outcomes

Upon completion of the course, participants will be able to:

- Select the appropriate pavement preservation treatment(s) after analyzing given pavement and traffic conditions.

Target Audience

The primary audience for the Pavement Preservation Treatment Construction WBT course is Federal, State, and local highway construction and maintenance teams, specifically the highway workers and inspectors involved in the placement of pavement preservation treatments. Although not in the primary audience, design engineers will also benefit from the online guide and the associated training. The training course is primarily targeted at individuals unfamiliar with pavement preservation policy and technical information.

Training Level: Intermediate

Fee: 2016: $25 Per Person; 2017: N/A

Length: .5 HOURS (CEU: 0 UNITS)

Class Size: MINIMUM: 1; MAXIMUM: 1

NHI Customer Service: (877) 558-6873 • nhicustomerservice@dot.gov

COURSE NUMBER

FHWA-NHI-131117

COURSE TITLE

Basic Materials for Highway and Structure Construction and Maintenance - WEB-BASED

This training is provided by the Transportation Curriculum Coordination Council (TCCC) in partnership with NHI to review basic materials for highway and structure construction and maintenance. The training was prepared by State DOT personnel for State DOT personnel. It contains good practices from various agencies. Each State agency/company has their own specifications, which the viewer needs to review and follow. This course is primarily intended for inspectors and technicians.

Although there are a number of materials used in the construction and maintenance process for both highways and structures, this course is focused on the three basic materials. They are Aggregate, Portland Cement Concrete (referred to as PCC), and Hot Mix Asphalt (referred to as HMA).

This training is directed toward the entry level technician, to give them a general view of the basic materials used in construction and maintenance. The course modules will address the procedures used in the production and sampling of aggregates.

Module 1 is called Basic Aggregates and includes quarry inspection, sand operation, stockpiling, and sampling. Module 2 covers Portland cement, including the production of Portland Cement, the hydration process, as well as other cementing materials used in concrete such as water, admixtures, and aggregates. Module 3 reviews Hot Mix Asphalt, including the asphalt binder and aggregates used in the production.

NHI is hosting this and other TCCC Web-based developments to serve a critical need for training. We need your feedback to determine whether we should continue posting other Web-based trainings like this one. Please take the time to complete the evaluation form provided at the end of the training, or email nhimarketing@dot.gov.

OUTCOMES

Upon completion of the course, participants will be able to:

- Identify aggregate production and sampling procedures
- Recognize the ingredients of PCC and the part each plays in concrete production
- Recognize the ingredients of HMA and the part each plays in hot mix asphalt production

TARGET AUDIENCE

This training is designed for Level I and Level II State/local public agency personnel and their industry counterparts involved in the construction, maintenance and testing process for highways and structures. Level I or Entry refers to employees/trainees with little to no experience in the subject area and perform his/her activities under direct supervision. Level II or Intermediate refers to employees that understand and demonstrate skills in one or more areas of the entry level and perform specific tasks under general supervision.

TRAINING LEVEL: Basic

FEE: 2016: $25 Per Person; 2017: N/A

LENGTH: 3 HOURS (CEU: 0 UNITS)

CLASS SIZE: MINIMUM: 1; MAXIMUM: 1

NHI Customer Service: (877) 558-6873 • nhicustomerservice@dot.gov

Course Number

FHWA-NHI-131121

Course Title

Construction of Portland Cement Concrete Pavements - WEB-BASED

Improving and maintaining the quality of concrete is an important aspect of keeping pavements safe and long lasting. This training provides participants with an overview of the entire Portland cement concrete (PCC) paving and restoration process: setting forms, mixing, hauling, curing and applicable repair techniques. This training is presented in several modules:

1. Construction Quality
2. PCC Production Overview
3. Slipform Paving
4. Fixed Form Paving
5. Pavement Curing, Sawing, and Joint Sealing Operations
6. Concrete Pavement Restoration

This self-paced, Web-based training is designed for participants to progress at their own pace. The training focuses on the proper methods for construction of concrete paving and pavement restoration techniques with an emphasis on cause and effect.

Outcomes

Upon completion of the course, participants will be able to:

- Describe the differences between truck-mixed and ready-mixed concrete
- Identify factors in production and paving operations that contribute to achieving a smooth ride
- Describe the differences between slip-form and fixed-form paving
- Identify the factors that impact saw timing and crack control
- Recognize the importance and key factors in placing joint sealant materials
- Identify the components of concrete pavement restoration application and construction techniques
- Describe the purpose and appropriate use of full depth and partial depth repairs
- Indentify critical factors for curing and sawing operations that affect pavement performance
- Describe the purpose of grinding and dowel bar retrofit
- Identify applicable repair techniques for concrete pavement restoration
- Describe purpose of slab stabilization and joint and crack resealing

Target Audience

This training is designed for contractors, technicians, and inspectors who are involved in daily pavement operations for the placement and restoration of PCC pavements. Participants should have some working knowledge of concrete pavement construction.

Training Level: Intermediate

Fee: 2016: $50 Per Person; 2017: $50 Per Person

Length: 10 HOURS (CEU: 0 UNITS)

Class Size: MINIMUM: 1; MAXIMUM: 1

NHI Customer Service: (877) 558-6873 • nhicustomerservice@dot.gov

COURSE NUMBER

FHWA-NHI-131122

COURSE TITLE

Portland Cement Concrete Paving Inspection - WEB-BASED

This training is provided by the Transportation Curriculum Coordination Council (TCCC) in partnership with NHI to review inspection practices for Portland cement concrete paving projects. The training was originally developed by the Iowa Department of Transportation and more currently updated and reviewed by the TCCC and NHI. This course is recommended for the Transportation Curriculum Coordination Council levels I and II. This course is primarily intended for inspectors and technicians.

This training course has been prepared to provide guidance and instruction to inspectors involved in the construction of Portland cement concrete (PCC) pavements. The important tasks involved in this work are explained and proper procedures are described. The material is targeted for those who have not had experience in PCC paving construction.

OUTCOMES

Upon completion of the course, participants will be able to:

- Identify the materials in a PCC mixture and the concrete properties
- Comprehend Design Project Plans and recognize the joints types and saw cuts
- Identify the safety requirements and recognize safe Traffic Control practices
- Recognize and comprehend the use of the equipment in a PCC Paving project
- Recognize various sub grade treatments
- Inspect project tasks for compliance with pre-paving requirements, i.e., survey stakes, proof rolling, subgrade, and dowel baskets
- Inspect project tasks for compliance with PCC Paving requirements, i.e., string line, place and consolidate, finish, and texture
- Perform post-construction checks

TARGET AUDIENCE

This training is designed for FHWA, State, and local agencies and their industry counterparts involved in the process of placement and inspection of Portland cement concrete paving. It is applicable to anyone desiring a better understanding of activities and inspection procedures on Portland cement concrete paving projects.

TRAINING LEVEL: Intermediate

FEE: 2016: $50 Per Person; 2017: $50 Per Person

LENGTH: 5 HOURS (CEU: 0 UNITS)

CLASS SIZE: MINIMUM: 1; MAXIMUM: 1

NHI Customer Service: (877) 558-6873 • nhicustomerservice@dot.gov

Course Number

FHWA-NHI-131126

Course Title

Concrete Pavement Preservation Series (Includes NHI-131126A-K)

The Transportation Curriculum Coordination Council (TCCC) in partnership with NHI is pleased to offer this comprehensive training series (FHWA-NHI-131126) for concrete pavement preservation. The training was developed by the National Concrete Pavement Technology Center at Iowa State University in cooperation with FHWA.

The Concrete Pavement Preservation Series presents current guidelines and recommendations for the design, construction, and selection of cost-effective concrete pavement preservation strategies. It concentrates primarily on strategies and methods that are applicable at the project level, and not at the network level, where pavement management activities function and address such issues as prioritizing and budgeting.

Registration in NHI-131126 enrolls you in all 11 courses in the Concrete Pavement Preservation Series (NHI-131126A-K) plus gives you access to a downloadable version of the FHWA Concrete Pavement Preservation Guide! You can take some or all of these courses when it best suits your schedule.

NHI-131126 includes:

- Introduction module with downloadable version of the FHWA Concrete Pavement Preservation Guide
- NHI-131126A: Pavement Preservation Concepts
- NHI-131126B: Concrete Pavement Evaluation
- NHI-131126C: Slab Stabilization
- NHI-131126D: Partial-depth Repairs
- NHI-131126E: Full-depth Repairs
- NHI-131126F: Retrofitted Edge Drains
- NHI-131126G: Dowel Bar Retrofit
- NHI-131126H: Diamond Grinding and Grooving
- NHI-131126I: Joint Resealing and Crack Sealing
- NHI-131126J: Concrete Overlays
- NHI-131126K: Strategy Selection

Outcomes

Upon completion of the course, participants will be able to:

- Define pavement preservation
- List the major components of a pavement evaluation and the types of information gained from each
- Identify the purpose and suitable application of various concrete pavement preservation treatments
- Describe recommended materials and construction/installation practices for each treatment
- List factors to consider in the selection of concrete pavement preservation treatments

Target Audience

The Concrete Pavement Preservation Series meets the needs of a diverse audience to include design engineers, quality control personnel, contractors, suppliers, technicians, and trades people. While the course is aimed at those who have some familiarity with concrete pavements and pavement preservation, it should also be of value to those that are new to the field. This course is recommended for the Transportation Curriculum Coordination Council levels I - IV.

TRAINING LEVEL: Intermediate

FEE: 2016: $50 Per Person; 2017: N/A

LENGTH: 11 HOURS (CEU: 0 UNITS)

CLASS SIZE: MINIMUM: 1; MAXIMUM: 1

NHI Customer Service: (877) 558-6873 • nhicustomerservice@dot.gov

Course Number

FHWA-NHI-131126A

Course Title

Concrete Pavement Preservation Series: Pavement Preservation Concepts

This training was prepared by the Transportation Curriculum Coordination Council (TCCC) in partnership with NHI to provide guidance on critical concrete pavement preservation issues. The training was developed by the National Concrete Pavement Technology Center at Iowa State University in cooperation with FHWA.

This module discusses how preventative maintenance impacts pavement preservation, good candidates for preservation, and the benefits to pavement preservation.

This module is part of the curriculum from the Concrete Pavement Preservation Series (FHWA-NHI-131126) which presents current guidelines and recommendations for the design, construction, and selection of cost-effective concrete pavement preservation strategies. The other Web-based training modules are:

- NHI-131126 Concrete Pavement Preservation Series with downloadable version of the FHWA Concrete Pavement Preservation Guide
- NHI-131126A: Pavement Preservation Concepts
- NHI-131126B: Concrete Pavement Evaluation
- NHI-131126C: Slab Stabilization
- NHI-131126D: Partial-depth Repairs
- NHI-131126E: Full-depth Repairs
- NHI-131126F: Retrofitted Edge Drains
- NHI-131126G: Dowel Bar Retrofit
- NHI-131126H: Diamond Grinding and Grooving
- NHI-131126I: Joint Resealing and Crack Sealing
- NHI-131126J: Concrete Overlays
- NHI-131126K: Strategy Selection

Outcomes

Upon completion of the course, participants will be able to:

- Define pavement preservation and preventive maintenance
- Describe characteristics of suitable pavements for preventive maintenance
- Describe the importance of selecting and placing the "right" treatment and placing it at the "right" time
- List the benefits of pavement preservation

Target Audience

The intended audience is quite diverse, and includes design engineers, quality control personnel, contractors, suppliers, technicians, and trades people. While the course is aimed at those who have some familiarity with concrete pavements and pavement preservation, it should also be of value to those that are new to the field. This course is recommended for the Transportation Curriculum Coordination Council levels I - IV.

TRAINING LEVEL: Intermediate

FEE: 2016: $25 Per Person; 2017: $25 Per Person

LENGTH: 1 HOURS (CEU: 0 UNITS)

CLASS SIZE: MINIMUM: 1; MAXIMUM: 1

NHI Customer Service: (877) 558-6873 • nhicustomerservice@dot.gov

Course Number

FHWA-NHI-131126B

Course Title

Concrete Pavement Preservation Series: Concrete Pavement Evaluation

This training was prepared by the Transportation Curriculum Coordination Council (TCCC) in partnership with NHI to provide guidance on critical concrete pavement preservation issues. The training was sponsored by the FHWA and developed by the National Concrete Pavement Technology Center at Iowa State University in cooperation with FHWA.

This module discusses how preventative maintenance impacts pavement preservation, good candidates for preservation, and the benefits to pavement preservation. This module also describes the common procedures associated with conducting thorough pavement evaluations.

This module is part of the curriculum from the Concrete Pavement Preservation Series (FHWA-NHI-131126) which presents current guidelines and recommendations for the design, construction, and selection of cost-effective concrete pavement preservation strategies. The other Web-based training modules are:

- NHI-131126 Concrete Pavement Preservation Series with downloadable version of the FHWA Concrete Pavement Preservation Guide
- NHI-131126A: Pavement Preservation Concepts
- NHI-131126B: Concrete Pavement Evaluation
- NHI-131126C: Slab Stabilization
- NHI-131126D: Partial-depth Repairs
- NHI-131126E: Full-depth Repairs
- NHI-131126F: Retrofitted Edge Drains
- NHI-131126G: Dowel Bar Retrofit
- NHI-131126H: Diamond Grinding and Grooving
- NHI-131126I: Joint Resealing and Crack Sealing
- NHI-131126J: Concrete Overlays
- NHI-131126K: Strategy Selection

Outcomes

Upon completion of the course, participants will be able to:

- Describe the need for a thorough pavement evaluation
- Name the common pavement evaluation components
- Describe what information is obtained from each pavement evaluation component

Target Audience

The intended audience is quite diverse, and includes design engineers, quality control personnel, contractors, suppliers, technicians, and trades people. While the course is aimed at those who have some familiarity with concrete pavements and pavement preservation, it should also be of value to those that are new to the field. This course is recommended for the Transportation Curriculum Coordination Council levels I - IV.

TRAINING LEVEL: Intermediate

FEE: 2016: $25 Per Person; 2017: $25 Per Person

LENGTH: 2 HOURS (CEU: 0 UNITS)

CLASS SIZE: MINIMUM: 1; MAXIMUM: 1

NHI Customer Service: (877) 558-6873 • nhicustomerservice@dot.gov

Course Number

FHWA-NHI-131126C

Course Title

Concrete Pavement Preservation Series: Slab Stabilization

This training was prepared by the Transportation Curriculum Coordination Council (TCCC) in partnership with NHI to provide guidance on critical concrete pavement preservation issues. The training was developed by the National Concrete Pavement Technology Center at Iowa State University in cooperation with FHWA.

This module covers the use of slab stabilization (also known as undersealing) and slab jacking of concrete pavements. Slab stabilization restores support beneath slabs where voids have been detected, and slab jacking is used to raise depressed or settled slabs.

This module is part of the curriculum from the Concrete Pavement Preservation Series (FHWA-NHI-131126) which presents current guidelines and recommendations for the design, construction, and selection of cost-effective concrete pavement preservation strategies. The other Web-based training modules are:

- NHI-131126 Concrete Pavement Preservation Series with downloadable version of the FHWA Concrete Pavement Preservation Guide
- NHI-131126A: Pavement Preservation Concepts
- NHI-131126B: Concrete Pavement Evaluation
- NHI-131126C: Slab Stabilization
- NHI-131126D: Partial-depth Repairs
- NHI-131126E: Full-depth Repairs
- NHI-131126F: Retrofitted Edge Drains
- NHI-131126G: Dowel Bar Retrofit
- NHI-131126H: Diamond Grinding and Grooving
- NHI-131126I: Joint Resealing and Crack Sealing
- NHI-131126J: Concrete Overlays
- NHI-131126K: Strategy Selection

Outcomes

Upon completion of the course, participants will be able to:

- List benefits of slab stabilization and slab jacking
- Describe recommended materials and mixtures
- Describe recommended construction steps for both procedures
- Identify typical construction problems and remedies for slab stabilization

Target Audience

The intended audience is quite diverse, and includes design engineers, quality control personnel, contractors, suppliers, technicians, and trades people. While the course is aimed at those who have some familiarity with concrete pavements and pavement preservation, it should also be of value to those that are new to the field.This course is recommended for the Transportation Curriculum Coordination Council levels I - IV.

TRAINING LEVEL: Intermediate

FEE: 2016: $25 Per Person; 2017: $25 Per Person

LENGTH: 1 HOURS (CEU: 0 UNITS)

CLASS SIZE: MINIMUM: 1; MAXIMUM: 1

NHI Customer Service: (877) 558-6873 • nhicustomerservice@dot.gov

Course Number

FHWA-NHI-131126D

Course Title

Concrete Pavement Preservation Series: Partial-depth Repairs

This training was prepared by the Transportation Curriculum Coordination Council (TCCC) in partnership with NHI to provide guidance on critical concrete pavement preservation issues. The training was developed by the National Concrete Pavement Technology Center at Iowa State University in cooperation with FHWA.

This module covers the procedures for partial-depth repairs (PDR) on PCC pavements. PDR is the removal and replacement of small, shallow areas of deteriorated PCC at spalled or distressed joints.

This module is part of the curriculum from the Concrete Pavement Preservation Series (FHWA-NHI-131126) which presents current guidelines and recommendations for the design, construction, and selection of cost-effective concrete pavement preservation strategies. The other Web-based training modules are:

- NHI-131126 Concrete Pavement Preservation Series with downloadable version of the FHWA Concrete Pavement Preservation Guide
- NHI-131126A: Pavement Preservation Concepts
- NHI-131126B: Concrete Pavement Evaluation
- NHI-131126C: Slab Stabilization
- NHI-131126D: Partial-depth Repairs
- NHI-131126E: Full-depth Repairs
- NHI-131126F: Retrofitted Edge Drains
- NHI-131126G: Dowel Bar Retrofit
- NHI-131126H: Diamond Grinding and Grooving
- NHI-131126I: Joint Resealing and Crack Sealing
- NHI-131126J: Concrete Overlays
- NHI-131126K: Strategy Selection

Outcomes

Upon completion of the course, participants will be able to:

- List benefits and appropriateness of partial-depth repairs
- List the advantages and disadvantages of different available repair materials
- Describe recommended construction procedures
- Identify typical construction problems and appropriate remedies

Target Audience

The intended audience is quite diverse, and includes design engineers, quality control personnel, contractors, suppliers, technicians, and trades people. While the course is aimed at those who have some familiarity with concrete pavements and pavement preservation, it should also be of value to those that are new to the field. This course is recommended for the Transportation Curriculum Coordination Council levels I - IV.

TRAINING LEVEL: Intermediate

FEE: 2016: $25 Per Person; 2017: $25 Per Person

LENGTH: 1 HOURS (CEU: 0 UNITS)

CLASS SIZE: MINIMUM: 1; MAXIMUM: 1

NHI Customer Service: (877) 558-6873 • nhicustomerservice@dot.gov

Course Number

FHWA-NHI-131126E

Course Title

Concrete Pavement Preservation Series: Full-depth Repairs

This training was prepared by the Transportation Curriculum Coordination Council (TCCC) in partnership with NHI to provide guidance on critical concrete pavement preservation issues. The training was developed by the National Concrete Pavement Technology Center at Iowa State University in cooperation with FHWA.

This module covers the procedures for cast-in-place Portland cement concrete (PCC) full-depth repair (FDR) of jointed concrete pavements (JCP) including jointed plain (JPCP) and jointed reinforced concrete pavements (JRCP). FDR techniques for continuously reinforced concrete pavements (CRCP) are discussed separately toward the end of the presentation. FDR is the cast-in-place concrete repairs that extend the full-depth of the existing slab.

This module is part of the curriculum from the Concrete Pavement Preservation Series (FHWA-NHI-131126) which presents current guidelines and recommendations for the design, construction, and selection of cost-effective concrete pavement preservation strategies. The other Web-based training modules are:

- NHI-131126 Concrete Pavement Preservation Series with downloadable version of the FHWA Concrete Pavement Preservation Guide
- NHI-131126A: Pavement Preservation Concepts
- NHI-131126B: Concrete Pavement Evaluation
- NHI-131126C: Slab Stabilization
- NHI-131126D: Partial-depth Repairs
- NHI-131126E: Full-depth Repairs
- NHI-131126F: Retrofitted Edge Drains
- NHI-131126G: Dowel Bar Retrofit
- NHI-131126H: Diamond Grinding and Grooving
- NHI-131126I: Joint Resealing and Crack Sealing
- NHI-131126J: Concrete Overlays
- NHI-131126K: Strategy Selection

Outcomes

Upon completion of the course, participants will be able to:

- List the benefits of full-depth repairs
- Describe primary design considerations in terms of dimensions, load transfer, and materials
- Describe recommended construction activities
- Identify typical construction problems and remedies

Target Audience

The intended audience is quite diverse, and includes design engineers, quality control personnel, contractors, suppliers, technicians, and trades people. While the course is aimed at those who have some familiarity with concrete pavements and pavement preservation, it should also be of value to those that are new to the field.This course is recommended for the Transportation Curriculum Coordination Council levels I - IV.

TRAINING LEVEL: Intermediate

FEE: 2016: $25 Per Person; 2017: $25 Per Person

LENGTH: 2 HOURS (CEU: 0 UNITS)

CLASS SIZE: MINIMUM: 1; MAXIMUM: 1

NHI Customer Service: (877) 558-6873 • nhicustomerservice@dot.gov

COURSE NUMBER

FHWA-NHI-131126F

COURSE TITLE

Concrete Pavement Preservation Series: Retrofitted Edge Drains

This training was prepared by the Transportation Curriculum Coordination Council (TCCC) in partnership with NHI to provide guidance on critical concrete pavement preservation issues. The training was developed by the National Concrete Pavement Technology Center at Iowa State University in cooperation with FHWA.

This module presents design and construction information on retrofitted edge drains. This treatment is not as widely used as it once was, largely because it has limited applicability. Specifically, it must be targeted to those pavements that are 1) in good structural condition and 2) have bases with some degree of permeability that would allow water to be drained from beneath the pavement and to the edge drain.

This module is part of the curriculum from the Concrete Pavement Preservation Series (FHWA-NHI-131126) which presents current guidelines and recommendations for the design, construction, and selection of cost-effective concrete pavement preservation strategies. The other Web-based training modules are:

- NHI-131126 Concrete Pavement Preservation Series with downloadable version of the FHWA Concrete Pavement Preservation Guide
- NHI-131126A: Pavement Preservation Concepts
- NHI-131126B: Concrete Pavement Evaluation
- NHI-131126C: Slab Stabilization
- NHI-131126D: Partial-depth Repairs
- NHI-131126E: Full-depth Repairs
- NHI-131126F: Retrofitted Edge Drains
- NHI-131126G: Dowel Bar Retrofit
- NHI-131126H: Diamond Grinding and Grooving
- NHI-131126I: Joint Resealing and Crack Sealing
- NHI-131126J: Concrete Overlays
- NHI-131126K: Strategy Selection

OUTCOMES

Upon completion of the course, participants will be able to:

- List benefits of drainage
- List components of edge drain systems
- Describe recommended installation procedures
- Identify typical construction problems and remedies

TARGET AUDIENCE

The intended audience is quite diverse, and includes design engineers, quality control personnel, contractors, suppliers, technicians, and trades people. While the course is aimed at those who have some familiarity with concrete pavements and pavement preservation, it should also be of value to those that are new to the field.This course is recommended for the Transportation Curriculum Coordination Council levels I - IV.

TRAINING LEVEL: Intermediate

FEE: 2016: $25 Per Person; 2017: N/A

LENGTH: 1 DAYS (CEU: 0 UNITS)

CLASS SIZE: MINIMUM: 1; MAXIMUM: 1

NHI Customer Service: (877) 558-6873 • nhicustomerservice@dot.gov

COURSE NUMBER

FHWA-NHI-131126G

COURSE TITLE

Concrete Pavement Preservation Series: Dowel Bar Retrofit

This training was prepared by the Transportation Curriculum Coordination Council (TCCC) in partnership with NHI to provide guidance on critical concrete pavement preservation issues. The training was developed by the National Concrete Pavement Technology Center at Iowa State University in cooperation with FHWA.

This module presents design and construction information on load transfer restoration (LTR), sometimes referred to as retrofitted load transfer. In the introduction we will describe the difference between load transfer restoration (generic term) and dowel bar retrofitting (DBR) which is a specific means of achieving LTR. There are other methods available, but DBR is the most proven.

This module is part of the curriculum from the Concrete Pavement Preservation Series (FHWA-NHI-131126) which presents current guidelines and recommendations for the design, construction, and selection of cost-effective concrete pavement preservation strategies. The other Web-based training modules are:

- NHI-131126 Concrete Pavement Preservation Series with downloadable version of the FHWA Concrete Pavement Preservation Guide
- NHI-131126A: Pavement Preservation Concepts
- NHI-131126B: Concrete Pavement Evaluation
- NHI-131126C: Slab Stabilization
- NHI-131126D: Partial-depth Repairs
- NHI-131126E: Full-depth Repairs
- NHI-131126F: Retrofitted Edge Drains
- NHI-131126G: Dowel Bar Retrofit
- NHI-131126H: Diamond Grinding and Grooving
- NHI-131126I: Joint Resealing and Crack Sealing
- NHI-131126J: Concrete Overlays
- NHI-131126K: Strategy Selection

OUTCOMES

Upon completion of the course, participants will be able to:

- List benefits and applications of load transfer restoration
- Describe recommended materials and mixtures
- Describe recommended construction procedures
- Identify typical construction problems and remedies

TARGET AUDIENCE

The intended audience is quite diverse, and includes design engineers, quality control personnel, contractors, suppliers, technicians, and trades people. While the course is aimed at those who have some familiarity with concrete pavements and pavement preservation, it should also be of value to those that are new to the field.This course is recommended for the Transportation Curriculum Coordination Council levels I - IV.

TRAINING LEVEL: Intermediate

FEE: 2016: $25 Per Person; 2017: $25 Per Person

LENGTH: 1 HOURS (CEU: 0 UNITS)

CLASS SIZE: MINIMUM: 1; MAXIMUM: 1

NHI Customer Service: (877) 558-6873 • nhicustomerservice@dot.gov

COURSE NUMBER

FHWA-NHI-131126H

COURSE TITLE

Concrete Pavement Preservation Series: Diamond Grinding and Grooving

This training was prepared by the Transportation Curriculum Coordination Council (TCCC) in partnership with NHI to provide guidance on critical concrete pavement preservation issues. The training was developed by the National Concrete Pavement Technology Center at Iowa State University in cooperation with FHWA.

This module describes recommended procedures for surface restoration of Portland cement concrete (PCC) pavements, specifically diamond grinding and diamond grooving operations.

This module is part of the curriculum from the Concrete Pavement Preservation Series (FHWA-NHI-131126) which presents current guidelines and recommendations for the design, construction, and selection of cost-effective concrete pavement preservation strategies. The other Web-based training modules are:

- NHI-131126 Concrete Pavement Preservation Series with downloadable version of the FHWA Concrete Pavement Preservation Guide
- NHI-131126A: Pavement Preservation Concepts
- NHI-131126B: Concrete Pavement Evaluation
- NHI-131126C: Slab Stabilization
- NHI-131126D: Partial-depth Repairs
- NHI-131126E: Full-depth Repairs
- NHI-131126F: Retrofitted Edge Drains
- NHI-131126G: Dowel Bar Retrofit
- NHI-131126H: Diamond Grinding and Grooving
- NHI-131126I: Joint Resealing and Crack Sealing
- NHI-131126J: Concrete Overlays
- NHI-131126K: Strategy Selection

OUTCOMES

Upon completion of the course, participants will be able to:

- Differentiate between diamond grinding and diamond grooving and list the benefits of each
- Identify appropriate blade spacing dimensions for grinding and grooving
- Describe recommended construction procedures
- Identify typical construction problems and remedies

TARGET AUDIENCE

The intended audience is quite diverse, and includes design engineers, quality control personnel, contractors, suppliers, technicians, and trades people. While the course is aimed at those who have some familiarity with concrete pavements and pavement preservation, it should also be of value to those that are new to the field. This course is recommended for the Transportation Curriculum Coordination Council levels I - IV.

TRAINING LEVEL: Intermediate

FEE: 2016: $25 Per Person; 2017: $25 Per Person

LENGTH: 1 HOURS (CEU: 0 UNITS)

CLASS SIZE: MINIMUM: 1; MAXIMUM: 1

NHI Customer Service: (877) 558-6873 • nhicustomerservice@dot.gov

Course Number

FHWA-NHI-131126I

Course Title

Concrete Pavement Preservation Series: Joint Sealing and Crack Resealing

This training was prepared by the Transportation Curriculum Coordination Council (TCCC) in partnership with NHI to provide guidance on critical concrete pavement preservation issues. The training was developed by the National Concrete Pavement Technology Center at Iowa State University in cooperation with FHWA.

This module covers joint resealing and crack sealing for concrete pavements. Joint resealing and crack sealing is defined as placement of an approved sealant material in an existing joint or crack to reduce moisture infiltration and prevent intrusion of incompressibles.

This module is part of the curriculum from the Concrete Pavement Preservation Series (FHWA-NHI-131126) which presents current guidelines and recommendations for the design, construction, and selection of cost-effective concrete pavement preservation strategies. The other Web-based training modules are:

- NHI-131126 Concrete Pavement Preservation Series with downloadable version of the FHWA Concrete Pavement Preservation Guide
- NHI-131126A: Pavement Preservation Concepts
- NHI-131126B: Concrete Pavement Evaluation
- NHI-131126C: Slab Stabilization
- NHI-131126D: Partial-depth Repairs
- NHI-131126E: Full-depth Repairs
- NHI-131126F: Retrofitted Edge Drains
- NHI-131126G: Dowel Bar Retrofit
- NHI-131126H: Diamond Grinding and Grooving
- NHI-131126I: Joint Resealing and Crack Sealing
- NHI-131126J: Concrete Overlays
- NHI-131126K: Strategy Selection

Outcomes

Upon completion of the course, participants will be able to:

- List the benefits of joint resealing
- Describe desirable sealant properties and characteristics
- Describe recommended installation procedures
- Identify typical construction problems and appropriate remedies

Target Audience

The intended audience is quite diverse, and includes design engineers, quality control personnel, contractors, suppliers, technicians, and trades people. While the course is aimed at those who have some familiarity with concrete pavements and pavement preservation, it should also be of value to those that are new to the field.This course is recommended for the Transportation Curriculum Coordination Council levels I - IV.

TRAINING LEVEL: Intermediate

FEE: 2016: $25 Per Person; 2017: $25 Per Person

LENGTH: 1 HOURS (CEU: 0 UNITS)

CLASS SIZE: MINIMUM: 1; MAXIMUM: 1

NHI Customer Service: (877) 558-6873 • nhicustomerservice@dot.gov

Course Number

FHWA-NHI-131126J

Course Title

Concrete Pavement Preservation Series: Concrete Overlays

This training was prepared by the Transportation Curriculum Coordination Council (TCCC) in partnership with NHI to provide guidance on critical concrete pavement preservation issues. The training was developed by the National Concrete Pavement Technology Center at Iowa State University in cooperation with FHWA.

This module provides guidance on the selection of concrete pavement preservation strategies. Based on a collective review of a number of recent published documents, this module covers the seven step process that can be used to determine the most appropriate treatment (or combination of treatments) for a PCC pavement.

This module is part of the curriculum from the Concrete Pavement Preservation Series (FHWA-NHI-131126) which presents current guidelines and recommendations for the design, construction, and selection of cost-effective concrete pavement preservation strategies. The other Web-based training modules are:

- NHI-131126 Concrete Pavement Preservation Series with downloadable version of the FHWA Concrete Pavement Preservation Guide
- NHI-131126A: Pavement Preservation Concepts
- NHI-131126B: Concrete Pavement Evaluation
- NHI-131126C: Slab Stabilization
- NHI-131126D: Partial-depth Repairs
- NHI-131126E: Full-depth Repairs
- NHI-131126F: Retrofitted Edge Drains
- NHI-131126G: Dowel Bar Retrofit
- NHI-131126H: Diamond Grinding and Grooving
- NHI-131126I: Joint Resealing and Crack Sealing
- NHI-131126J: Concrete Overlays
- NHI-131126K: Strategy Selection

Outcomes

Upon completion of the course, participants will be able to:

- Describe the treatment selection process
- List the components of a life-cycle cost analysis
- List other factors that may enter the selection process

Target Audience

The intended audience is quite diverse, and includes design engineers, quality control personnel, contractors, suppliers, technicians, and trades people. While the course is aimed at those who have some familiarity with concrete pavements and pavement preservation, it should also be of value to those that are new to the field. This course is recommended for the Transportation Curriculum Coordination Council levels I - IV.

TRAINING LEVEL: Intermediate

FEE: 2016: $25 Per Person; 2017: $25 Per Person

LENGTH: 1 HOURS (CEU: 0 UNITS)

CLASS SIZE: MINIMUM: 1; MAXIMUM: 1

NHI Customer Service: (877) 558-6873 • nhicustomerservice@dot.gov

Course Number

FHWA-NHI-131126K

Course Title

Concrete Pavement Preservation Series: Strategy Selection

This training was prepared by the Transportation Curriculum Coordination Council (TCCC) in partnership with NHI to provide guidance on critical concrete pavement preservation issues. The training was developed by the National Concrete Pavement Technology Center at Iowa State University in cooperation with FHWA.

This module provides guidance on the selection of concrete pavement preservation strategies. Based on a collective review of a number of recent published documents, this module covers the seven step process that can be used to determine the most appropriate treatment (or combination of treatments) for a PCC pavement.

This module is part of the curriculum from the Concrete Pavement Preservation Series (FHWA-NHI-131126) which presents current guidelines and recommendations for the design, construction, and selection of cost-effective concrete pavement preservation strategies. The other Web-based training modules are:

- NHI-131126 Concrete Pavement Preservation Series with downloadable version of the FHWA Concrete Pavement Preservation Guide
- NHI-131126A: Pavement Preservation Concepts
- NHI-131126B: Concrete Pavement Evaluation
- NHI-131126C: Slab Stabilization
- NHI-131126D: Partial-depth Repairs
- NHI-131126E: Full-depth Repairs
- NHI-131126F: Retrofitted Edge Drains
- NHI-131126G: Dowel Bar Retrofit
- NHI-131126H: Diamond Grinding and Grooving
- NHI-131126I: Joint Resealing and Crack Sealing
- NHI-131126J: Concrete Overlays
- NHI-131126K: Strategy Selection

Outcomes

Upon completion of the course, participants will be able to:

- Describe the treatment selection process
- List factors that might enter into the selection process
- Describe pavement deficiencies addressed by the different preservation treatments
- Describe how the benefits and costs of alternative treatment strategies are computed in a cost-effectiveness analysis
- Describe a process used to select the preferred treatment strategy

Target Audience

The intended audience is quite diverse, and includes design engineers, quality control personnel, contractors, suppliers, technicians, and trades people. While the course is aimed at those who have some familiarity with concrete pavements and pavement preservation, it should also be of value to those that are new to the field.This course is recommended for the Transportation Curriculum Coordination Council levels I - IV.

TRAINING LEVEL: Intermediate

FEE: 2016: $25 Per Person; 2017: N/A

LENGTH: .3 HOURS (CEU: 0 UNITS)

CLASS SIZE: MINIMUM: 1; MAXIMUM: 1

NHI Customer Service: (877) 558-6873 • nhicustomerservice@dot.gov

COURSE NUMBER

FHWA-NHI-131127

COURSE TITLE

Concrete Series - WEB-BASED

The Transportation Curriculum Coordination Council (TCCC) in partnership with NHI is pleased to offer this comprehensive training series (FHWA-NHI-131127) for any engineer or supervisor working with Portland cement. The training was developed by the National Concrete Pavement Technology Center at Iowa State University. It is the first training of its kind offered by NHI, and we would like to give special recognition to the TCCC for their efforts. This course is recommended for the Transportation Curriculum Coordination Council levels II - IV.

The TCCC Concrete Series is part of a curriculum from the "Integrated Materials and Construction Practices for Concrete Pavement" manual developed through the National Concrete Pavement Technology Center at Iowa State University.

To streamline registration and enable you to take some or all of these courses when it best suits your schedule, we have created this new series option which automatically registers you for all 11 modules-it's that easy. They are as follows:

Module 1 - TCCC Design of Pavement (FHWA-NHI-134101)

Module 2 - TCCC Fundamentals of Materials Used for Concrete Pavements (FHWA-NHI-134084)

Module 3 - TCCC Mix Design Principles (FHWA-NHI-134087)

Module 4 - TCCC Fresh Concrete Properties (FHWA-NHI-134097)

Module 5 - TCCC Basics of Cement Hydration (FHWA-NHI-134096)

Module 6 - TCCC Incompatibility in Concrete Pavement Systems (FHWA-NHI-134085)

Module 7 - TCCC Early Age Cracking (FHWA-NHI-134095)

Module 8 - TCCC Hardened Concrete Properties- Durability (FHWA-NHI-134075)

Module 9 - TCCC Construction of Concrete Pavements (FHWA-NHI-134098)

Module 10 - TCCC QCQA for Concrete Pavements (FHWA-NHI-134100)

Module 11 - TCCC Troubleshooting for Concrete Pavements (FHWA-NHI-134102)

OUTCOMES

Upon completion of the course, participants will be able to:

- Explain concrete pavement construction as a complex, integrated system involving several discrete practices that interrelate and affect one another in various ways
- Recognize and implement technologies, tests, and best practices to identify materials, concrete properties, and construction practices that are known to optimize concrete performance
- Identify factors that lead to premature distress in concrete, and learn how to avoid or reduce those factors
- Apply appropriate how-to and troubleshooting information

TARGET AUDIENCE

This training is intended as both a training tool and a reference to help concrete paving engineers, quality control personnel, specifiers, contractors, suppliers, technicians, and tradespeople bridge the gap between recent research and practice regarding optimizing the performance of concrete for pavements.

TRAINING LEVEL: Intermediate

FEE: 2016: $50 Per Person; 2017: $50 Per Person

LENGTH: 12 HOURS (CEU: 0 UNITS)

CLASS SIZE: MINIMUM: 1; MAXIMUM: 1

NHI Customer Service: (877) 558-6873 • nhicustomerservice@dot.gov

Course Number

FHWA-NHI-131128

Course Title

Testing Self-Consolidating Concrete - WEB-BASED

This training was prepared by the Transportation Curriculum Coordination Council (TCCC) in partnership with NHI to review the properties and applications of self-consolidating concrete as well as the test methods used for measuring SCC properties according to ASTM test methods. This training is recommended for the Transportation Curriculum Coordination Council levels I, II, and III. This course is primarily intended for inspectors and technicians.

This training includes an overview of the fresh properties of self-consolidating concrete including terminology, target guidelines and quality control. In addition, ASTM test methods for slump flow and flow rate, passing ability using the j-ring, column segregation, static segregation and making self consolidated concrete test cylinders are reviewed.

Outcomes

Upon completion of the course, participants will be able to:

- Define self-consolidating concrete
- Understand the terminology associated with self-consolidating concrete
- Perform the tests associated with SCC
- Report the test results

Target Audience

This course is designed for anyone who would like to understand more about self consolidating concrete, including personnel running self-consolidating concrete tests in the field along with supervisors in charge of field testing technicians.

Training Level: Basic

Fee: 2016: $25 Per Person; 2017: $25 Per Person

Length: 1 HOURS (CEU: 0 UNITS)

Class Size: MINIMUM: 1; MAXIMUM: 1

NHI Customer Service: (877) 558-6873 • nhicustomerservice@dot.gov

Course Number

FHWA-NHI-131129

Course Title

HMA Paving Field Inspection - WEB-BASED

This training was prepared by the Transportation Curriculum Coordination Council (TCCC) in partnership with NHI to provide guidance and instruction to inspectors involved in the construction of hot mix asphalt (HMA) pavements. The important tasks involved in this work are explained and proper procedures are described. This training is recommended for the Transportation Curriculum Coordination Council levels I, II, and III. This course is primarily intended for inspectors and technicians.

This training is arranged in a fashion to help the inspector first learn the various aspects of what is involved in a HMA paving operation and then become familiar with the duties that are a part of the HMA pavement grade inspection responsibilities. It also explains how to recognize the mix properties of a HMA mixture. The information included will assist the inspector in recognizing problems during a project and offering solutions to the problems. This training is not intended to cover every aspect of HMA paving.

Outcomes

Upon completion of the course, participants will be able to:

- Know various aspects of what is involved in a HMA paving operation
- Understand the duties of a HMA paving inspector
- Recognize the mix properties of a HMA mixture
- Recognize the problems that may occur on HMA paving projects
- Understand the product and project so solutions can be recommended

Target Audience

This training would be beneficial to anyone that is involved with an HMA paving project, but focuses on technicians/ inspectors that are involved with the production, placement, and inspection of HMA paving projects.

Training Level: Intermediate

Fee: 2016: $50 Per Person; 2017: $50 Per Person

Length: 4.5 HOURS (CEU: 0 UNITS)

Class Size: MINIMUM: 1; MAXIMUM: 1

NHI Customer Service: (877) 558-6873 • nhicustomerservice@dot.gov

Course Number

FHWA-NHI-131130

Course Title

Advanced Self-Consolidating Concrete - WEB-BASED

.

This training was prepared by the Transportation Curriculum Coordination Council (TCCC) in partnership with NHI to reviews advanced concepts, properties, and applications of self-consolidating concrete. This training is recommended for the Transportation Curriculum Coordination Council levels II, III, and IV. This course is primarily intended for inspectors and technicians.

This training will cover the basic characteristics of self-consolidating concrete as well as advantages of using SCC as compared to conventional concrete. In addition, it will discuss SCC's composition and proportioning as well as fresh and hardened properties. Finally, we will review specific examples where SCC has been used as well as the details of SCC use in slipform paving.

Outcomes

Upon completion of the course, participants will be able to:

- Define self-consolidating concrete
- List procedures for creating SCC
- Identify SCC performance characteristics
- Compare SCC and conventional concrete
- Recognize SCC applications

Target Audience

Anyone who would like to understand more about self consolidating concrete, including personnel running self-consolidating concrete tests in the field along with supervisors in charge of field testing technicians.

Training Level: Intermediate

Fee: 2016: $25 Per Person; 2017: $25 Per Person

Length: 1.5 HOURS (CEU: 0 UNITS)

Class Size: MINIMUM: 1; MAXIMUM: 1

NHI Customer Service: (877) 558-6873 • nhicustomerservice@dot.gov

Course Number

FHWA-NHI-131132

Course Title

Chip Seal Best Practices - WEB-BASED

The Chip Seal Best Practices course presents ways to assist in the development and implementation of pavement preservation programs by identifying the benefits of using chip seal as part of a preventive maintenance program.

This course has six modules. Module 1 is an introduction into chip seals, module 2 covers designing chip seal mixes, module 3 is selecting the proper materials for the chip seal mix, module 4 focuses on the use of the equipment, module 5 covers proper construction practices, and module 6 rounds out the course with performance measures of chip seals. The combination of all this information provides an excellent overview of successful chip seal practices worldwide.

Outcomes

Upon completion of the course, participants will be able to:

- Define chip seal
- Describe how chip seals are used as a preventive maintenance treatment for pavement
- Identify materials used in chip seals
- Describe the characteristics of chip seal design
- Identify types of chip seal
- Identify the important considerations of aggregate and binder selection
- Describe aggregate-binder compatibility
- Describe equipments used in chip seal practices
- Identify important variables in construction practice
- Define the measures of control implemented over the quality of materials and construction
- Identify construction best practices
- Describe the components of engineering-based performance measures
- Identify qualitative performance indicators for chip seal
- Define common visible chip seal distresses

Target Audience

This training is recommended for the Transportation Curriculum Coordination Council levels I, II and III. This training would benefit entry level construction inspectors, maintenance employees and contractor personnel as well as serve as refresher training for those already well versed in the selection and application of a chip seal as a preventive maintenance treatment.

Training Level: Basic

Fee: 2016: $25 Per Person; 2017: N/A

Length: 3 HOURS (CEU: 0 UNITS)

Class Size: MINIMUM: 1; MAXIMUM: 1

NHI Customer Service: (877) 558-6873 • nhicustomerservice@dot.gov

COURSE NUMBER

FHWA-NHI-131133

COURSE TITLE

Roller Compacted Concrete Pavements - WEB-BASED

The Roller Compacted Concrete (RCC) Pavements course provides detailed overviews of RCC properties and materials, mixture proportioning, structural design issues, and production and construction considerations, plus troubleshooting guidelines and an extensive reference list for more comprehensive information.

This course contains six modules. Module 1 is an introduction in RCC covering the characteristics, benefits, limitations, selection considerations, and typical uses. Module 2 discusses the property differences between RCC and conventional mixes, material requirements and testing. Module 3 covers mix proportioning of RCC, while Module 4 gets into structural design of RCC pavements. Module 5 acquaints the student with production and the proper handling and storage of materials, mixing and batching, and production planning. Module 6 covers the actual construction of a RCC pavement. All of the modules for this training were developed from the August 2010 "Guide for Roller-Compacted Concrete Pavements" which is available from the Portland Cement Association website www.cement.org/pavements.

OUTCOMES

Upon completion of the course, participants will be able to:

- Define RCC key elements and common uses
- Define RCC properties and materials
- Describe RCC mix proportioning
- Describe structural design of RCC pavement
- Identify RCC production
- Identify RCC pavement construction

TARGET AUDIENCE

This training provides agencies, contractors, materials suppliers, and others with a thorough introduction to and updated review of RCC and its many paving applications. This training is recommended for the Transportation Curriculum Coordination Council levels II through IV.

TRAINING LEVEL: Basic

FEE: 2016: $50 Per Person; 2017: N/A

LENGTH: 6 HOURS (CEU: 0 UNITS)

CLASS SIZE: MINIMUM: 1; MAXIMUM: 1

NHI Customer Service: (877) 558-6873 • nhicustomerservice@dot.gov

COURSE NUMBER

FHWA-NHI-131134

COURSE TITLE

Superpave for Construction - WEB-BASED

The Superpave for Construction Course contains information for field construction personnel on the Superpave mix design system and the control of field produced Hot Mix Asphalt.

There are two modules in this course. The first module introduces the Superpave Hot Mix Asphalt design testing and analysis. It will cover design testing procedures, design analysis methods, and will include calculations to analyze the volumetrics of paving samples. Module two includes relevant volumetric examples including the use of phase diagrams to calculate volumetric properties. Example problems are included. This course is an excellent learning tool to assist in understanding corrective actions for volumetric parameters.

OUTCOMES

Upon completion of the course, participants will be able to:

- Describe the benefits of Superpave over previous mix design methodologies
- Understand Superpave mix design procedures and testing
- Understand mix design analysis methods
- Perform the calculation necessary to analyze the volumetrics of paving samples for comparison
- Describe how to use phase diagrams to calculate volumetric properties
- Describe factors which can influence key mass-volume relationships and calculations
- Understand corrective action for volumetric parameters
- Calculate and evaluate volumetric properties through example problems

TARGET AUDIENCE

This training is targeted to intermediate and advanced technicians from both contractor and agency employment, which will be involved in construction of pavements using Superpave. This training is recommended for the Transportation Curriculum Coordination Council levels II and III.

TRAINING LEVEL: Basic

FEE: 2016: $25 Per Person; 2017: $25 Per Person

LENGTH: 3.5 HOURS (CEU: 0 UNITS)

CLASS SIZE: MINIMUM: 1; MAXIMUM: 1

NHI Customer Service: (877) 558-6873 • nhicustomerservice@dot.gov

Course Number

FHWA-NHI-131135

Course Title

Aggregate Sampling Basics - WEB-BASED

The Aggregate Sampling Basics course will cover the importance of proper sampling, why we need to sample aggregate, and why we need special procedures to do so. We will cover how to obtain a proper sample that will accurately represent the materials by utilizing sampling principles and preferred methods.

The specifications covered in the course are from the American Association of State Highway and Transportation Officials or AASHTO. The course starts at the beginning with what are aggregates, what are aggregate uses, and continues through proper sampling. It also has information on aggregate processing and sieving. The course contains interaction with the student and quizzes to make sure the material was understood.

Outcomes

Upon completion of the course, participants will be able to:

- Define aggregates
- Describe aggregate processing
- Describe aggregate sampling

Target Audience

This training is targeted to the beginning technician that will be obtaining aggregate samples for testing during production or on a project for agency, industry or consultant. This training is recommended for the Transportation Curriculum Coordination Council levels I and II.

Training Level: Basic

Fee: 2016: $25 Per Person; 2017: $25 Per Person

Length: 1 HOURS (CEU: 0 UNITS)

Class Size: MINIMUM: 1; MAXIMUM: 1

NHI Customer Service: (877) 558-6873 • nhicustomerservice@dot.gov

Course Number

FHWA-NHI-131136

Course Title

Materials Testing: Reducing Aggregate Samples - WEB-BASED

The Materials Testing and Reducing Aggregate Samples course will cover the two methods for splitting a sample; using a mechanical splitter and quartering. The purpose of these procedures is to reduce large samples of aggregate to the appropriate size for testing. The end product should be a sample that is representative of the source.

The American Association of State Highway and Transportation Officials or AASHTO procedures and specifications are used throughout the course. The course covers two methods used for splitting, the mechanical method and the quartering method. Both of these processes are covered in detail. There are questions for the students as a review of the material. References are given for further information.

Outcomes

Upon completion of the course, participants will be able to:

- Define aggregate reducing
- Describe the aggregate reducing method using mechanical splitter
- Describe the aggregate reducing method using quartering

Target Audience

This training is targeted to the beginning technician that will be reducing samples for testing using mechanical spitting and/or quartering for a contractor, producer, agency, or consultant. This training is recommended for the Transportation Curriculum Coordination Council levels I and II.

Training Level: Basic

Fee: 2016: $25 Per Person; 2017: $25 Per Person

Length: 1 HOURS (CEU: 0 UNITS)

Class Size: MINIMUM: 1; MAXIMUM: 1

NHI Customer Service: (877) 558-6873 • nhicustomerservice@dot.gov

WBT

Course Number

FHWA-NHI-131137

Course Title

Special Mixture Design Considerations and Methods for Warm Mix Asphalt - WEB-BASED

Highway transportation agencies are exploring the use of warm mix asphalt (WMA) for pavement projects. Because of the potential environmental and engineering benefits that WMA provides, agency and industry personnel want to learn the proper design considerations for a quality WMA mixture design. Mixture design technicians and engineers are particularly interested in design differences between WMA and HMA.

The Special Mixture Design Considerations and Methods for Warm Mix Asphalt course explains the key differences between WMA and HMA design procedures. Participants in this course compare important elements of the mixtures and review the effects of those elements on the final WMA product. Learners also have an opportunity to apply AASHTO R35 standard practice to a WMA design modification, converting an HMA mixture design to WMA.

Outcomes

Upon completion of the course, participants will be able to:

- Describe differences between warm mix asphalt (WMA) and hot mix asphalt (HMA) mixture design processes.
- Convert HMA mixtures to WMA mixtures.

Target Audience

This training was developed for experienced HMA mixture design technicians and engineers who are interested in using WMA. Participants should have basic computer skills, such as manipulating windows, using directories, and opening Web browsers.

Training Level: Basic

Fee: 2016: $25 Per Person; 2017: $25 Per Person

Length: 2 HOURS (CEU: 0 UNITS)

Class Size: MINIMUM: 1; MAXIMUM: 1

NHI Customer Service: (877) 558-6873 • nhicustomerservice@dot.gov

Course Number

FHWA-NHI-131138

Course Title

AASHTO Designation: T 308 - WEB-BASED

The TCCC AASHTO Designation: T308 course explains the importance of asphalt content, describes the equipment needed to perform the test procedure, shows how to perform the ignition furnace test procedure (both Method A - internal balance and Method B - external balance), and instructs how to calculate and apply the correction factors.

Some of the topics covered in this training include, background and purpose of asphalt content, apparatus, correction factors determination, test procedure, calculations, and wrap-up of the test procedure which includes reporting.

Upon completion of this course, participants will know why performing AASHTO T-308 is necessary, will know how to perform the test procedure, and can accurately calculate and apply the correction factors. This course is an excellent learning tool to demonstrate Asphalt Content by Ignition Oven to new technicians.

Outcomes

Upon completion of the course, participants will be able to:

- Explain the impact that asphalt binder content can have on a pavement
- Define the purpose of the ignition method, as well as the benefits and limitations of the test procedure
- Understand the basic concepts behind the test procedure
- Identify the equipment needed to perform the test procedure for both Method A and Method B
- Understand why correction factors must be determined
- Explain how to determine the asphalt binder correction factor
- Explain how to determine the aggregate correction factor
- Describe how the ignition test is performed for either Method A, Internal Balance Method or Method B, External Balance Method
- Calculate the measured (corrected) asphalt binder content percent for both Method A and Method B
- Reporting the test results
- Preparing sample for a gradation analysis according to AASHTO T 30

Target Audience

This training is designed for plant technicians, private lab, or contractor employees who are qualified to sample hot mix, aggregate or asphalt cement, and perform acceptance tests including Asphalt Content by Ignition Oven (AASHTO Designation: T 308-10). It is also useful for laboratory and personnel assessment technicians.

Training Level: Basic

Fee: 2016: $25 Per Person; 2017: $25 Per Person

Length: 2 HOURS (CEU: 0 UNITS)

Class Size: MINIMUM: 1; MAXIMUM: 1

NHI Customer Service: (877) 558-6873 • nhicustomerservice@dot.gov

Course Number

FHWA-NHI-131140

Course Title

Hot In-place Recycling

This training was developed by the Transportation Curriculum Coordination Council (TCCC) in partnership with AASHTO and NHI. Hot in-place recycling (HIR) is a pavement preservation and corrective maintenance technique that consists of heating and softening the existing asphalt pavement. When combined with an asphalt overlay, HIR can be classified as structural rehabilitation.

The HIR techniques described inthis training provide owner agencies with cost-effective and sustainable methods to repair their aging pavements. HIR processes have been used on all functional classes of roadways. When properly designed, specified, and constructed, HIR methods can result in significant cost savings as compared to conventional maintenance operations, while reducing carbon dioxide emissions.

This course contains three modules:

1. Introduction to Hot In-Place Recycling
2. Pre-Production Inspection
3. Full Production Pavement Recycling

Outcomes

Upon completion of the course, participants will be able to:

- Explain the purpose, benefits, and use of HIR;
- Identify the purpose and use of HIR designs and the equipment used for its applications;
- Identify the preparation and planning steps necessary for an HIR application; and
- Describe the production, evaluation, steps necessary for an HIR application.

Target Audience

This course is intended for local, county, and State owner agency technicians andinspectors. It is also useful for individuals who need awareness or basicunderstanding of hot in-place recycling. Training level: This training is recommended for the Transportation Curriculum Coordination Council levels I, II, III, and IV.

Training Level: Basic

Fee: 2016: $25 Per Course; 2017: N/A

Length: 2.5 HOURS (CEU: 0 UNITS)

Class Size: MINIMUM: 1; MAXIMUM: 1

NHI Customer Service: (877) 558-6873 • nhicustomerservice@dot.gov

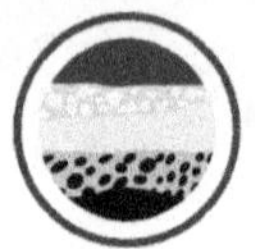

COURSE NUMBER

FHWA-NHI-132036

COURSE TITLE

Earth Retaining Structures

The goal of this course is to provide agencies with state-of-the-practice design tools and construction techniques to expand implementation of safe and cost-effective earth retention technologies. This course addresses the selection, design, construction, and performance of earth retaining structures used for support of fills and excavations or cut slopes. Instructors cover factors that affect wall selection, including contracting approaches with an emphasis on required bidding documents for each approach. Class discussions will include design procedures and case histories, demonstrating the selection, design, and performance of various earth retaining structures. Detailed information on subsurface investigation, soil and rock property design parameter selection, lateral earth pressures for wall system design, and load and resistance factor design (LRFD) for retaining walls are provided.

OUTCOMES

Upon completion of the course, participants will be able to:

- Describe potential applications for Earth Retaining Structures (ERS)
- Select a technically appropriate and cost-effective ERS
- Select appropriate material properties, soil design parameters, and earth pressure diagrams
- Perform design analysis and prepare conceptual designs
- Review contractor submitted documents
- Discuss contracting methods
- Describe construction and inspection activities for ERS

TARGET AUDIENCE

The primary audience for this course is agency and consultant bridge/structures, geotechnical, and roadway design engineers; engineering geologists; and consultant review specialists. In addition, management, specification, and contracting specialists and construction engineers involved in design and contracting aspects of retaining structures are encouraged to attend. Attendees should have a basic knowledge of soil mechanics and structural engineering, including some understanding of LRFD concepts.

TRAINING LEVEL: Intermediate

FEE: 2016: $925 Per Person; 2017: N/A

LENGTH: 3 DAYS (CEU: 1.8 UNITS)

CLASS SIZE: MINIMUM: 20; MAXIMUM: 30

NHI Customer Service: (877) 558-6873 • nhicustomerservice@dot.gov

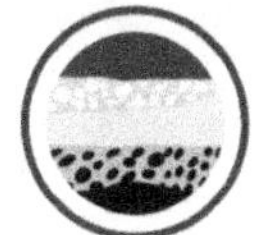

Course Number

FHWA-NHI-132040

Course Title

Geotechnical Aspects of Pavements

This course covers the latest methods and procedures to address the geotechnical issues in pavement design, construction, and performance for new construction, reconstruction, and rehabilitation pavement projects. The course content includes geotechnical exploration and characterization of in-place and constructed subgrades; design and construction of subgrades and unbound layers for paved and unpaved roads, with emphasis on the American Association of State Highway Transportation Officials (AASHTO) 1993 empirical design procedure and on the new Mechanistic-Empirical Pavement Design Guide (MEPDG); drainage of bases, subbases, and subgrades and its impact on providing safe, cost-effective, and durable pavements; problematic soils, soil improvement, stabilization, and other detailed geotechnical issues in pavement design and construction; and construction methods, specifications, and QC/QA (quality control/quality assurance) inspection for pavement projects.

The goal of the course is for each participant to recognize the importance of the geotechnical aspects relevant to the design, construction, and performance of a pavement system. Participants will develop an appreciation for the importance of adequate subsurface exploration and laboratory characterization of subgrade soils as well as the requisite pavement design parameters for subgrades, unbound base and subbase layers, including drainage features. The course is designed to elicit maximum input from participants, particularly regarding an understanding of the impact of geotechnical features on the long-term performance of pavement systems.

NOTE TO PARTICIPANT: Please bring a calculator that can perform trigonometric, log, and other engineering calculations, a note pad, and a pencil.

NOTE TO HOST: For this course, the host is asked to identify a state speaker to conduct a host state presentation. The presentation is usually on the first day of the class and lasts approximately 25 minutes with an additional 15 minutes of discussion. The objective of the presentation is to communicate the state's current practices and experience to the course participants. The state representative should have experience in geotechnical pavement activities. A detailed list of issues to be addressed in the host presentation will be provided. Also for this course, the host is asked to secure at least 6 laptop computers to be used during team exercises. The host can request that at least 6 participants bring their laptops to the course. The machines must have Microsoft Excel (Office 97 or later) and the optional Solver add-in tool installed. Lastly, the host state is asked to complete a "Questionnaire on Geotechnical Practices in Pavement Design" and provide policies and special provisions for (1.) obtaining subsurface information and laboratory testing in relation to pavement design, (2.) pavement design along with any agency design guides, (3.) field construction monitoring for subgrade approval and pavement component approval as well as contractors QC requirements for pavement component construction.

Outcomes

Upon completion of the course, participants will be able to:

- Explain the geotechnical parameters of interest in pavement design and their effects on the performance of different types of pavements
- Explain the influence of climate, moisture, and drainage on pavement performance
- Identify and explain the impact of unsuitable subgrades on pavement performance
- Determine the geotechnical inputs needed for design of pavements, both for the AASHTO 93 empirical design procedure and the new MEPDG
- Evaluate and select appropriate remediation measures for pavement subgrades
- Explain the geotechnical aspects of construction specifications and inspection requirements
- Identify subgrade problems during construction and develop recommended solutions

Target Audience

Many groups within an agency are involved with different aspects of definition, design, use, and construction verification of pavement geomaterials. These groups include pavement design engineers, geotechnical engineers, materials engineers, specification writers, and construction engineers who are or will be involved in the design, evaluation, and construction (or reconstruction or rehabilitation) of pavements. This course was developed as a forum for these various personnel to work together to enhance current procedures for building and maintaining more cost-efficient pavement

structures.

TRAINING LEVEL: Basic

FEE: 2016: $925 Per Person; 2017: N/A

LENGTH: 3 DAYS (CEU: 1.8 UNITS)

CLASS SIZE: MINIMUM: 20; MAXIMUM: 30

NHI Customer Service: (877) 558-6873 • nhicustomerservice@dot.gov

Course Number

FHWA-NHI-134001

Course Title

Principles and Applications of Highway Construction Specifications

Well-written highway construction specifications are those that can be interpreted accurately to minimize confusion and reduce owner-contractor disputes. Across the country, current practices, standards, and requirements for writing specifications are changing. Agencies also are using effective specifications to manage risk and support alternative contracting methods.

NHI 134001 Principles of Writing Highway Construction Specifications is a highly engaging, two-day, instructor-led training session. It includes content that highlights the role of specifications as contract documents and tools for assigning risk. Course participants engage in lessons and practice sessions to identify types of specifications, select the most appropriate type for a given project, and generate an original, effective highway construction specification.

This is not a grammar course; however, adequate course content emphasizes the use of basic grammar and writing style so that the learners can generate specifications that are correct, consistent, clear, complete, and concise.

Outcomes

Upon completion of the course, participants will be able to:

- Explain the purposes of a specification.
- Explain how specifications are used to assign risk and influence the behavior of different parties, within a given a scenario.
- Compare the functions of Standard and Supplemental Specifications with the functions of Special Provisions.
- Explain how the "order of precedence" affects writing specifications and preparing plans.
- Describe the purpose of the General Provisions.
- Explain how a consistent writing style can affect the interpretation of specifications.
- Complete a checklist of the information needed before writing or revising a specification.
- Explain the potential benefits of writing in the active voice.
- Rewrite passive voice sentences into the active voice.
- Evaluate specifications to determine the need for imperative or indicative mood.
- State the five Cs used in specification writing. (Note: the five Cs include: correct; consistent; clear; complete; concise.)
- Explain each element of the AASHTO five-part format.
- Identify potential ambiguities in the wording, given a sample specification.
- Identify the potential benefits of each of the five Cs, given a sample specification.
- Apply the five Cs and the host agency's preferred format to revise the specification, given a sample specification.
- Write a new specification to a given set of criteria using the five Cs and the host agency's preferred format, given a sample specification.
- Compare method versus end-result specifications.
- Relate the type of specification to the allocation of risk.
- Write an end-result specification to replace a method specification, given an excerpt from a method specification.

Target Audience

This course is designed primarily for individuals who write, review, and implement an agency's contract specifications. Participants might represent Federal, State, and local transportation agencies; other public agencies; contractors; and consultant firms.Individuals who do not write specifications but may contribute to their development, as well as those who use specifications, could also benefit from this course and the interaction with their classmates. Such participants might include personnel from environmental, materials, or construction sections or units; legal departments; work zone and safety professionals; contractor personnel; and any others involved with the design and construction of transportation facilities.

TRAINING LEVEL: Intermediate

FEE: 2016: $475 Per Person; 2017: $475 Per Person

LENGTH: 2 DAYS (CEU: 1.2 UNITS)

CLASS SIZE: MINIMUM: 20; MAXIMUM: 30

NHI Customer Service: (877) 558-6873 • nhicustomerservice@dot.gov

Course Number

FHWA-NHI-134061

Course Title

Construction Program Management and Inspection

The Federal Highway Administration's (FHWA) responsibilities for construction project and program oversight has changed considerably throughout the years. Today, the FHWA field engineers are typically involved in a diverse array of issues that were not common in construction projects of decades past. Changes in legislation, declines in staffing resources and expertise, and increased complexity of the Federal-aid construction program have all had an impact on how the FHWA conducts construction program management and oversight. Today's FHWA field engineers must have a more focused and programmatic approach in fulfilling construction stewardship and oversight responsibilities.

This 2-day training workshop highlights the FHWA roles and resources to assist the State in delivering a quality construction program. The training will assist the FHWA field engineers in maintaining and improving technical competence and in selecting a balanced program of construction management techniques.

The workshop uses the "Construction Program Management and Inspection Guide" as instructional material. While the workshop is focused primarily at FHWA's staff and FHWA oversight activities, participation by State partners and other relevant entities is highly encouraged to further educate and train Federal Aide partners to "act on FHWA's behalf in line with the Divisions/State DOT Stewardship Agreement.

Outcomes

Upon completion of the course, participants will be able to:

- Manage and oversee Federal-aid construction programs.

Target Audience

This training is targeted at FHWA Division field engineers and State agencies, and will provide staff with the background and knowledge they need for managing and overseeing their Federal-aid construction programs. The training is geared towards the new FHWA generalist employee but is also intended as a refresher for the veteran FHWA engineer.

Training Level: Basic

Fee: 2016: $335 Per Person; 2017: N/A

Length: 2 DAYS (CEU: 0 UNITS)

Class Size: MINIMUM: 15; MAXIMUM: 40

NHI Customer Service: (877) 558-6873 • nhicustomerservice@dot.gov

COURSE NUMBER

FHWA-NHI-134075

COURSE TITLE

Hardened Concrete Properties - Durability - WEB-BASED

This training is provided by the Transportation Curriculum Coordination Council (TCCC) in partnership with NHI to review integrated materials and construction practices for concrete pavement. The training was developed by the National Concrete Pavement Technology Center at Iowa State University. This training is recommended for the Transportation Curriculum Coordination Council levels II - IV. This course is primarily intended for inspectors and technicians.

Durability as a property of hardened concrete is essential for long-lasting pavements. This workshop discusses factors that contribute to durable concrete and covers permeability, frost resistance, sulfate resistance, alkali silica attack, and a brief look at abrasion resistance.

This module is part of a curriculum from the "Integrated Materials and Construction Practices for Concrete Pavement" manual developed through the National Concrete Pavement Technology Center at Iowa State University. The other Web-based training modules include:

FHWA-NHI-134084 TCCC Fundamentals of Materials Used for Concrete Pavements

FHWA-NHI-134085 TCCC Incompatibility in Concrete Pavement Systems

FHWA-NHI-134087 TCCC Mix Design Principles

FHWA-NHI-134095 TCCC Early Age Cracking

FHWA-NHI-134096 TCCC Basics of Cement Hydration

FHWA-NHI-134097 TCCC Fresh Concrete Properties

FHWA-NHI-134098 TCCC Construction of Concrete Pavements

FHWA-NHI-134100 TCCC QCQA for Concrete Pavements

FHWA-NHI-134101 TCCC Design of Pavement

FHWA-NHI-134102 TCCC Troubleshooting for Concrete Pavements

OUTCOMES

Upon completion of the course, participants will be able to:

- Recognize factors contribute to durable concrete
- Explain the importance of permeability, alkali-silica reaction, abrasion resistance and, in certain regions in the country, frost resistance and sulfate resistance of hardened concrete
- Identify tests that can be performed to determine the variables affecting the durability of hardened concrete

TARGET AUDIENCE

This training is designed for FHWA, State, and local agencies and their industry counterparts involved in the process to assure that concrete meets all the requirements for durability. It is applicable to anyone desiring a better understanding of the factors of durability.

TRAINING LEVEL: Basic

FEE: 2016: $25 Per Person; 2017: $25 Per Person

LENGTH: 1 HOURS (CEU: 0 UNITS)

CLASS SIZE: MINIMUM: 1; MAXIMUM: 1

NHI Customer Service: (877) 558-6873 • nhicustomerservice@dot.gov

Course Number

FHWA-NHI-134084

Course Title

Fundamentals of Materials Used for Concrete Pavements - WEB-BASED

This training is provided by the Transportation Curriculum Coordination Council (TCCC) in partnership with NHI to review integrated materials and construction practices for concrete pavement. It is the first training of its kind offered by NHI, and we would like to give special recognition to the TCCC for their efforts. This training is recommended for the Transportation Curriculum Coordination Council levels II - IV. This course is primarily intended for inspectors and technicians.

The materials used in Portland cement concrete play an extremely valuable role in the performance of the concrete. This training covers both the non-reactive and reactive materials used in Portland cement concrete. This would include the aggregates, curing compound, reinforcement, and the materials that are chemically reactive.

This module is part of a curriculum from the "Integrated Materials and Construction Practices for Concrete Pavement" manual developed through the National Concrete Pavement Technology Center at Iowa State University. The other Web-based training modules include:

FHWA-NHI-134075 TCCC Hardened Concrete Properties - Durability

FHWA-NHI-134085 TCCC Incompatibility in Concrete Pavement Systems

FHWA-NHI-134087 TCCC Mix Design Principles

FHWA-NHI-134095 TCCC Early Age Cracking

FHWA-NHI-134096 TCCC Basics of Cement Hydration

FHWA-NHI-134097 TCCC Fresh Concrete Properties

FHWA-NHI-134098 TCCC Construction of Concrete Pavements

FHWA-NHI-134100 TCCC QCQA for Concrete Pavements

FHWA-NHI-134101 TCCC Design of Pavement

FHWA-NHI-134102 TCCC Troubleshooting for Concrete Pavements

Outcomes

Upon completion of the course, participants will be able to:

- Identify materials used in Portland cement concrete
- Describe the importance of each material and the role it plays in the performance of the concrete
- Describe how each material reacts with the other materials to obtain strength, permeability, workability, etc.

Target Audience

This training is designed for FHWA, State, and local agencies and their industry counterparts involved in the process to assure that the materials used in Portland cement concrete meet specification requirements and are compatible to provide good, durable concrete. It is applicable to anyone desiring a better understanding of the materials used in Portland cement concrete.

Training Level: Intermediate

Fee: 2016: $25 Per Person; 2017: $25 Per Person

Length: 2 HOURS (CEU: 0 UNITS)

Class Size: MINIMUM: 1; MAXIMUM: 1

NHI Customer Service: (877) 558-6873 • nhicustomerservice@dot.gov

Course Number

FHWA-NHI-134085

Course Title

Incompatibility in Concrete Pavement Systems - WEB-BASED

This training is provided by the Transportation Curriculum Coordination Council (TCCC) in partnership with NHI to review integrated materials and construction practices for concrete pavement. It is the first training of its kind offered by NHI, and we would like to give special recognition to the TCCC for their efforts. This training is recommended for the Transportation Curriculum Coordination Council levels II - IV.

The materials used in Portland cement concrete play an extremely valuable role in the performance of the concrete. This training covers the incompatibilities of materials used in Portland cement concrete. Although certain materials may be perfectly acceptable on their own, when they are combined they are not compatible with each other. This can cause early stiffening, retardation, cracking, and the lack of a quality of air void system.

This module is part of a curriculum from the "Integrated Materials and Construction Practices for Concrete Pavement" manual developed through the National Concrete Pavement Technology Center at Iowa State University. The other Web-based training modules include:

FHWA-NHI-134075 TCCC Hardened Concrete Properties - Durability

FHWA-NHI-134084 TCCC Fundamentals of Materials Used for Concrete Pavements

FHWA-NHI-134087 TCCC Mix Design Principles

FHWA-NHI-134095 TCCC Early Age Cracking

FHWA-NHI-134096 TCCC Basics of Cement Hydration

FHWA-NHI-134097 TCCC Fresh Concrete Properties

FHWA-NHI-134098 TCCC Construction of Concrete Pavements

FHWA-NHI-134100 TCCC QCQA for Concrete Pavements

FHWA-NHI-134101 TCCC Design of Pavement

FHWA-NHI-134102 TCCC Troubleshooting for Concrete Pavements

Outcomes

Upon completion of the course, participants will be able to:

- Identify the causes of incompatible conditions leading to early stiffening or setting and occasional early age cracking
- Recognize the importance to use the correct air void system
- Describe test methods used to identify incompatibilities

Target Audience

This training is designed for FHWA, State, and local agencies and their industry counterparts involved in the process to assure that the materials used in Portland cement concrete meet specification requirements and are compatible to provide good, durable concrete. It is applicable to anyone desiring a better understanding of the materials used in Portland cement concrete.

Training Level: Intermediate

Fee: 2016: $25 Per Person; 2017: N/A

Length: 1 HOURS (CEU: 0 UNITS)

Class Size: MINIMUM: 1; MAXIMUM: 1

NHI Customer Service: (877) 558-6873 • nhicustomerservice@dot.gov

Course Number

FHWA-NHI-134087

Course Title

Mix Design Principles - WEB-BASED

This training is provided by the Transportation Curriculum Coordination Council (TCCC) in partnership with NHI to review integrated materials and construction practices for concrete pavement. This training is recommended for the Transportation Curriculum Coordination Council levels II - IV.

This module discusses mix design and mix proportioning. Mix design is the process of choosing the characteristics we are looking for in the concrete mixture. Mix proportioning, on the other hand, involves taking the information provided by the mix design process and using that information to determine the actual proportions of ingredients in the mixture. This course discusses theoretical, laboratory, and field testing to determine the Portland cement concrete mix that will achieve the best possible durability, strength, constructability, economy, and uniformity.

This module is part of a curriculum from the "Integrated Materials and Construction Practices for Concrete Pavement" manual developed through the National Concrete Pavement Technology Center at Iowa State University. The other Web-based training modules include:

FHWA-NHI-134075 TCCC Hardened Concrete Properties - Durability

FHWA-NHI-134084 TCCC Fundamentals of Materials Used for Concrete Pavements

FHWA-NHI-134085 TCCC Incompatibility in Concrete Pavement Systems

FHWA-NHI-134095 TCCC Early Age Cracking

FHWA-NHI-134096 TCCC Basics of Cement Hydration

FHWA-NHI-134097 TCCC Fresh Concrete Properties

FHWA-NHI-134098 TCCC Construction of Concrete Pavements

FHWA-NHI-134100 TCCC QCQA for Concrete Pavements

FHWA-NHI-134101 TCCC Design of Pavement

FHWA-NHI-134102 TCCC Troubleshooting for Concrete Pavements

Outcomes

Upon completion of the course, participants will be able to:

- Describe the overall goal of mix design
- Define the difference between mix design and mix proportioning
- Recognize field and laboratory testing plans
- Describe test methods used to identify incompatibilities

Target Audience

This training is designed for FHWA, State, and local agencies and their industry counterparts involved in the process to assure that the mix design and proportioning of Portland cement concrete materials meet specification requirements and provide good, durable concrete. It is applicable to anyone desiring a better understanding of the mix design of Portland cement concrete.

Training Level: Intermediate

Fee: 2016: $25 Per Person; 2017: $25 Per Person

Length: 1 HOURS (CEU: 0 UNITS)

Class Size: MINIMUM: 1; MAXIMUM: 1

NHI Customer Service: (877) 558-6873 • nhicustomerservice@dot.gov

COURSE NUMBER

FHWA-NHI-134095

COURSE TITLE

Early Age Cracking - WEB-BASED

This training is provided by the Transportation Curriculum Coordination Council (TCCC) in partnership with NHI to review integrated materials and construction practices for concrete pavement. The training was developed by the National Concrete Pavement Technology Center at Iowa State University. It is the first training of its kind offered by NHI, and we would like to give special recognition to the TCCC for their efforts. This training is recommended for the Transportation Curriculum Coordination Council levels II - IV. This course is primarily intended for inspectors and technicians.

Cracks are not a problem as long as they are controlled through jointing; ideally the concrete will crack below the saw joint to relieve the stress. Uncontrolled random cracks are not aesthetically acceptable and can reduce ride quality, durability, and particularly load transfer. Early cracking in this module is defined as those cracks that occur before the concrete is open to public traffic. In this module, we will be talking about early age cracking. Primarily, why does it occur and how can it be eliminated or at least controlled?

This module is part of a curriculum from the "Integrated Materials and Construction Practices for Concrete Pavement" manual developed through the National Concrete Pavement Technology Center at Iowa State University. The other Web-based training modules include:

FHWA-NHI-134075 TCCC Hardened Concrete Properties - Durability

FHWA-NHI-134084 TCCC Fundamentals of Materials Used for Concrete Pavements

FHWA-NHI-134085 TCCC Incompatibility in Concrete Pavement Systems

FHWA-NHI-134087 TCCC Mix Design Principles

FHWA-NHI-134096 TCCC Basics of Cement Hydration

FHWA-NHI-134097 TCCC Fresh Concrete Properties

FHWA-NHI-134098 TCCC Construction of Concrete Pavements

FHWA-NHI-134100 TCCC QCQA for Concrete Pavements

FHWA-NHI-134101 TCCC Design of Pavement

FHWA-NHI-134102 TCCC Troubleshooting for Concrete Pavements

OUTCOMES

Upon completion of the course, participants will be able to:

- Describe the various mechanisms that can lead to early age cracking
- Define and understand why curling and warping occur
- Recognize how curling and warping affect early age cracking
- Recognize the proper use of the materials and maintaining good construction practices can control early age cracking
- Describe how certain material properties and construction methods can affect early age cracking and can help prevent the cracking from occurring

TARGET AUDIENCE

This training is designed for FHWA, State, and local agencies and their industry counterparts involved in the process to assure that concrete meets all the requirements to prevent early age cracking. It is applicable to anyone desiring a better understanding of the causes and prevention of early age cracking.

TRAINING LEVEL: Intermediate

FEE: 2016: $25 Per Person; 2017: $25 Per Person

LENGTH: 1 HOURS (CEU: 0 UNITS)

CLASS SIZE: MINIMUM: 1; MAXIMUM: 1

NHI Customer Service: (877) 558-6873 • nhicustomerservice@dot.gov

COURSE NUMBER

FHWA-NHI-134096

COURSE TITLE

Basics of Cement Hydration - WEB-BASED

This training is provided by the Transportation Curriculum Coordination Council (TCCC) in partnership with NHI to review integrated materials and construction practices for concrete pavement. The training was developed by the National Concrete Pavement Technology Center at Iowa State University. This training is recommended for the Transportation Curriculum Coordination Council levels III and IV. This course is primarily intended for inspectors and technicians.

This module covers how a concrete mixture changes from a plastic state to become a solid concrete slab in a relatively short period of time. Central to this transformation is a complex process called hydration, an irreversible series of chemical reactions between water and cement.

This module is part of a curriculum from the "Integrated Materials and Construction Practices for Concrete Pavement" manual developed through the National Concrete Pavement Technology Center at Iowa State University. The other Web-based training modules include:

FHWA-NHI-134075 TCCC Hardened Concrete Properties - Durability

FHWA-NHI-134084 TCCC Fundamentals of Materials Used for Concrete Pavements

FHWA-NHI-134085 TCCC Incompatibility in Concrete Pavement Systems

FHWA-NHI-134087 TCCC Mix Design Principles

FHWA-NHI-134095 TCCC Early Age Cracking

FHWA-NHI-134097 TCCC Fresh Concrete Properties

FHWA-NHI-134098 TCCC Construction of Concrete Pavements

FHWA-NHI-134100 TCCC QCQA for Concrete Pavements

FHWA-NHI-134101 TCCC Design of Pavement

FHWA-NHI-134102 TCCC Troubleshooting for Concrete Pavements

OUTCOMES

Upon completion of the course, participants will be able to:

- Knowledge of physical and chemical occurrences during cement hydration
- Identify various factors that can adversely affect these occurrences
- Recognize the different temperature changes during particular stages of hydration

TARGET AUDIENCE

This training is designed for FHWA, State, and local agencies and their industry counterparts involved in the process to assure that the mix design and proportioning of Portland cement concrete materials meet specification requirements and provide good, durable concrete. It is applicable to anyone desiring a better understanding of the mix design of Portland cement concrete.

TRAINING LEVEL: Intermediate

FEE: 2016: $25 Per Person; 2017: N/A

LENGTH: 1 HOURS (CEU: 0 UNITS)

CLASS SIZE: MINIMUM: 1; MAXIMUM: 1

NHI Customer Service: (877) 558-6873 • nhicustomerservice@dot.gov

Course Number

FHWA-NHI-134097

Course Title

Fresh Concrete Properties - WEB-BASED

This training is provided by the Transportation Curriculum Coordination Council (TCCC) in partnership with NHI to review integrated materials and construction practices for concrete pavement. The training was developed by the National Concrete Pavement Technology Center at Iowa State University. This training is recommended for the Transportation Curriculum Coordination Council levels III and IV. This course is primarily intended for inspectors and technicians.

This module covers the properties of fresh concrete needed to produce high-quality, long lasting pavements and how to monitor these properties.

This module is part of a curriculum from the "Integrated Materials and Construction Practices for Concrete Pavement" manual developed through the National Concrete Pavement Technology Center at Iowa State University. The other Web-based training modules include:

FHWA-NHI-134075 TCCC Hardened Concrete Properties - Durability

FHWA-NHI-134084 TCCC Fundamentals of Materials Used for Concrete Pavements

FHWA-NHI-134085 TCCC Incompatibility in Concrete Pavement Systems

FHWA-NHI-134087 TCCC Mix Design Principles

FHWA-NHI-134095 TCCC Early Age Cracking

FHWA-NHI-134096 TCCC Basics of Cement Hydration

FHWA-NHI-134098 TCCC Construction of Concrete Pavements

FHWA-NHI-134100 TCCC QCQA for Concrete Pavements

FHWA-NHI-134101 TCCC Design of Pavement

FHWA-NHI-134102 TCCC Troubleshooting for Concrete Pavements

Outcomes

Upon completion of the course, participants will be able to:

- List the main properties of fresh concrete
- Describe what affects each property
- Recognize how to monitor these properties through concrete testing

Target Audience

This training is designed for FHWA, State, and local agencies and their industry counterparts involved in the process to assure that the properties of a concrete mixture provide ease in placement, ease of consolidation, and long lasting pavement. It is applicable to anyone desiring a better understanding of the properties of Portland cement concrete.

Training Level: Intermediate

Fee: 2016: $25 Per Person; 2017: N/A

Length: 1 HOURS (CEU: 0 UNITS)

Class Size: MINIMUM: 1; MAXIMUM: 1

NHI Customer Service: (877) 558-6873 • nhicustomerservice@dot.gov

COURSE NUMBER

FHWA-NHI-134101

COURSE TITLE

Design of Pavement - WEB-BASED

This training is provided by the Transportation Curriculum Coordination Council (TCCC) in partnership with NHI to review integrated materials and construction practices for concrete pavement. The training was developed by the National Concrete Pavement Technology Center at Iowa State University. This training is recommended for the Transportation Curriculum Coordination Council levels III and IV. This course is primarily for inspectors and technicians.

This module covers pavement design and subgrade concepts as they relate to materials and construction. It does not provide sufficient detail to actually design or evaluate a design. It covers the primary goal of pavement design, which is to provide a pavement with the following characteristics: safe, long lasting, cost effective, low maintenance, and constructible.

This module is part of a curriculum from the "Integrated Materials and Construction Practices for Concrete Pavement" manual developed through the National Concrete Pavement Technology Center at Iowa State University. The other Web-based training modules include:

FHWA-NHI-134075 TCCC Hardened Concrete Properties - Durability

FHWA-NHI-134084 TCCC Fundamentals of Materials Used for Concrete Pavements

FHWA-NHI-134085 TCCC Incompatibility in Concrete Pavement Systems

FHWA-NHI-134087 TCCC Mix Design Principles

FHWA-NHI-134095 TCCC Early Age Cracking

FHWA-NHI-134096 TCCC Basics of Cement Hydration

FHWA-NHI-134097 TCCC Fresh Concrete Properties

FHWA-NHI-134098 TCCC Construction of Concrete Pavements

FHWA-NHI-134100 TCCC QCQA for Concrete Pavements

FHWA-NHI-134102 TCCC Troubleshooting for Concrete Pavements

OUTCOMES

Upon completion of the course, participants will be able to:

- Identify pavement types and design features
- Recognize what design variables are controlled by field operations
- Discuss the two primary types of pavement distresses (performance measures)
- Recognize how subgrades and bases effect construction operations and long-term pavement performance

TARGET AUDIENCE

This training is designed for FHWA, State, and local agencies and their industry counterparts involved in designing, constructing, and inspecting Portland cement concrete pavements.

TRAINING LEVEL: Intermediate

FEE: 2016: $25 Per Person; 2017: $25 Per Person

LENGTH: 1 HOURS (CEU: 0 UNITS)

CLASS SIZE: MINIMUM: 1; MAXIMUM: 1

NHI Customer Service: (877) 558-6873 • nhicustomerservice@dot.gov

Course Number

FHWA-NHI-134109B

Course Title

Maintenance Training Series: Shaping and Shoulders - WEB-BASED

Shoulders play an important role in both pavement performance and roadway safety. Maintaining shoulders in a proper and timely manner is a primary goal of transportation agencies. In an effort to assist agencies in meeting this goal, the Shaping and Shoulders training provides information on the maintenance of both paved and unpaved shoulders, including specific details on the maintenance of gravel shoulders. This course is primarily intended for inspectors and technicians.

In addition to a discussion of the various types of shoulders, project selection considerations, and key maintenance issues, this training places shoulders and shaping into the context of an overall maintenance and pavement preservation program.

This training was developed as part of the Maintenance Training Series. To access all the trainings in the series, enroll in the 134109 course.

Outcomes

Upon completion of the course, participants will be able to:

- Identify desirable characteristics of various types of shoulders
- Identify project selection considerations for shaping and shoulders
- Describe shoulder shaping and blading activities, including equipment requirements and construction activities
- Describe how a shoulder and ditching program forms the core of the overall maintenance and pavement preservation program

Target Audience

This course is designed for State, regional, and county personnel who manage operations programs and deal with oversight and quality assurance across broad geographic areas. This target audience also is involved with handling materials, scheduling, budgeting, and planning.

Training Level: Basic

Fee: 2016: $25 Per Person; 2017: N/A

Length: 1.5 HOURS (CEU: 0 UNITS)

Class Size: MINIMUM: 0; MAXIMUM: 0

NHI Customer Service: (877) 558-6873 • nhicustomerservice@dot.gov

COURSE NUMBER

FHWA-NHI-134109C

COURSE TITLE

Maintenance Training Series: Thin HMA Overlays and Leveling - WEB-BASED

Thin HMA overlays and leveling are common pavement treatments and can be a central part of a maintenance crew's activities. During the Thin HMA Overlays and Leveling training, participants will be introduced to the characteristics and purposes of thin HMA overlays as well as the placement of leveling courses. Each of these techniques is capable of improving the functionality of an otherwise structurally sound pavement.

The training also covers information on the materials, personnel, and equipment needed for thin HMA overlays; items that should be considered when making project selection decisions; and guidance on proper mixture compaction. This information is designed to help participants improve project planning and execution for thin HMA overlays and leveling treatments.

This training was developed as part of the Maintenance Training Series. To access all the courses in the series, enroll in the 134109 course.

OUTCOMES

Upon completion of the course, participants will be able to:

- Determine the purpose of thin HMA overlays and leveling courses
- Identify material components of HMA overlays
- Identify personnel and equipment needed for HMA overlays and leveling construction
- Identify project selection considerations for thin HMA overlays and leveling
- Identify how this treatment can be incorporated into an overall system preservation program

TARGET AUDIENCE

This course is designed for State, regional, and county personnel who manage operations programs and deal with oversight and quality assurance across broad geographic areas. This target audience also is involved with handling materials, scheduling, budgeting, and planning.

TRAINING LEVEL: Basic

FEE: 2016: $25 Per Person; 2017: N/A

LENGTH: 1.5 HOURS (CEU: 0 UNITS)

CLASS SIZE: MINIMUM: 0; MAXIMUM: 0

NHI Customer Service: (877) 558-6873 • nhicustomerservice@dot.gov

Course Number

FHWA-NHI-134109D

Course Title

Maintenance Training Series: Base and Subbase Stabilization and Repair - WEB-BASED

Before preservation treatments can be applied, localized repairs may be necessary for a pavement's base or subbase. The Base and Subbase Stabilization and Repair course gives participants the knowledge they need to determine if the base or subbase must be stabilized or repaired, to select the appropriate stabilization and repair methods for a given project, and to ensure the repair is performed properly.

This training reviews the failures and distresses that indicate structural deterioration exists in a roadway. The course also covers project selection and trade-off considerations through example roadway projects that give participants the opportunity to evaluate a roadway and determine if it is a candidate for reconstruction or repair. Participants can use this information, as well as guidance on design and construction, to make sound project planning decisions.

This training was developed as part of the Maintenance Training Series. To access all the courses in the series, enroll in the 134109 course.

Outcomes

Upon completion of the course, participants will be able to:

- Identify the symptoms of a localized base or subbase problem, which require greater depth of stabilization and repair than a hot-mix asphalt (HMA) or portland cement concrete (PCC) surface repair patch
- Determine when it is appropriate to employ base or subbase repair on a preventive maintenance project
- Identify the most appropriate repair methods if base or subbase failures are identified in a project

Target Audience

This course is designed for State, regional, and county personnel who manage operations programs and deal with oversight and quality assurance across broad geographic areas. This target audience also is involved with handling materials, scheduling, budgeting, and planning.

Training Level: Basic

Fee: 2016: $25 Per Person; 2017: N/A

Length: 1 HOURS (CEU: 0 UNITS)

Class Size: MINIMUM: 0; MAXIMUM: 0

NHI Customer Service: (877) 558-6873 • nhicustomerservice@dot.gov

Course Number

FHWA-NHI-134109E

Course Title

Maintenance Training Series: Roadway Drainage - WEB-BASED

Shoulder, ditch, and pipe or culvert maintenance activities are performed frequently throughout the year. These activities are critical for avoiding hazardous roadway conditions and extending the life of pavements by controlling water flow along maintainable pathways. This course, Roadway Drainage, provides information on the purpose, function, and components of roadway drainage systems.

This course reviews the components of shoulders and ditches, the purpose of a roadway drainage inventory, and the permits used in roadway drainage maintenance. Examples of existing drainage inventories are provided. In addition, the benefits of proper water removal are discussed through examples of drainage system issues, such as ponding and washouts, in order to emphasize the connection between good drainage and roadway safety.

This training was developed as part of the Maintenance Training Series. To access all the courses in the series, enroll in the 134109 course.

Outcomes

Upon completion of the course, participants will be able to:

- Identify the purpose and function of roadway drainage systems
- Identify eight components of roadway drainage systems
- Identify the purpose of a roadway drainage inventory
- Identify the purpose of permits in roadway drainage maintenance
- Identify the components of shoulders and ditches

Target Audience

This course is designed for State, regional, and county personnel who manage operations programs and deal with oversight and quality assurance across broad geographic areas. This target audience also is involved with handling materials, scheduling, budgeting, and planning.

Training Level: Basic

Fee: 2016: $25 Per Person; 2017: N/A

Length: 1 HOURS (CEU: 0 UNITS)

Class Size: MINIMUM: 0; MAXIMUM: 0

NHI Customer Service: (877) 558-6873 • nhicustomerservice@dot.gov

Course Number

FHWA-NHI-132010A

WBT

Course Title

Earthquake Engineering Fundamentals (Web-based)

This 4-hour NHI training course 132010A entitled "Earthquake Engineering Fundamentals" is a Web-Based pre-requisite to the 2-day 132094A "LRFD Seismic Analysis and Design of Transportation Geotechnical Features" and 132094B "LRFD Seismic Analysis and Design of Structural Foundations and Earth Retaining Structures" courses. The participants will generally be notified to take the WBT course about 1 month before the ILT sessions and must complete this course before the start of the 132094A or 132094B course. The course consists of 6 lessons including: Earthquake Fundamentals (L1); Intro to LRFD Seismic Design (L2); Earthquake Ground Motions (L3); Seismic Hazard Analysis (L4); AASHTO Design Ground Motion Characterization (L5); and Intro to Geotechnical Hazards (L6).

Outcomes

Upon completion of the course, participants will be able to:

- Describe basic earthquake concepts
- Explain basic concepts of LRFD Seismic Design
- Describe earthquake ground motions
- Describe aspects of seismic hazard analysis
- Explain AASHTO design ground motion characterization
- Describe basic aspects of geotechnical hazards

Target Audience

This course is intended to engage a target audience of bridge and geotechnical engineers with zero and up to 20 years of experience that are preparing to attend the 132094A and 132094B Instructor-Led Training courses.

Training Level: Intermediate

Fee: 2016: $0 Per Person; 2017: N/A

Length: 4 HOURS (CEU: .4 UNITS)

Class Size: MINIMUM: 0; MAXIMUM: 0

NHI Customer Service: (877) 558-6873 • nhicustomerservice@dot.gov

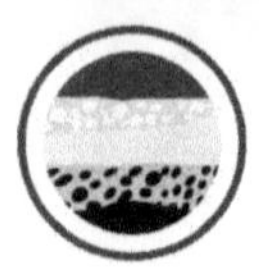

Course Number

FHWA-NHI-132010B

Course Title

Introduction to LRFD for Foundation Design

NHI-132010B Introduction to LRFD for Foundation and Substructure Design is a web-based training (WBT) course covering fundamental and basic principles related to Load and Resistance Factor Design (LRFD) for Highway Bridge Foundations. The course is developed to assist engineers in understanding the transition from Allowable Stress Design (ASD) to LRFD for structural foundations. Topics in this course include basic elements of LRFD development and implementation, principles of limit state design, loads and load combinations, soil and rock properties, and shallow and deep foundation design.

This WBT is designed to be both a stand-alone course that provides introductory information on LRFD for bridge foundations, and a prerequisite for attending NHI-132082 LRFD for Highway Bridge Substructure, a 4 day instructor-led course. NHI-132010B is a recommended prerequisite to NHI-132082 as well as other foundation design courses in the geotechnical curriculum.

Outcomes

Upon completion of the course, participants will be able to:

- Describe the development of LRFD and relationship to AASHTO.
- Identify the LRFD equation, limit states, LRFD design objectives, and foundation materials associated with LRFD.
- List loads, load combinations, and load factors associated with LRFD.
- Categorize soil and rock properties to provide a basis for determination of geotechnical resistance of soil and rock.
- Recognize shallow and deep foundation design by LRFD.

Target Audience

The target audience for NHI-132010B Introduction to LRFD for Foundation Design is individuals responsible for, or involved with, the design and construction of bridge substructures on surface transportation projects. Typically, the individuals will include an audience that is novice to LRFD, but has a background in bridge foundation design on surface transportation facilities such as geotechnical engineers, bridge and transportation engineers, geologists, and managers. This course is intended for those with general knowledge and/or skills with the bridge and/or geotechnical foundation and substructure design who desire to become familiar with LRFD.

Training Level: Basic

Fee: 2016: $50 Per Person; 2017: N/A

Length: 8 HOURS (CEU: .8 UNITS)

Class Size: MINIMUM: 0; MAXIMUM: 0

NHI Customer Service: (877) 558-6873 • nhicustomerservice@dot.gov

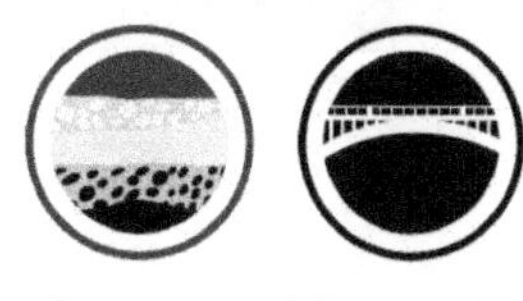

Course Number

FHWA-NHI-132012

Course Title

Soils and Foundations Workshop

This course is geared toward practicing design and construction engineers who routinely deal with soil and foundation problems but have little theoretical background in soil mechanics or foundation engineering. The course takes a project-oriented approach whereby the soils input to a bridge project is followed from conception to completion. In each phase of the project, the soil concepts will be developed into specific foundation designs and recommendations. The classroom presentation includes a variety of exercises to verify achievement of learning objectives. Each participant will take away a comprehensive reference manual on soils and foundations and a participant workbook containing a copy of all slides presented and completed exercises.

NOTE TO PARTICIPANT: All participants should bring calculators that perform trigonometric calculations, a note pad, and a pencil.

NOTE TO HOST: In addition to the typical host requirements of NHI courses, for this course the host is asked to arrange for the state's geotechnical engineering group to conduct a short presentation (usually on the second day of the course) summarizing the administrative and technical procedures followed by the host state.

Outcomes

Upon completion of the course, participants will be able to:

- Identifying the minimum level of geotechnical input in various project phases of a highway project
- Recalling the equipment and procedures used to implement a subsurface investigation of soil and rock conditions
- Demonstrating basic skills in visual description of soils native to the host state
- Recalling geotechnical facilities and personnel in the host state
- Recalling the basic soil test procedures and how the results of the various soil tests are applied results to highway projects
- Listing procedures used for both settlement and stability analysis, and recalling design solutions to stability and settlement problems for approach roadway embankments
- Listing procedures used for determining bearing capacity and settlement of shallow foundations such as spread footings
- Identifying the basic skills needed in the design and construction management of driven pile and drilled shaft foundations
- Recalling the driven pile and drilled shaft foundation construction equipment and construction inspection procedures
- Description static load testing and recalling the basic skills needed to interpret static load test results
- Recalling the basic skills needed in the design and construction of earth retaining structures
- Discussing the format and minimum content of an adequate foundation report

Target Audience

Personnel from the following units at the transportation agency could benefit from this workshop: geotechnical, bridge design, roadway design, materials, construction, and maintenance. The personnel who will benefit the most are the first-line supervisors involved in the design of highway structures and embankments. The greatest impact will be achieved by convincing structural, design, and construction engineers to use procedures from this course as a guide for routine geotechnical work. All attendees should be encouraged to attend the entire course, not just sections that are in their specialty. One of the major benefits of this course is to give engineers an appreciation of activities outside their specialties that influence, or are influenced by, the work of the geotechnical engineer.

TRAINING LEVEL: Basic

FEE: 2016: $1075 Per Person; 2017: N/A

LENGTH: 4 DAYS (CEU: 2.4 UNITS)

CLASS SIZE: MINIMUM: 20; MAXIMUM: 30

NHI Customer Service: (877) 558-6873 • nhicustomerservice@dot.gov

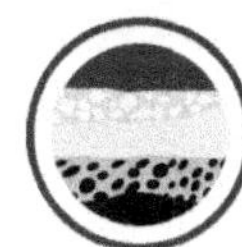

Course Number

FHWA-NHI-132013

Course Title

Geosynthetics Engineering Workshop (1-Day)

This course (1-day) provides training on construction with geosynthetics in transportation applications. The course examines the use of geotextiles, geogrids, pavement edge drains, drainage composites, erosion control materials and sediment control materials. Construction of filtration, drainage, temporary and permanent erosion control, sediment control, roadway separation, roadway reinforcement, roadway subgrade improvement, pavement overlays, embankments over soft foundations, mechanically stabilized earth walls, mechanically stabilized earth slopes applications are reviewed.

This 1-day construction summary course provides an introduction to geosynthetic installations.

Outcomes

Upon completion of the course, participants will be able to:

- Recognize geosynthetic applications for transportation facilities, construction and maintenance
- Identify types of geosynthetics and the functions they perform
- State and review general construction procedures and inspection items for geosynthetic installations
- Locate references on geosynthetic materials and geosynthetic applications

Target Audience

Federal, State and local transportation personnel (bridge, hydraulic, pavement, geotechnical, construction, and maintenance engineers, and construction inspectors and technicians) involved with construction and maintenance of transportation facilities that include earthwork construction.

Training Level: Basic

Fee: 2016: $625 Per Person; 2017: N/A

Length: 1 DAYS (CEU: .6 UNITS)

Class Size: MINIMUM: 20; MAXIMUM: 30

NHI Customer Service: (877) 558-6873 • nhicustomerservice@dot.gov

Course Number

FHWA-NHI-132013A

Course Title

Geosynthetics Engineering Workshop (3-Day)

This 3-day course, provides training on the appropriate, cost-effective utilization of geosynthetics in transportation applications. The course examines the use of geotextiles, geogrids, pavement edge drains, drainage composites, erosion control materials, sediment control materials, and geomembranes. In addition, instructors also cover applications of filtration, drainage, temporary and permanent erosion control, sediment control, roadway separation, roadway reinforcement, roadway subgrade improvement, pavement overlays, embankments over soft foundations, mechanically stabilized earth walls, mechanically stabilized earth slopes, geomembrane containment ponds, and geomembrane pavement encapsulation.

Outcomes

Upon completion of the course, participants will be able to:

- List six geosynthetic applications for transportation facilities, construction and maintenance
- Identify types of geosynthetics, and the functions they perform
- Discuss if geosynthetics are a feasible, cost-effective option for construction or maintenance of transportation works
- State and locate general construction procedures and inspection items for geosynthetic installations
- Locate references on geosynthetic materials and geosynthetic applications
- Prepare basic designs for filtration, drainage, temporary and permanent erosion control, sediment control, roadways, pavement overlays, embankments over soft foundations, mechanically stabilized earth walls, and reinforced earth slope transportation applications
- Select appropriate test methods for material properties and design parameters for specific geosynthetic projects, and differentiate between index and performance tests/properties
- Locate and review appropriate materials and construction specifications for geosynthetic projects
- Discuss the need for site specific monitoring or special inspection schemes

Target Audience

Federal, State and local transportation personnel (bridge, hydraulic, pavement, geotechnical, construction, and maintenance engineers, and construction inspectors and technicians) involved with design and/or construction of transportation facilities that include earthwork. In addition, public agency and private sector construction engineers and project inspectors responsible for installation, construction monitoring and inspection of geosynthetics installations can attend either course. There are no prerequisites, although prior attendance in FHWA-NHI-132012 Soils and Foundations Workshop is recommended.

Training Level: Basic

Fee: 2016: $925 Per Person; 2017: N/A

Length: 3 DAYS (CEU: 1.8 UNITS)

Class Size: MINIMUM: 20; MAXIMUM: 30

NHI Customer Service: (877) 558-6873 • nhicustomerservice@dot.gov

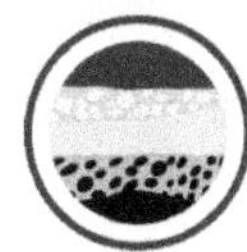

COURSE NUMBER

FHWA-NHI-132013B

COURSE TITLE

Geosynthetics Engineering Workshop - Hydraulics and Drainage (1.5-Day)

This 1.5-day provides training on the appropriate, cost-effective utilization of geosynthetics in hydraulic applications for transportation works. The course examines the use of geotextiles, pavement edge drains, drainage composites, erosion control materials, and sediment control materials. Applications of filtration, drainage, temporary and permanent erosion control, and sediment control are addressed. Geomembrane applications of flow barriers, containment ponds and pavement are summarized.

OUTCOMES

Upon completion of the course, participants will be able to:

- List four geosynthetic hydraulic applications for transportation works
- Identify three types of erosion control geosynthetics and the functions they perform
- Discuss feasibility and cost-effectiveness of geosynthetics in drainage and filtration applications
- State and review construction procedures for geosynthetic drainage, filtration, and erosion control installations
- Review design concepts and determine the basic design requirements for geosynthetics in conventional drains and erosion control applications
- Explain difference between temporary and permanent erosion control geosynthetics
- Select appropriate material property and design parameter test methods and prepare specification requirement for hydraulic applications of geosynthetics

TARGET AUDIENCE

Federal, State and local transportation personnel (hydraulic, erosion control, geotechnical, construction, and maintenance engineers, and construction inspectors and technicians) involved with design and/or construction and/or maintenance of transportation facilities that incorporate drainage and/or erosion control features. In addition, public agency and private sector construction engineers and project inspectors responsible for installation, construction monitoring and inspection of geosynthetic drainage and/or erosion control installations can attend either course. There are no prerequisites, although prior attendance of NHI course 132012 - Soils and Foundations Workshop is recommended.

TRAINING LEVEL: Basic

FEE: 2016: $700 Per Person; 2017: N/A

LENGTH: 1.5 DAYS (CEU: .9 UNITS)

CLASS SIZE: MINIMUM: 20; MAXIMUM: 30

NHI Customer Service: (877) 558-6873 • nhicustomerservice@dot.gov

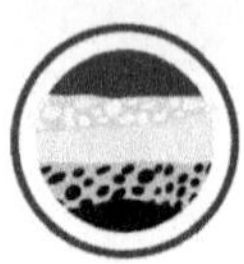

GEOTECHNICAL

Course Number

FHWA-NHI-132013C

Course Title

Geosynthetics Engineering Workshop - Roadways (1.5-Day)

Course 132013 C (1 1/2 day) provides training on the appropriate, cost-effective utilization of geosynthetics in roadway applications. The course examines the use of geotextiles, geogrids, and pavement edge drains in pavement structures. Applications of filtration, drainage, roadway separation, roadway reinforcement, roadway subgrade improvement, and pavement overlays are addressed.

Outcomes

Upon completion of the course, participants will be able to:

- List four geosynthetic pavement applications for transportation works
- Identify four types of geosynthetics used in pavement systems, and identify functions they perform
- Calculate if geosynthetics are a feasible, cost-effective option for construction or maintenance of pavements
- State and review construction procedures for geosynthetic pavement installations
- Review design concepts and determine the basic design requirements for geosynthetics in pavement systems
- Select appropriate material property and design parameter test methods and prepare specification requirement for geosynthetic reinforcement

Target Audience

Federal, State and local transportation personnel (pavement, geotechnical, construction, and maintenance engineers, and construction inspectors and technicians) involved with design and/or construction and/or maintenance of pavement systems. In addition, public agency and private sector construction engineers and project inspectors responsible for installation, construction monitoring and inspection of geosynthetics installations can attend either course. There are no prerequisites, although prior attendance in FHWA-NHI-132012 Soils and Foundations Workshop and FHWA-NHI-132040 Geotechnical Aspects of Pavements are recommended.

Training Level: Basic

Fee: 2016: $700 Per Person; 2017: N/A

Length: 1.5 DAYS (CEU: .9 UNITS)

Class Size: MINIMUM: 20; MAXIMUM: 30

NHI Customer Service: (877) 558-6873 • nhicustomerservice@dot.gov

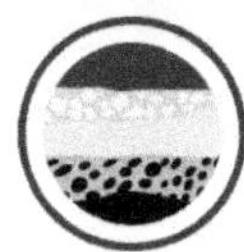

COURSE NUMBER

FHWA-NHI-132013D

COURSE TITLE

Geosynthetics Engineering Workshop - Reinforcement (1.5-Day)

This 1.5-day course provides training on the appropriate, cost-effective utilization of geosynthetics in transportation applications. The course examines the use of geotextiles and geogrids as soil reinforcement. Embankments over soft foundations, geosynthetic reinforced MSE walls, reinforced soil slopes, and geosynthetic reinforced load transfer platforms are addressed.

OUTCOMES

Upon completion of the course, participants will be able to:

- List three geosynthetic reinforcement applications for transportation works
- Identify four types of geosynthetic reinforcements, and discuss relative strengths and cost
- State and review construction procedures for geosynthetic reinforcement installations
- Identify primary design references for geosynthetic reinforcement applications for transportation works
- Review design concepts and determine the basic design requirements for embankments over soft foundations, mechanically stabilized earth walls and earth slopes transportation applications
- Select appropriate material property and design parameter test methods and prepare specification requirement for geosynthetic reinforcement

TARGET AUDIENCE

Federal, State and local transportation personnel (bridge, geotechnical, construction engineers, and construction inspectors and technicians) involved with design and/or construction of transportation facilities that incorporate reinforced soil earthworks. In addition, public agency and private sector construction engineers and project inspectors responsible for installation, construction monitoring and inspection of reinforced soil installations can attend either course. There are no prerequisites, although prior attendance in FHWA-NHI-132012 Soils and Foundations Workshop is recommended.

TRAINING LEVEL: Basic

FEE: 2016: $700 Per Person; 2017: N/A

LENGTH: 1.5 DAYS (CEU: .9 UNITS)

CLASS SIZE: MINIMUM: 20; MAXIMUM: 30

NHI Customer Service: (877) 558-6873 • nhicustomerservice@dot.gov

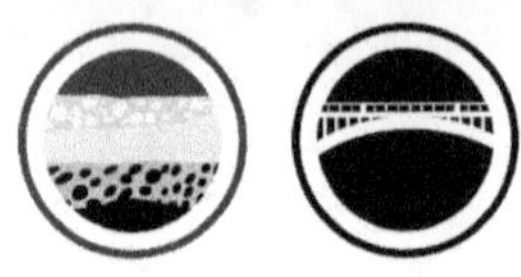

GEOTECHNICAL

COURSE NUMBER

FHWA-NHI-132014

COURSE TITLE

Drilled Shafts

Drilled shafts are an alternate type of deep foundation that may be more cost effective and perform better than other types of deep foundations in bridge piers at river crossings and in retrofit operations, high-mast lighting, earth retaining structures, single-column piers, and similar applications. This course provides participants with specific technical guidance on all aspects of designing, installing, and monitoring the construction of drilled shafts. The lessons address the following topics: applications, advantages, and disadvantages of drilled shafts for transportation structure foundations; general requirements for subsurface investigations; construction methods; construction case histories; construction specifications; principles of designing drilled shafts for axial and lateral loading; expansive soils, downdrag, and similar effects; load testing; inspection; integrity testing; repair and retrofit of defective shafts; and cost estimation. The participants will receive a comprehensive reference manual on drilled shaft construction and design used by engineers who perform detailed designs of drilled shafts, write construction specifications, and evaluate the performance of contractors through a comprehensive inspection program.

OUTCOMES

Upon completion of the course, participants will be able to:

- Describe the various drilling rigs and tools that are available to construct drilled shafts under varied subsurface soil and rock conditions
- Recognize the basic features of drilling aids, such as casings and drilling slurries, and the reasons for certain fundamental requirements for these aids
- Design drilled shafts for axial loading in simple soil and rock profiles
- Demonstrate a general understanding of the elements of designing drilled shafts for lateral loads
- Demonstrate an understanding of the need for load tests and available methods for performing the tests
- Formulate the basic elements of construction specifications for drilled shafts
- Demonstrate an understanding of integrity testing, repair, and retrofit of defective shafts
- Estimate costs for drilled shafts

TARGET AUDIENCE

The target audience for this course includes geotechnical engineers, bridge designers, and resident engineers. The course embraces both construction and design, and it is important that all participants attend all lessons, not just those in their immediate areas of interest. A key issue is how the details of construction affect the way in which a drilled shaft should be designed and how the intent of the design affects inspection. Participants are expected to have a degree in engineering for which they have passed an undergraduate course in soil mechanics and/or have successfully completed NHI course FHWA-NHI-132012 Soils and Foundations Workshop.

TRAINING LEVEL: Intermediate

FEE: 2016: $925 Per Person; 2017: N/A

LENGTH: 3 DAYS (CEU: 1.8 UNITS)

CLASS SIZE: MINIMUM: 20; MAXIMUM: 30

NHI Customer Service: (877) 558-6873 • nhicustomerservice@dot.gov

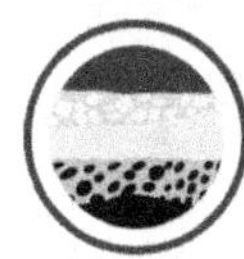

Course Number

FHWA-NHI-132033

Course Title

Soil Slope and Embankment Design and Construction

This course covers important aspects associated with the design and construction of soil slopes and embankments. It is intended to provide transportation earthwork professionals with knowledge to recognize potential problems with soil slope/embankment stability and deformation in transportation projects. Participants will develop the skills necessary to design and evaluate soil slopes and embankments and learn about the implications for construction and inspections. The course embraces both design and construction.

Participants will receive a comprehensive reference manual, used by practicing highway and geotechnical engineers covering investigation, design, construction, and mitigation of soil slopes and embankments. The participant workbook contains copies of visual aids and student exercises that closely follow the PowerPoint slide presentations. The participant exercises promote interaction in the classroom and illustrate the basic principles and analyses.

Outcomes

Upon completion of the course, participants will be able to:

- Recognize potential failure modes or deformation types for soil slopes and embankments
- Identify the potential failure modes for soil slopes and the type of analysis required to evaluate stability of the slope
- Determine the stability of a slope using slope stability charts
- Recognize the major design consideration for embankments constructed using earth fill, rock fill, and lightweight fill
- List the steps necessary for designing an embankment over compressible foundation soil
- List the common causes/triggering mechanisms for landslides/slope instabilities
- List appropriate stabilization methods

Target Audience

The target audience is bridge, geotechnical, or transportation engineers with 0 to 20 years of experience and responsible for the design, analysis, and construction maintenance or remediation of soil slopes and embankments on surface transportation facilities.

Training Level: Intermediate

Fee: 2016: $850 Per Person; 2017: N/A

Length: 2.5 DAYS (CEU: 1.5 UNITS)

Class Size: MINIMUM: 20; MAXIMUM: 30

NHI Customer Service: (877) 558-6873 • nhicustomerservice@dot.gov

Course Number

FHWA-NHI-132035

Course Title

Rock Slopes

This course presents geological investigation techniques, shear strength theories for determining rock strength, and design methods for rock slopes with different failure mechanisms. Other topics include rock blasting, rock slope stabilization methods, and contracting issues. Classroom instruction includes the discussion of sample problems and case histories involving rock slope analyses and designs.

Participants will receive a comprehensive reference manual (FHWA-NHI-99-007) and the accompanying exercises (FHWA-NHI-99-036). The reference manual covers investigation, design, and construction of rock slopes for highway/geotechnical engineers. It is geared towards practicing engineers who are involved with rock slope design and stabilization, but may not have the complete theoretical background. The exercises (FHWA-NHI-99-036) are designed to promote interaction in the classroom and to illustrate the basic principles and analyses. Solutions are included with each exercise.

Outcomes

Upon completion of the course, participants will be able to:

- Describe the basic principles of rock slope design
- Plan and execute a geological investigation, including geologic mapping
- Perform appropriate in situ and laboratory strength tests
- Determine rational design parameters by proper evaluation of in situ and laboratory test data along with appropriate rock strength correlations
- Identify the failure mechanisms associated with rock slopes and apply appropriate design methodologies
- Design effective rockfall protection and slope stabilization measures
- Design a monitoring program for cut slopes

Target Audience

The target audience for this course includes FHWA, State, and local highway agency employees; college and university faculty; and consultant engineers/geologists who are or will be involved in the design, excavation, and stabilization of rock slopes. An undergraduate degree in geology, engineering geology, civil engineering, or equivalent engineering experience in the highway/transportation field is desirable.

Training Level: Intermediate

Fee: 2016: $775 Per Person; 2017: N/A

Length: 2 DAYS (CEU: 1.2 UNITS)

Class Size: MINIMUM: 20; MAXIMUM: 30

NHI Customer Service: (877) 558-6873 • nhicustomerservice@dot.gov

Course Number

FHWA-NHI-132036

Course Title

Earth Retaining Structures

The goal of this course is to provide agencies with state-of-the-practice design tools and construction techniques to expand implementation of safe and cost-effective earth retention technologies. This course addresses the selection, design, construction, and performance of earth retaining structures used for support of fills and excavations or cut slopes. Instructors cover factors that affect wall selection, including contracting approaches with an emphasis on required bidding documents for each approach. Class discussions will include design procedures and case histories, demonstrating the selection, design, and performance of various earth retaining structures. Detailed information on subsurface investigation, soil and rock property design parameter selection, lateral earth pressures for wall system design, and load and resistance factor design (LRFD) for retaining walls are provided.

Outcomes

Upon completion of the course, participants will be able to:

- Describe potential applications for Earth Retaining Structures (ERS)
- Select a technically appropriate and cost-effective ERS
- Select appropriate material properties, soil design parameters, and earth pressure diagrams
- Perform design analysis and prepare conceptual designs
- Review contractor submitted documents
- Discuss contracting methods
- Describe construction and inspection activities for ERS

Target Audience

The primary audience for this course is agency and consultant bridge/structures, geotechnical, and roadway design engineers; engineering geologists; and consultant review specialists. In addition, management, specification, and contracting specialists and construction engineers involved in design and contracting aspects of retaining structures are encouraged to attend. Attendees should have a basic knowledge of soil mechanics and structural engineering, including some understanding of LRFD concepts.

Training Level: Intermediate

Fee: 2016: $925 Per Person; 2017: N/A

Length: 3 DAYS (CEU: 1.8 UNITS)

Class Size: MINIMUM: 20; MAXIMUM: 30

NHI Customer Service: (877) 558-6873 • nhicustomerservice@dot.gov

GEOTECHNICAL

COURSE NUMBER

FHWA-NHI-132037

COURSE TITLE

Spread Footings: LRFD Design and Construction

This updated course (January 2012) replaces NHI training 132037 Shallow Foundations, which was developed in 2001. Designed in accordance with the AASHTO Bridge LRFD Specification, 5th Ed., 2010, the course describes basic principles and state-of-the-practice analysis, load and resistance factor design (LRFD) procedures, and construction procedures for shallow foundations in soil and rock with particular application to transportation facilities.

The main topics covered are LRFD procedures for spread footings, vertical stress distribution, tolerable settlement criteria, settlement criteria in coarse-grained soils, settlement in fine-grained soils, time rate of consolidation settlement in fine-grained soils, bearing resistance in soil and rock, sliding resistance in soil and rock, problematic soils and ground improvement techniques, and inspection and construction monitoring methods for spread footings. Group exercises are interspersed throughout the course, enabling participants to be actively involved in the learning experience. This course provides FHWA recommended technical guidance in accordance with standard of practice for design and construction of spread footings for LRFD.

OUTCOMES

Upon completion of the course, participants will be able to:

- List the key steps in the design of spread footings
- Identify common significant construction points for an abutment footing
- Compute settlement of spread footings at service limit states
- Use the bearing capacity equation to evaluate the nominal bearing resistance of spread footings at strength limit state
- Recognize the impact of each parameter on the calculated bearing resistance
- Estimate nominal bearing resistance for spread footings on rock
- List the ground improvement techniques that may be used to improve problematic soils
- Identify key geotechnical construction activities as they relate to spread footings

TARGET AUDIENCE

This course is intended for geotechnical professionals, foundation designers, highway/bridge engineers, and project engineers who are involved in the analysis, design, construction, and maintenance of spread footings for surface transportation facilities. The target audience may also include FHWA, State, and local highway agency employees, college and university faculty, and consultant engineers/geologists who are or will be involved in the research, design, construction and maintenance of spread footings/shallow foundations.

TRAINING LEVEL: Intermediate

FEE: 2016: $775 Per Person; 2017: N/A

LENGTH: 2 DAYS (CEU: 1.2 UNITS)

CLASS SIZE: MINIMUM: 20; MAXIMUM: 30

NHI Customer Service: (877) 558-6873 • nhicustomerservice@dot.gov

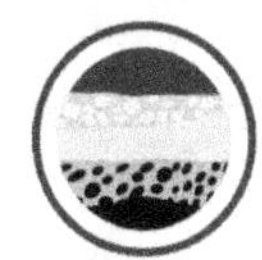  

COURSE NUMBER

FHWA-NHI-132040

COURSE TITLE

Geotechnical Aspects of Pavements

This course covers the latest methods and procedures to address the geotechnical issues in pavement design, construction, and performance for new construction, reconstruction, and rehabilitation pavement projects. The course content includes geotechnical exploration and characterization of in-place and constructed subgrades; design and construction of subgrades and unbound layers for paved and unpaved roads, with emphasis on the American Association of State Highway Transportation Officials (AASHTO) 1993 empirical design procedure and on the new Mechanistic-Empirical Pavement Design Guide (MEPDG); drainage of bases, subbases, and subgrades and its impact on providing safe, cost-effective, and durable pavements; problematic soils, soil improvement, stabilization, and other detailed geotechnical issues in pavement design and construction; and construction methods, specifications, and QC/QA (quality control/quality assurance) inspection for pavement projects.

The goal of the course is for each participant to recognize the importance of the geotechnical aspects relevant to the design, construction, and performance of a pavement system. Participants will develop an appreciation for the importance of adequate subsurface exploration and laboratory characterization of subgrade soils as well as the requisite pavement design parameters for subgrades, unbound base and subbase layers, including drainage features. The course is designed to elicit maximum input from participants, particularly regarding an understanding of the impact of geotechnical features on the long-term performance of pavement systems.

NOTE TO PARTICIPANT: Please bring a calculator that can perform trigonometric, log, and other engineering calculations, a note pad, and a pencil.

NOTE TO HOST: For this course, the host is asked to identify a state speaker to conduct a host state presentation. The presentation is usually on the first day of the class and lasts approximately 25 minutes with an additional 15 minutes of discussion. The objective of the presentation is to communicate the state's current practices and experience to the course participants. The state representative should have experience in geotechnical pavement activities. A detailed list of issues to be addressed in the host presentation will be provided. Also for this course, the host is asked to secure at least 6 laptop computers to be used during team exercises. The host can request that at least 6 participants bring their laptops to the course. The machines must have Microsoft Excel (Office 97 or later) and the optional Solver add-in tool installed. Lastly, the host state is asked to complete a "Questionnaire on Geotechnical Practices in Pavement Design" and provide policies and special provisions for (1.) obtaining subsurface information and laboratory testing in relation to pavement design, (2.) pavement design along with any agency design guides, (3.) field construction monitoring for subgrade approval and pavement component approval as well as contractors QC requirements for pavement component construction.

OUTCOMES

Upon completion of the course, participants will be able to:

- Explain the geotechnical parameters of interest in pavement design and their effects on the performance of different types of pavements
- Explain the influence of climate, moisture, and drainage on pavement performance
- Identify and explain the impact of unsuitable subgrades on pavement performance
- Determine the geotechnical inputs needed for design of pavements, both for the AASHTO 93 empirical design procedure and the new MEPDG
- Evaluate and select appropriate remediation measures for pavement subgrades
- Explain the geotechnical aspects of construction specifications and inspection requirements
- Identify subgrade problems during construction and develop recommended solutions

TARGET AUDIENCE

Many groups within an agency are involved with different aspects of definition, design, use, and construction verification of pavement geomaterials. These groups include pavement design engineers, geotechnical engineers, materials engineers, specification writers, and construction engineers who are or will be involved in the design, evaluation, and construction (or reconstruction or rehabilitation) of pavements. This course was developed as a forum for these various personnel to work together to enhance current procedures for building and maintaining more cost-efficient pavement

structures.

TRAINING LEVEL: Basic

FEE: 2016: $925 Per Person; 2017: N/A

LENGTH: 3 DAYS (CEU: 1.8 UNITS)

CLASS SIZE: MINIMUM: 20; MAXIMUM: 30

NHI Customer Service: (877) 558-6873 • nhicustomerservice@dot.gov

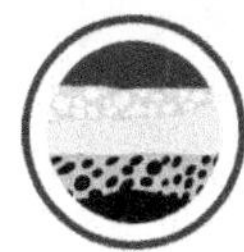

Course Number

FHWA-NHI-132041

Course Title

Geotechnical Instrumentation

The course is designed to provide participants with the necessary knowledge and skills to plan, select, and implement instrumentation programs in geotechnical features for construction monitoring and performance verification. The course will discuss measurement tools, including recommendations for a systematic and complete approach to planning monitoring programs. The course presents recommendations for selecting proper instrumentation for various types of construction. Tasks covered include calibration, maintenance and installation of instrumentation, collection of data, processing and presentation of collected data, interpretation of processed data, and reporting of results.

Outcomes

Upon completion of the course, participants will be able to:

- Recognize effective uses of geotechnical instrumentation in transportation projects
- Identify benefits of instrumentation
- Identify typical instrumentation programs for common transportation structures
- Recognize role of instrumentation and how it is used for answering key geotechnical questions
- Identify available instruments and where to find additional information and assistance
- Plan an instrumentation program in a systematic way
- Examine practical methods to collect and use data from instrumentation
- Perform an evaluation of the need for and potential benefits of geotechnical instrumentation on a project

Target Audience

The target audience is bridge, geotechnical, or transportation engineers with 0 to 20 years of experience and responsible for the planning, design, and implementation of instrumentation programs for geotechnical construction on surface transportation facilities.

Training Level: Intermediate

Fee: 2016: $775 Per Person; 2017: N/A

Length: 2 DAYS (CEU: 1.2 UNITS)

Class Size: MINIMUM: 20; MAXIMUM: 30

NHI Customer Service: (877) 558-6873 • nhicustomerservice@dot.gov

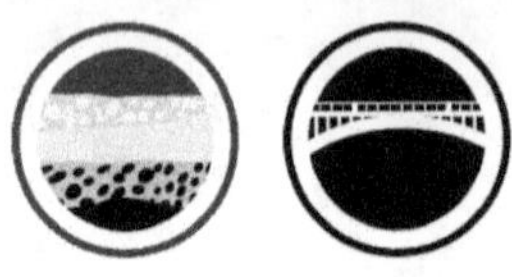

GEOTECHNICAL

Course Number

FHWA-NHI-132042

Course Title

Design of Mechanically Stabilized Earth Walls and Reinforced Soil Slopes

Mechanically stabilized earth walls (MSEWs) are commonly used on roadway projects and are typically cost effective and aesthetically pleasing. The basic concept behind MSEWs is to combine soil, reinforcing materials made of steel or polymers, and appropriate facing to produce a composite system with engineering properties that are ideal for most roadway applications. Reinforced soil slopes (RSS) utilize the same types of reinforcement for the construction of steep embankments. Both MSEWs and RSS structures can provide substantial savings in construction time and costs when compared with other types of earth retaining systems.

The goal of the course is to educate agencies about state-of-the-practice design tools. This includes comprehensive instruction on the design of MSEWs using load resistance factor design (LRFD). The course also presents construction practices to promote implementation of mechanically stabilized earth technology in cost effective earth retention structures. This course would most benefit persons who are involved in the design and construction of earth retention structures for surface transportation projects.

NOTE TO PARTICIPANT: Please bring a calculator that performs trigonometric calculations, a note pad, and a pencil.

NOTE TO HOST: In addition to the typical host requirements of NHI courses, for this course the host state technical contact is asked to bring 30 copies of the standard MSE wall and the RSS specifications (or special provisions), a complete set of applicable state DOT state construction specifications, standard plates, standard details, inspection guidelines, etc. pertaining to earth retaining structures. Copies should be forwarded to the instructors a month before the course. The host agency is also asked to provide approximately 20-25 pounds of dry sand. About 1/2 bag of "play" sand from a hardware store will suffice.

Outcomes

Upon completion of the course, participants will be able to:

- Recognize potential applications for MSEWs and RSS structures in transportation facilities
- Prepare conceptual and basic (i.e., for simple geometry) designs, and be able to check contractor-submitted designs for walls and slopes
- Examine and select appropriate material properties and parameters used in design
- Calculate the cost of conceptual MSEWs and RSS structures and determine if construction is a cost-effective option
- Select appropriate specification/contracting method(s) and prepare detailed specifications for materials and methods of construction
- Define and communicate major components of construction inspection of MSEWs and RSS structures to confirm compliance with design

Target Audience

The primary audience for this course is agency and consultant bridge/structures, geotechnical, and roadway design engineers; engineering geologists; and consultant review specialists. In addition, management, specification and contracting specialists, and construction engineers interested in design and contracting aspects of MSEWs and RSS structures are encouraged to attend. Attendees should have a basic knowledge of soil mechanics and structural engineering. (Note that NHI offers a 1-day course, FHWA-NHI-132043 Construction of MSEW and RSS.)

Training Level: Intermediate

Fee: 2016: $925 Per Person; 2017: N/A

Length: 3 DAYS (CEU: 1.8 UNITS)

Class Size: MINIMUM: 20; MAXIMUM: 30

NHI Customer Service: (877) 558-6873 • nhicustomerservice@dot.gov

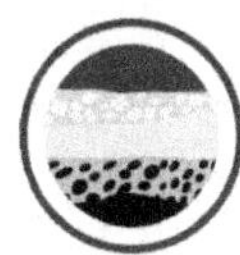

Course Number

FHWA-NHI-132043

Course Title

Construction of Mechanically Stabilized Earth Walls and Reinforced Soil Slopes

This course presents the concepts of mechanically stabilized earth wall (MSEW) and reinforced soil slope (RSS) systems and their application to roadways. The construction materials for both systems are described and guidance on acceptance for use is given. MSEW and RSS system construction steps are taught and typical construction practices and techniques are presented.

Outcomes

Upon completion of the course, participants will be able to:

- Recognize potential applications for MSEWs and RSS structures in transportation facilities
- Recognize differences between available systems and their components
- Understand the intent of specification/contracting method(s)
- Define and communicate major components of construction inspection of MSEWs and RSS structures to confirm compliance with design
- Understand the steps for MSEW and RSS construction and the corresponding points for inspection

Target Audience

The primary audience for this course is agency and consultant construction engineers, inspectors, and technicians. In addition, management; specification and contracting specialists; bridge/structures, geotechnical, and roadway design engineers; and engineering geologists interested in construction aspects of MSEWs and RSS structures are encouraged to attend. Attendees should have a basic knowledge of soil mechanics and structural engineering. (Note that NHI offers a 3-day course, FHWA-NHI-132042 Design of MSEWs and RSSs and a 3-day course, FHWA-NHI-132080 Inspection of MSEWs and RSSs.)

Training Level: Intermediate

Fee: 2016: $625 Per Person; 2017: N/A

Length: 1 DAYS (CEU: .6 UNITS)

Class Size: MINIMUM: 20; MAXIMUM: 35

NHI Customer Service: (877) 558-6873 • nhicustomerservice@dot.gov

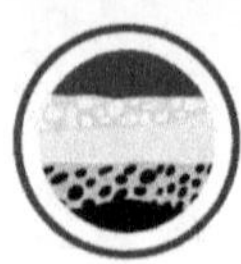

Course Number

FHWA-NHI-132069

Course Title

Driven Pile Foundation Inspection

This course provides Federal, State, and local inspectors with practical knowledge and standard industry practices for inspecting pile-driving operations at transportation construction sites.

To establish a national standard for transportation personnel, NHI developed the course based on a number of Federal and State sources: the course materials from the Florida Department of Transportation's Pile Driving Inspector's Qualification test, AASHTO's 2000 Bridge Construction Specifications, and the NHI courses Driven Pile Foundations - Design and Construction (FHWA-NHI-132021) and Driven Pile Foundations - Construction Monitoring (FHWA-NHI-132022). However, the local specifications, inspection reports, and plan sheets available from the hosting agency also will be discussed. The course includes a 3-hour qualification examination.

Outcomes

Upon completion of the course, participants will be able to:

- Explain the inspector's role, duties, and responsibilities
- Describe the pile-driving system components
- Recognize key inspection elements of the contract documents
- Identify proper communication and coordination with the engineer and contractor
- Identify the key elements of a pile installation plan
- Recognize and identify pile-driving system components and tools
- Verify tip elevations, cutoff elevations, pile penetration, and length driven for vertical and battered piles
- Perform inspection of pile-driving operations and verify compliance with construction tolerances
- Recognize when to stop driving based upon provided driving criteria, minimum tip or penetration, and refusal guidelines.
- Verify pile condition, labeling, and marking for compliance
- Recognize and explain the difference between test piles and production piles and the various types of pile testing
- Identify "driving" irregularities
- Identify and document pay quantities
- Interpret and apply applicable AASHTO specifications relating to foundation acceptance
- List potential problems and safety issues

Target Audience

The target audience for this course includes those who inspect pile-driving operations during construction of foundations and major structures. In addition, project management and construction engineers in charge of pile-driving construction inspections are encouraged to attend. Attendees should have completed courses in basic courses in reading construction plans as well as construction math and high school algebra.

Training Level: Intermediate

Fee: 2016: $850 Per Person; 2017: N/A

Length: 2.5 DAYS (CEU: 1.5 UNITS)

Class Size: MINIMUM: 20; MAXIMUM: 35

NHI Customer Service: (877) 558-6873 • nhicustomerservice@dot.gov

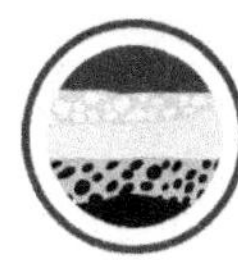

Course Number

FHWA-NHI-132070

Course Title

Drilled Shaft Foundation Inspection

Drilled Shaft Foundation Inspection is a stand-alone course developed to provide a basis for local, regional, or national qualification of drilled shaft foundation inspectors. The goal of this course is to provide inspectors with practical knowledge and standard industry practices for the inspection of drilled shaft foundation construction. A 2-hour qualification exam is administered on the third day of the course.

The course follows recommended FHWA specifications and practices for drilled shaft construction but may be modified to follow local agency specifications and practices.

NOTE TO PARTICIPANT: All participants should be advised by the local coordinator/session host that they are encouraged to complete NHI 132070B Drilled Shaft Inspector Tutorial (WBT). All participants should also be advised to bring a calculator that performs basic math, in particular works with negative numbers, a built in Pi function, and square root functions.

NOTE TO HOST: This course requires participation of a host agency technical representative. The objective of the host agency technical representative is to communicate the state's or region's current practices and experience to the course participants. The technical representative is asked to provide the instructor with a complete set of applicable state DOT construction specifications, standard plates, standard details, set of typical plans including boring sheets, a technical special provision, inspection guidelines and state standard Drilled Shaft Inspection Reporting and Recording forms. Also, the host agency is asked to provide seven duplicate sets of local soil and rock samples, one rock core, one set of rock cuttings, a full set of slurry testing equipment, a variety of spacers and standoffs used locally, and thirty sets of typical plans.

Outcomes

Upon completion of the course, participants will be able to:

- Identify and understand the role and duties of the inspector
- Recognize key inspection elements of the contract documents
- Identify proper communication and coordination with the engineer and contractor
- Interpret and verify contractor compliance with items in the drilled shaft installation plan
- Recognize and identify drilled shaft construction equipment and tools
- Perform visual field verification of soil/rock material for comparison to supplied soil boring data/logs
- Calculate percent recovery and rock quality designation (RQD)
- Recognize and identify the various types of drilled shaft construction
- Perform inspection of drilled shaft excavations for compliance with plans, construction tolerances, and cleanliness
- Verify reinforcing cage construction compliance including side spacers and cross-hole sonic logging (CSL) tubes
- Determine concrete volumes for theoretical shafts and develop concrete curves
- Identify shaft "concreting" irregularities
- Perform calculations for volume, area, circumference, and elevation
- Locate, explain, and apply applicable FHWA, AASHTO, and State DOT specifications relating to compliance

Target Audience

The target audience for this course includes agency and consultant personnel who inspect foundations or major structures. In addition, project management and construction engineers in charge of drilled shaft construction inspection are encouraged to attend. This course is designed to be most beneficial to foundation inspectors who are responsible for inspecting drilled shafts during construction.

Training Level: Intermediate

Fee: 2016: $850 Per Person; 2017: N/A

Length: 2.5 DAYS (CEU: 1.5 UNITS)

Class Size: MINIMUM: 20; MAXIMUM: 35

NHI Customer Service: (877) 558-6873 • nhicustomerservice@dot.gov

Course Number

FHWA-NHI-132070B

WBT

GEOTECHNICAL

Course Title

Drilled Shaft Inspector Tutorial - WEB-BASED

This training is a prerequisite of another NHI training and is offered at no cost.

132070B Drilled Shaft Inspector Tutorial provides training on the fundamental concepts of drilled shafts construction for those involved in the inspection of drilled shafts. This 4-hour Web-based training (WBT) includes the following topics related to drilled shafts: foundations drilled shaft types, methods of construction, construction equipment, and tools. 132070B was developed as a companion training aid for the Instructor-led training 132070 Drilled Shaft Foundation Inspection course. This course replaces 132070A and it is recommended that this WBT be completed prior to attending 132070 Drilled Shaft Foundation Inspection.

This course details the work of the inspector prior, during and after completion of the drilled shaft construction process. Areas of focus include: the inspector's roles, functions, responsibilities, and levels of involvement at different phases of construction. The drilled shaft construction process is covered from the inspector's viewpoint with regards to the documents and tools required for inspection, including equipment and site required checks. Also highlighted is the inspector's role during the drilled shaft excavation process; the rebar cage fabrication and positioning process; and during the placement of concrete. Theoretical and actual drilled shaft concrete volumes calculation, post installation, load, and integrity tests, and other types of tests are also addressed in this course. 132070B details the steps in the drilled shaft construction process and identifies specific responsibilities and methods that will assist inspectors in safely achieving project goals.

Outcomes

Upon completion of the course, participants will be able to:

- Describe the inspector's duties and responsibilities during drilled shaft construction
- Explain the inspector's role in the drilled shaft construction process
- Identify different construction methods
- List equipment and tools used by the inspector and at a drilled shaft construction site
- List the steps in the drilled shaft construction process
- Identify specific responsibilities and methods to assist the inspector in achieving their goal

Target Audience

Federal, State, and local highway agency employees and consultant personnel who inspect foundations or major structures, as well as project managers and construction engineers responsible for drilled shaft construction inspection may benefit from this course.

Training Level: Intermediate

Fee: 2016: $0 Per Person; 2017: N/A

Length: 4 HOURS (CEU: 0 UNITS)

Class Size: MINIMUM: 0; MAXIMUM: 0

NHI Customer Service: (877) 558-6873 • nhicustomerservice@dot.gov

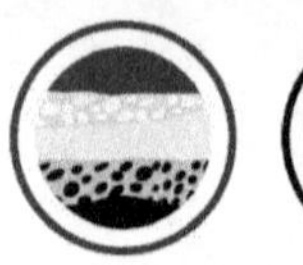

Course Number

FHWA-NHI-132078

Course Title

Micropile Design and Construction

The primary goal of this course is to provide the target audience with guidance on when and where it is appropriate to use micropiles, and educate engineers about the state of the practice in the design and construction of micropiles. The course covers stepwise procedures for the design of micropiles for structural support and for slope stability applications. Construction, inspection and integrity-testing aspects and issues are discussed as well. Classroom presentations include exercises that will lead participants through the technical and cost feasibility aspects of structural support and slope stability design with micropiles. Each participant will receive a workbook and reference manual containing detailed micropile design examples for various applications.

FHWA-NHI-132012 Soils and Foundations course is a recommended prerequisite.

Outcomes

Upon completion of the course, participants will be able to:

- Briefly describe the history and current status of the micropile industry
- Identify potential micropile applications
- Explain construction constraints, techniques, and performance
- Assess feasibility of micropiles for a given application
- Prepare conceptual and basic designs, and evaluate contractor-submitted designs
- Select appropriate specification/contracting method(s) and prepare contract documents
- Describe construction monitoring and inspection requirements

Target Audience

This course is directed toward practicing geotechnical, foundation, construction and bridge/structural engineers who have knowledge and experience in the design and construction of driven piles and drilled shaft foundations. Engineers involved with the design and construction of structure foundations will all benefit from this training, which builds upon the basic concepts presented in NHI courses FHWA-NHI-132012, FHWA-NHI-132014, and FHWA-NHI-132021.

Training Level: Intermediate

Fee: 2016: $775 Per Person; 2017: N/A

Length: 2 DAYS (CEU: 1.2 UNITS)

Class Size: MINIMUM: 20; MAXIMUM: 30

NHI Customer Service: (877) 558-6873 • nhicustomerservice@dot.gov

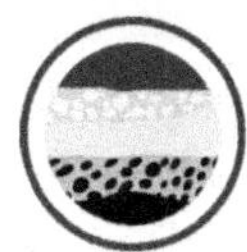

COURSE NUMBER

FHWA-NHI-132079

COURSE TITLE

Subsurface Investigation Qualification

This course is part of a series to develop a training and qualification/certification program for geotechnical inspectors and field personnel. The course follows FHWA guidelines and practices for subsurface investigations. Topics addressed in the course include exploration equipment and methods, safety, borehole sealing, drilling and sampling requirements and criteria, proper visual classification and description of soils and rocks, common drilling errors, and dealing with difficult subsurface site conditions. A 2-hour qualification exam is administered at the end of the course.

OUTCOMES

Upon completion of the course, participants will be able to:

- Explain the investigation specialist's general role and duties, as well as the importance of coordination and communication with the field personnel and engineers
- Explain the purpose of geotechnical subsurface investigations and why adequate, consistent, and quality investigations are essential
- Identify the major components of the typical subsurface investigation plan
- Identify common drilling rigs, uses, and components
- Explain the importance of accurate borehole logging and documentation
- Describe the importance of accurate groundwater investigations
- Discuss safety issues involving operation of a drill rig

TARGET AUDIENCE

The target audience for this course includes drillers, drilling inspectors, engineers, geologists, and technicians involved in field data collection and quality assurance of subsurface investigations.

TRAINING LEVEL: Intermediate

FEE: 2016: $925 Per Person; 2017: N/A

LENGTH: 3 DAYS (CEU: 1.8 UNITS)

CLASS SIZE: MINIMUM: 20; MAXIMUM: 35

NHI Customer Service: (877) 558-6873 • nhicustomerservice@dot.gov

GEOTECHNICAL

Course Number

FHWA-NHI-132080

Course Title

Inspection of Mechanically Stabilized Earth Walls and Reinforced Soil Slopes

This course is part of a series to develop a training and qualification/certification program for geotechnical field inspectors. Topics addressed in the course include the types and durability of mechanically stabilized earth walls (MSEWs) and reinforced soil slopes (RSS); construction methods and sequences; alignment control; methods of fill and compaction control; plans, specifications, and the geotechnical report; shop drawings; and safety. A 2-hour qualification exam is administered at the end of the course.

Outcomes

Upon completion of the course, participants will be able to:

- Identify the basic MSEW and RSS types and design philosophy
- Explain the role and duties of the MSEW and RSS inspector
- Identify current practices for constructing MSE structures
- Define key inspection elements for MSEWs and RSS contract documents to assure compliance
- Explain the logical steps to ensure proper communication with engineers and field personnel
- Understand the steps for MSEW and RSS construction

Target Audience

The target audience for this course includes inspectors (novice to senior level), engineers, geologists, and technicians involved in field data collection and quality assurance for MSEWs and RSS structures. In addition, managers; specification and contracting specialists; bridge/structure, geotechnical and roadway design engineers; and engineering geologists interested in construction aspects of MSEWs and RSS structures are encouraged to attend.

Training Level: Intermediate

Fee: 2016: $925 Per Person; 2017: N/A

Length: 3 DAYS (CEU: 1.8 UNITS)

Class Size: MINIMUM: 20; MAXIMUM: 35

NHI Customer Service: (877) 558-6873 • nhicustomerservice@dot.gov

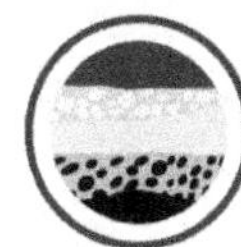

Course Number

FHWA-NHI-132081

Course Title

Highway Slope Maintenance and Slide Restoration

As focus changes toward the asset management of our existing infrastructure, the value of maintaining and managing our embankment and cut slopes becomes more apparent. This course provides the essentials to slope maintenance and slide restoration for transportation field personnel with an asset management perspective. This course is not meant to be highly technical, and explains, conceptually and in layman's terms, the conditions and factors affecting slope movement, stability and deterioration, and the cost considerations of maintenance, stabilization and of slope failures. The course also provides the fundamental aspects of slope management systems and discusses the rationale of slope management considering the legal implications of slope failures and rock fall.

Outcomes

Upon completion of the course, participants will be able to:

- Discuss common soil and rock slope movement and instability
- Describe common factors and conditions under which slopes deteriorate and become less stable
- Describe the affects of earth material properties on slope stability
- Discuss the influences of water on slope stability
- Identify failure-prone conditions
- Describe the importance of necessary communication and coordination with geotechnical specialists
- Discuss best maintenance practices
- Discuss methods of slope monitoring
- Describe key components of slope management systems
- Recognize common soil and rock slope stabilization techniques
- Compare cost differences between preventative measures for slope maintenance and slide restoration and costs associated with slope failures
- Discuss legal implications of slope failures, rock fall and management systems

Target Audience

The target audience for this course includes a wide range of transportation personnel consisting of Federal, State and local maintenance, geotechnical, operations and asset management engineers, geologists, managers, supervisors and personnel involved in assessing, maintaining, managing and repairing cut-slopes, fill-slopes and associated features. Although the potential audience of this course is wide-ranging, the course is primarily provided for the State maintenance specialists.

Training Level: Basic

Fee: 2016: $850 Per Person; 2017: N/A

Length: 2.5 DAYS (CEU: 1.5 UNITS)

Class Size: MINIMUM: 20; MAXIMUM: 35

NHI Customer Service: (877) 558-6873 • nhicustomerservice@dot.gov

GEOTECHNICAL

COURSE NUMBER

FHWA-NHI-132082

COURSE TITLE

LRFD for Highway Bridge Substructures

This course expands the suite of FHWA services to assist State and local governments in the successful implementation of load and resistance factor design (LRFD). The course promotes the philosophy of the LRFD design platform, establishes the motivation for LRFD as a reassurance that safe design practices are being applied, and applies these principles to geotechnical design for bridge foundations. The course's PowerPoint slides and Participant Workbook are regularly updated and the course follows the latest American Association of State Highway Transportation Officials (AASHTO) "LRFD Bridge Design Specifications".

Major topics in this course include: loads, load distribution, and load combinations; principles of limit state designs; geotechnical spread footing design (soil and rock); driven pile and drilled shaft design (soil and rock); and substructure design and detailing for a cantilever abutment and hammerhead pier.

Structured as a combination of Instructor-led discussions and workshop exercises, this course includes LRFD theory applied to design examples and illustrates step-by-step LRFD design procedures. The training includes the extensive use of student exercises and example problems to demonstrate overall design, detailing, and construction principles addressed in the reference materials. The course also provides hands-on experience in the AASHTO LRFD design and detailing of bridge abutment and pier elements, deep and shallow foundation design, and earth retaining structures. Exercise and example problems are based on components of overall comprehensive bridge design examples using AASHTO LRFD and provide comparisons between ASD, LFD, and LRFD design methods.

OUTCOMES

Upon completion of the course, participants will be able to:

- Define AASHTO LRFD limit states and compute structural and geotechnical design loads
- Apply AASHTO LRFD criteria for design
- Integrate the AASHTO LRFD specification provisions into the host agency's current practice
- Integrate the geotechnical aspects of LRFD foundation design into LRFD structural design

TARGET AUDIENCE

Bridge, geotechnical, and transportation engineers with 0-20 years of experience who are responsible for the design and construction of bridge substructures on surface transportation facilities may benefit from this course.

TRAINING LEVEL: Intermediate

FEE: 2016: $1075 Per Person; 2017: N/A

LENGTH: 4 DAYS (CEU: 2.4 UNITS)

CLASS SIZE: MINIMUM: 20; MAXIMUM: 40

NHI Customer Service: (877) 558-6873 • nhicustomerservice@dot.gov

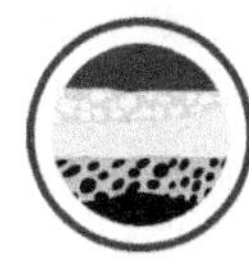

Course Number

FHWA-NHI-132083

GEOTECHNICAL

Course Title

Implementation of LRFD Geotechnical Design for Bridge Foundations

The Implementation of Load and Resistance Factor Design (LRFD): Geotechnical Design for Bridge Foundations is a Web-conference training (WCT) that is designed to assist transportation agencies with the successful development of LRFD Design Guidance for bridge foundations. The training follows the 2010 American Association of State Highway Transportation Officials (AASHTO) LRFD Bridge Design Specifications, Fifth Edition, while incorporating local experiences of design engineers and various transportation agencies. The course is presented in two sessions with a variety of interactive discussions and exercises to verify achievement of learning objectives.

The training introduces a step-by-step procedure for implementing the LRFD platform and provides recommendations to assist transportation agencies in implementing these steps. It also highlights principal changes in the AASHTO design specifications regarding the transition from allowable stress design (ASD) to LRFD and discusses options for selecting LRFD design methods.

Outcomes

Upon completion of the course, participants will be able to:

- Describe an LRFD implementation plan for development of LRFD guidance
- List and discuss the principal changes between ASD and LRFD
- Identify, implement, and compare three options for selection of LRFD Geotechnical Design Methods
- Calibrate resistance factors (f) by fitting to ASD methods
- Summarize the procedure for and evaluate results of reliability calibration of f
- List the conditions to adhere to when using AASHTO's LRFD design methods
- Select conditions for development of local LRFD geotechnical design methods
- Evaluate and address project site variability
- Recall the advantages of development of local LRFD geotechnical design methods through reliability calibration
- Describe a road map for the development of LRFD Design Guidance for bridge foundations

Target Audience

The Implementation of LRFD Geotechnical Design for Bridge Functions WCT is intended for State DOT engineers who are involved with the development of LRFD specifications for foundations, or who manage and conduct LRFD geotechnical research studies, or who are involved with LRFD design of bridge foundations. All intended learners are expected to have completed NHI Course 130082 or have equivalent knowledge of AASHTO LRFD Specifications for structural foundation, and are expected to be familiar with AASHTO Standard Specifications (ASD) design methods for foundations.

Training Level: Basic

Fee: 2016: $50 Per Person; 2017: N/A

Length: 4 HOURS (CEU: 0 UNITS)

Class Size: MINIMUM: 20; MAXIMUM: 0

NHI Customer Service: (877) 558-6873 • nhicustomerservice@dot.gov

COURSE NUMBER

FHWA-NHI-132084

COURSE TITLE

Geotechnical Subsurface Exploration - WEB-BASED

The Subsurface Explorations Web-based Training course will provide transportation engineers with a basic knowledge and understanding of subsurface exploration programs for design and construction of structure foundations, walls, and other geotechnical features. Properly conducted subsurface exploration programs are an essential part of geotechnical engineering, and are a critical step in understanding soil and rock properties necessary for design.

The course covers a range of topics related to subsurface exploration programs including earth materials, subsurface conditions, geophysical methods, drilling methods and equipment, soil and rock sampling methods, in-situ testing, and groundwater investigation. Upon completion of this course, participants will be able to apply basic geotechnical engineering principles and sound geotechnical methods to transportation projects.

OUTCOMES

Upon completion of the course, participants will be able to:

- Identify the key geotechnical considerations associated with typical transportation projects
- Describe the recommended process for characterizing subsurface conditions
- Identify the primary types of geophysical methods
- Identify types of drilling methods and equipment
- Identify types of soil and rock sampling methods
- Explain the purpose of in-situ tests and energy-efficiency parameters
- Explain the purpose of doing a groundwater investigation
- Describe minimum guidelines for the geotechnical investigation of both roadway and structure sites

TARGET AUDIENCE

The course is intended for transportation engineers and geotechnical specialists who are involved with the planning, design, and construction of surface transportation facilities. The course will be oriented toward those professionals who routinely deal with soils and foundations issues but who may have little theoretical background in soil mechanics or foundation engineering.

TRAINING LEVEL: Basic

FEE: 2016: $50 Per Person; 2017: N/A

LENGTH: 6 HOURS (CEU: 0 UNITS)

CLASS SIZE: MINIMUM: 0; MAXIMUM: 0

NHI Customer Service: (877) 558-6873 • nhicustomerservice@dot.gov

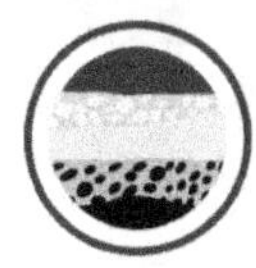

Course Number

FHWA-NHI-132091

Course Title

Earthwork Series: Grades and Grading - WEB-BASED

Grades and Grading is the third part of the Earthwork five part series. It is designed to prepare technical front-line workers for what they can expect to see during actual project inspection. Topics covered include an overview of the plans that pertain to earthwork and earthwork quantities, grade stakes that will be encountered and their meanings, how Global Positioning System (GPS) works and its functions in the field, and verifying and documenting grade information. This course is primarily intended for inspectors and technicians.

The introductory lesson covers an overview of the plan sheets that deal with earthwork and earthwork quantities, topographical images and their meaning, stationing and control points, and profile/section sheets. The second section covers the typical grade stakes used throughout a project and their meaning. The GPS section discusses the history of GPS in construction and how it relates to current projects. And the final section covers how to verify the grade and what information is needed in the documentation from the inspector.

This course provides the front-line technical inspector with the proper tools to assure that the project is built on a stable platform.

Outcomes

Upon completion of the course, participants will be able to:

- Describe the process of plan reading
- Identify the purpose of grade stakes
- Explain how Global Positioning System (GPS) works
- Describe requirements for grade verification and documentation

Target Audience

This training is designed for intermediate to advanced technicians who perform site preparation inspection on earthwork projects. The training was developed by the Transportation Curriculum Coordination Council (TCCC) in partnership with NHI and is recommended for TCCC levels II through IV.

Training Level: Basic

Fee: 2016: $0 Per Person; 2017: N/A

Length: 3.5 HOURS (CEU: 0 UNITS)

Class Size: MINIMUM: 1; MAXIMUM: 1

NHI Customer Service: (877) 558-6873 • nhicustomerservice@dot.gov

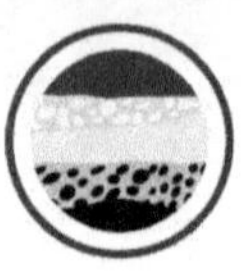

Course Number

FHWA-NHI-132092

Course Title

Earthwork Series: Excavation - WEB-BASED

Excavation is the fourth part of the Earthwork five-part series. Excavations of soil and rock are an integral part of highway construction due to the associated costs, safety concerns, engineering considerations, and short and long-term performance expectations. The Earthwork Series: Excavation course is an overview of the basic principles related to the requirements for proper excavation during a project.

This training consists of four modules, which cover the equipment used to excavate soils, and the procedures, requirements, and special considerations for mass excavation, permanent cut slopes, and temporary trench excavations. The course also covers some common problems and safety concerns associated with excavation.

Outcomes

Upon completion of the course, participants will be able to:

- Explain considerations and requirements for excavation
- Recall excavation safety procedures
- Relate common issues and solutions associated with excavation

Target Audience

This training is designed for state and local government employees, as well as private industry technicians and inspectors who work within or around excavations, are responsible for documenting excavation operations, or are responsible for verifying foundation materials and proper earthwork construction on highway projects. The course is a beneficial overview for all those working on an earthwork project, but intermediate and advanced technicians and inspectors are the primary target audience. This training was developed by the Transportation Curriculum Coordination Council (TCCC) in partnership with NHI and is recommended for TCCC Levels II through IV.

Training Level: Basic

Fee: 2016: $0 Per Person; 2017: N/A

Length: 4.5 HOURS (CEU: 0 UNITS)

Class Size: MINIMUM: 1; MAXIMUM: 1

NHI Customer Service: (877) 558-6873 • nhicustomerservice@dot.gov

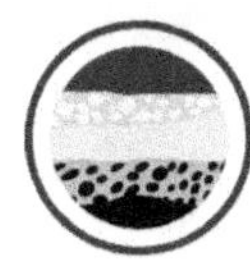

Course Number

FHWA-NHI-132093

Course Title

Earthwork Series: Fill Placement - WEB-BASED

Fill Placement is the fifth part of the Earthwork five-part series. Embankment construction; structural and utility bedding and backfilling; and the construction of drainage and filter systems are fundamental examples of highway earthwork - where the control of the material and how it is placed significantly influences engineering performance. The Earthwork Series: Fill Placement course is an overview of the basic applications where fill materials are to be used, and some common problems and safety considerations that you will need to know.

This training consists of four modules which cover culvert bedding and backfill, drainage filters and fabrics, embankment construction, key-ways, and benching. The course discusses material and placement requirements, methods used to control and assure placement, special construction considerations, common problems, and safety related issues.

Outcomes

Upon completion of the course, participants will be able to:

- Explain fill placement;
- Recall fill placement safety procedures; and
- Identify steps for addressing obstacles associated with fill placement.

Target Audience

This training is designed for State and local government employees as well as private industry technicians and inspectors who provide quality control/quality assurance testing, document fill placement activities, verify that earthwork has been constructed according to contract documents, or inspect earthwork activities on highway projects. The course is a beneficial overview for all those working on an earthwork project, but intermediate and advanced technicians and inspectors are the primary target audience. This training was developed by the Transportation Curriculum Coordination Council (TCCC) in partnership with AASHTO NHI and is recommended for TCCC Levels II through IV.

Training Level: Basic

Fee: 2016: $0 Per Person; 2017: N/A

Length: 5.5 HOURS (CEU: 0 UNITS)

Class Size: MINIMUM: 1; MAXIMUM: 1

NHI Customer Service: (877) 558-6873 • nhicustomerservice@dot.gov

Course Number

FHWA-NHI-132094

Course Title

LRFD Seismic Analysis and Design of Transportation Structures, Features, and Foundations

This course is a comprehensive and practical training course for analysis and design of transportation geotechnical features including soil and rock slopes, earth embankments, retaining walls, MSE walls, and buried structures; and bridge structural foundations including shallow and deep foundations, and abutment walls. It is developed in consideration of the requirements and recommendations of the seismic provisions in both the 2009 AASHTO LRFD Bridge Design Specifications and the AASHTO Guide Specifications for LRFD Seismic Bridge Design, the Final Report from NCHRP Project 12-70 "Seismic Analysis and Design of Retaining Walls, Buried Structures, Slopes, and Embankments", and 2006 FHWA Seismic Retrofitting Manual for Highway Structures.

In addition, the course reviews the fundamental principles including engineering seismology, earthquake hazard analysis, site characterization, ground motion characterization, and site response analysis, and highlight updated topics such as the 1000-yr USGS hazard map; updated AASHTO site classes/factors and spectral shapes; the "3-Point" Design Spectrum Construction method; derivation of the relative displacement spectrum; and regional differences in ground motion characteristics (i.e. western US (WUS) characteristics versus central and eastern US (CEUS). It addresses geotechnical hazards which can adversely impact bridges and other transportation structures and features during seismic event including slope instability, soil liquefaction, ground settlement, and fault Rupture. Liquefaction-induced lateral spread failures are also addressed.

Outcomes

Upon completion of the course, participants will be able to:

- Recognize sources of primary and secondary damage due to earthquakes
- Describe the AASHTO seismic design philosophy
- Describe the input for a seismic hazard analysis and interpret the output for a bedrock site condition
- Develop an AASHTO acceleration response spectra and adjust it for local site conditions
- Estimate the residual undrained shearing resistance of liquefied sand
- Develop the input for an equivalent linear seismic site response analysis
- Determine the appropriate seismic coefficient for a pseudo static slope stability analysis and calculate the permanent seismic displacement of an unstable soil slope
- Evaluate the potential for liquefaction triggering and consequences
- Identify potential mitigation measures for slope instability, liquefaction and lateral spreading
- Evaluate external stability of gravity and semi-gravity walls subject to seismic loading
- Discuss types of soil-foundation-structure interaction and how its effects are modeled
- Evaluate the geotechnical and structural capacity of a spread footing
- Identify the primary capacity considerations for deep foundations under seismic loading
- Develop the abutment spring stiffness relationship

Target Audience

This course is intended to engage a target audience of bridge and geotechnical engineers with zero and up to 20 years of experience through instructor-led presentations, discussions, Q&A, group activities, walkthrough examples, hands-on student exercises, and demonstrations.

TRAINING LEVEL: Basic

FEE: 2016: $1225 Per Person; 2017: N/A

LENGTH: 5 DAYS (CEU: 3 UNITS)

CLASS SIZE: MINIMUM: 20; MAXIMUM: 30

NHI Customer Service: (877) 558-6873 • nhicustomerservice@dot.gov

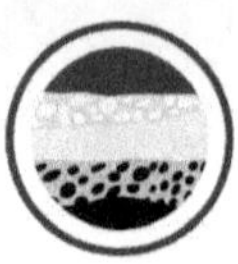

GEOTECHNICAL

COURSE NUMBER

FHWA-NHI-132094A

COURSE TITLE

LRFD Seismic Analysis and Design of Transportation Geotechnical Features

The Instructor-Led 132094A Course has a prerequisite Web-Based Training (WBT), NHI-132010A Earthquake Engineering Fundamentals, that participants must complete before the start of the 132094A course. The WBT prerequisite course consists of 6 lessons including: Earthquake Fundamentals (L1); Intro to LRFD Seismic Design (L2); Earthquake Ground Motions (L3); Seismic Hazard Analysis (L4); AASHTO Design Ground Motion Characterization (L5); and Intro to Geotechnical Hazards (L6).

This 2-day NHI training course 132094A entitled "LRFD Seismic Analysis and Design of Transportation Geotechnical Features" is a shortened version of the NHI training course 132094 "LRFD Seismic Design of Transportation Geotechnical Features and Structural Foundations" focusing specifically on the geotechnical earthquake engineering aspects. It is a comprehensive and practical training course that addresses seismic analysis and design of transportation geotechnical features including ground motion characterization, development of the AASHTO acceleration response spectrum for structural design using the 1000-yr USGS hazard map for reference site conditions, and evaluation of AASHTO site class and application of AASHTO soil factors to account for local soil conditions; site characterization for geotechnical seismic analysis; equivalent linear site response analysis; identification of geotechnical seismic hazards; seismic stability and deformation analysis of embankments and slopes; analysis procedures for liquefaction and liquefaction-induced lateral spread or flow failures; seismic settlement analysis; and geotechnical hazard mitigation measures. The 132094A course also focuses on interactions between the geotechnical specialist and the bridge design engineer in the seismic design process.

OUTCOMES

Upon completion of the course, participants will be able to:

- Describe the AASHTO seismic design performance criteria and develop an AASHTO acceleration response spectra for reference site (weak rock) conditions.
- Calculate fundamental period of the site and peak ground velocity from a spectral acceleration.
- Identify key soil properties necessary for seismic analysis and methods for evaluating them.
- Identify conditions warranting, establish input parameters, and conduct a one-dimensional equivalent linear site response analysis.
- Assess seismic slope stability and deformation potential in accordance with the AASHTO specifications and national state of art analysis and design guidance.
- Evaluate the potential for earthquake-induced liquefaction and its impacts on geotechnical transportation features in accordance with AASHTO specifications and national state-of-practice analysis and design guidance.
- Identify common mitigation methods for geotechnical seismic hazards.

TARGET AUDIENCE

This course is intended to engage a target audience of bridge and geotechnical engineers with zero and up to 20 years of experience, through instructor-led presentations, discussions, Q&A, group activities, walkthrough examples, and hands-on student exercises. At the end of design lessons, participants will have the opportunity to undertake a group design exercise to reinforce learning and enhance the transfer of new skills and knowledge to the workplace.

TRAINING LEVEL: Intermediate

FEE: 2016: $775 Per Person; 2017: N/A

LENGTH: 2 DAYS (CEU: 1.4 UNITS)

CLASS SIZE: MINIMUM: 20; MAXIMUM: 30

NHI Customer Service: (877) 558-6873 • nhicustomerservice@dot.gov

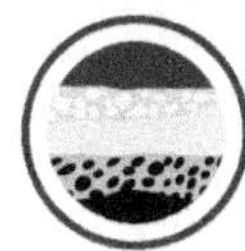

Course Number

FHWA-NHI-132094B

Course Title

LRFD Seismic Analysis and Design of Structural Foundations and Earth Retaining Structures

This 2-day NHI training course 132094B entitled "LRFD Seismic Analysis and Design of Structural Foundations and Earth Retaining Structures" is a shortened version of the NHI training course 132094 "LRFD Seismic Design of Transportation Geotechnical Features and Structural Foundations" focusing specifically on the seismic design of retaining wall and structural foundations aspects. It is a comprehensive and practical training course that addresses seismic analysis and design of transportation geotechnical features including ground motion characterization using the AASHTO acceleration response spectrum developed based upon the AASHTO or USGS hazard maps adjusted for local site conditions using AASHTO soil site factors to account for local soil conditions or upon a site specific analysis; identification and evaluation of geotechnical seismic hazards; soil-foundation-structure interaction; shallow foundation design; deep foundation design; and design or earth retaining structures, including free standing retaining walls and abutment walls. It is developed generally in consideration of the requirements and recommendations of the seismic provisions in both the AASHTO LRFD Bridge Design Specifications and the AASHTO Guide Specifications for LRFD Seismic Bridge Design, the Final Report from NCHRP Project 12-70 "Seismic Analysis and Design of Retaining Walls, Buried Structures, Slopes, and Embankments", and 2006 FHWA Seismic Retrofitting Manual for Highway Structures. The 132094B course also focuses on interactions between the geotechnical specialist and the bridge design engineer in the seismic design process.

Outcomes

Upon completion of the course, participants will be able to:

- Describe the AASHTO seismic design performance criteria and develop an AASHTO acceleration response spectra for reference site (weak rock) conditions.
- Calculate peak ground velocity and relative displacement from spectral acceleration.
- Identify the potential impacts of geotechnical hazards on foundations and earth retaining structures.
- Describe the two types of SFSI and recognize the importance of the interaction between structural designers and geotechnical engineers in the bridge design process.
- Evaluate the seismic capacity and stiffness of a shallow foundation.
- Evaluate the seismic capacity and stiffness of a deep foundation.
- Evaluate global and internal stability of earth retaining systems.
- Calculate bi-linear force-deformation relationship for seismic design and analysis of bridge abutment-backfill interaction.

Target Audience

This course is intended to engage a target audience of bridge and geotechnical engineers with zero and up to 20 years of experience, through instructor-led presentations, discussions, Q&A, group activities, walkthrough examples, and hands-on student exercises. At the end of design lessons, participants will have the opportunity to undertake a group design exercise to reinforce learning and enhance the transfer of new skills and knowledge to the workplace.

Training Level: Intermediate

Fee: 2016: $775 Per Person; 2017: N/A

Length: 2 DAYS (CEU: 1.3 UNITS)

Class Size: MINIMUM: 20; MAXIMUM: 30

NHI Customer Service: (877) 558-6873 • nhicustomerservice@dot.gov

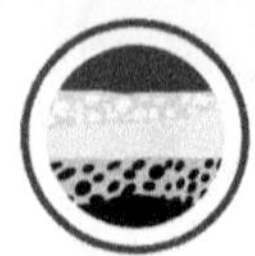

Course Number

FHWA-NHI-132097A

Course Title

Integrating GeoTechTools into Project Planning and Delivery (1-Day ILT)

NHI-132097A, Integrating GeoTechTools into Project Planning and Delivery, is a 1-day instructor-led course designed to promote the integration of GeoTechTools into your agency's practice, in both program development and project delivery. This course will teach you how to use GeoTechTools to make better informed and innovative decisions with geotechnologies, factor project constraints into geotechnology selection, and mitigate project and geotechnical risks with geotechnology selection.

NOTE TO HOST: For use in the afternoon session, the host must provide each participant a computer with internet access.

Outcomes

Upon completion of the course, participants will be able to:

- State what is GeoTechTools (GTT)
- Identify who (agency and role) can benefit from the use of GTT
- Identify the types of projects where using GTT would be appropriate
- Identify at which stage(s) of the project development and delivery process GTT should be employed
- Explain how and why GTT was developed
- Identify the scope and limitations of GTT
- Demonstrate how to navigate and explore within GTT
- State how your agency will incorporate GTT into project planning and delivery

Target Audience

The target audience for this 1-day course includes agency program managers, geotechnical engineers, pavement engineers, bridge engineers, project planners, and project managers.

Training Level: Basic

Fee: 2016: $625 Per Person; 2017: N/A

Length: 1 DAYS (CEU: .6 UNITS)

Class Size: MINIMUM: 20; MAXIMUM: 40

NHI Customer Service: (877) 558-6873 • nhicustomerservice@dot.gov

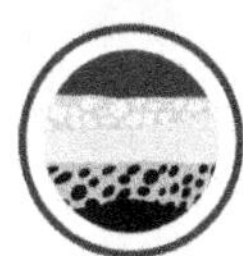

GEOTECHNICAL

Course Number

FHWA-NHI-132097B

Course Title

Integrating GeoTechTools into Project Planning and Delivery - 1 Day Technical Assistance

This course will provide technical assistance to State Departments of Transportation for GeoTechTools.

Outcomes

Upon completion of the course, participants will be able to:

- The instructor will work with the host to determine the areas of technical assistance.

Target Audience

The target audience for this technical assistance includes agency program managers, geotechnical engineers, pavement engineers, bridge engineers, project planners, and project managers

Training Level: Basic

Fee: 2016: $625 Per Person; 2017: N/A

Length: 1 DAYS (CEU: 0 UNITS)

Class Size: MINIMUM: 20; MAXIMUM: 30

NHI Customer Service: (877) 558-6873 • nhicustomerservice@dot.gov

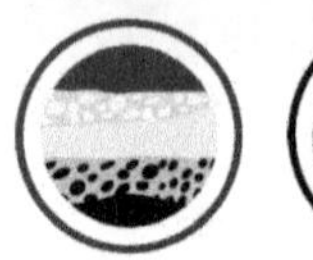

Course Number

FHWA-NHI-135046

Course Title

Stream Stability and Scour at Highway Bridges

The National Highway Institute's (NHI) 3-day Stream Stability and Scour at Highway Bridges course provides participants with comprehensive training in the prevention of hydraulic-related bridge failures. Course participants will receive training in conducting a stream stability classification and qualitative analysis of stream response and make estimates of scour at a bridge opening.

Material for the course comes primarily from two Hydraulic Engineering Circulars (HEC), "Evaluating Scour at Bridges" (HEC-18), 5th Edition (2012), and "Stream Stability at Highway Structures" (HEC-20), 4th Edition (2012). The effects of stream instability, scour, erosion, and stream aggradation and degradation are covered. Quantitative techniques are provided for estimating long-term degradation and for calculating the magnitude of contraction scour in a bridge opening. Procedures for estimating local scour at bridge piers and abutments for simple and complex substructures are also provided. A comprehensive workshop integrates qualitative analysis and analytical techniques to determine the need for a Scour Plan of Action for correcting stream instability and scour problems. For this 3-day course, the host agency will need to select 3 optional topics (out of 8 possible topics). Course instructors will contact the host prior to the course to complete a pre-course questionnaire, determine optional topics to be taught, and discuss the course schedule.

This comprehensive training provides preventive techniques for identifying, analyzing, and calculating various hydraulic factors that impact bridge stability. Public and private sector engineers responsible for maintaining the integrity of highway bridges will find it invaluable.

Prior to the beginning of the course, participants are strongly encouraged to enroll in the following Web-based training (WBT) courses: 135091 Basic Hydraulic Principles Review, 135086 Stream Stability Factors and Concepts, and 135087 Scour at Highway Bridges: Concepts and Definitions. Mastery of the concepts covered in these WBTs will enhance participation in the Instructor-led training.

Outcomes

Upon completion of the course, participants will be able to:

- Identify indicators of stream instability that can threaten bridges
- Identify stream types and their potential for instability problems
- Describe open-channel hydraulics concepts in bridge scour and stream instability analyses
- Define types of scour that can occur at bridge crossings
- Describe aggradation, degradation, and contraction scour
- Calculate contraction scour for live bed and clear water conditions
- Describe factors that influence scour at piers
- Calculate pier scour for three typical case studies
- Describe the factors that influence scour at abutments
- Describe how HEC-18, HEC-20, and HEC-23 provide analysis procedures for stream instability and bridge scour
- Perform Level I and II analyses
- Classify a stream using two different classification systems
- Conduct a qualitative analysis of stream responses
- Apply the HEC-18 scour equations to determine total scour at a bridge
- Determine the need for a Scour Plan of Action at a scour-critical bridge

Target Audience

Federal, State, and local highway hydraulic, structural, and geotechnical engineers as well as bridge inspectors responsible for maintaining the integrity of highway bridges against possible hydraulic-related problems. Consultants who perform bridge engineering work are encouraged to attend.

TRAINING LEVEL: Intermediate

FEE: 2016: $975 Per Person; 2017: N/A

LENGTH: 3 DAYS (CEU: 2 UNITS)

CLASS SIZE: MINIMUM: 20; MAXIMUM: 30

NHI Customer Service: (877) 558-6873 • nhicustomerservice@dot.gov

Course Number

FHWA-NHI-135048

Course Title

Countermeasure Design for Bridge Scour and Stream Instability (2.5-Day)

This course provides an overview of countermeasures to highway related failures from the effects of stream instability, scour, erosion, and stream aggradation and degradation problems. Material for the 2.5-day course comes primarily from Hydraulic Engineering Circular (HEC) "Bridge Scour and Stream Instability Countermeasures - Experience, Selection, and Design Guidance" (HEC-23).

Given a stream instability and scour problem, participants will select appropriate countermeasures to correct the problem. The course provides training in recommended strategies for developing a plan that includes appropriate countermeasures, including alternatives to conventional riprap and filter design.

Participants will apply hydraulics analysis techniques to countermeasure design for seven design guideline workshops. The course provides an introduction to fixed and portable instrumentation for scour monitoring using slides and video demonstrations. Participants will receive training in designing a monitoring program to reduce the risk from scour.

NHI Course 135046 provides training in identifying and analyzing stream instability and scour problems at highway bridges and is recommended as a prerequisite for this course.

NHI Courses #135086 and #135087 are Web-based training module and are prerequisites for NHI Hydraulics courses 135047 and 135048.

Outcomes

Upon completion of the course, participants will be able to:

- Develop a plan of action for a scour critical bridge
- Propose countermeasures for stream instability and scour problems
- Identify countermeasures for bridge scour and stream instability using the HEC-23 countermeasures matrix
- Design selected countermeasures with HEC-23 design guidelines

Target Audience

Federal, State, and local highway hydraulic, structural, and geotechnical engineers and bridge inspectors responsible for maintaining the integrity of highway bridges against possible hydraulic-related problems. Consultants who do bridge engineering work are also encouraged to attend.

Training Level: Intermediate

Fee: 2016: $900 Per Person; 2017: N/A

Length: 2.5 DAYS (CEU: 1.5 UNITS)

Class Size: MINIMUM: 20; MAXIMUM: 30

NHI Customer Service: (877) 558-6873 • nhicustomerservice@dot.gov

COURSE NUMBER

FHWA-NHI-133078

COURSE TITLE

Access Management, Location and Design

This course covers the complex technical issues that underlie effective access management practices on streets and highways and provides the technical rationale for proper signal spacing, driveway spacing and design, the application and design of auxiliary lanes. "Before" and "after" case studies illustrate the impacts of projects to improve traffic safety and operations. In addition, the course addresses the issues involved in developing and administering an effective access management program. The course references the state-of-the-practice as presented in the Transportation Research Board's Access Management Manual, the latest edition of AASHTO's A Policy on Geometric Design of Highways and Streets (Green Book), and pertinent NCHRP reports. In summary, this training provides a lasting reference and specific applications of techniques and practices that will enable transportation engineering and planning personnel to implement successful access management strategies and programs. All participants will receive the class notebook and a copy of the TRB Access Management Manual.

OUTCOMES

Upon completion of the course, participants will be able to:

- Discuss the impact of access on highway safety and operations
- Choose access management techniques to mitigate challenges
- Identify practices needed for implementing access management programs

TARGET AUDIENCE

This course targets transportation and planning professionals involved in traffic operations, roadway design, the planning of circulation systems, and land development. Specifically, the course is designed for those individuals directly involved in implementing access management solutions in their jurisdictions, as it focuses heavily on resources and solutions to reduce the impact of access points on traffic flow.

TRAINING LEVEL: Basic

FEE: 2016: $875 Per Person; 2017: N/A

LENGTH: 3 DAYS (CEU: 1.8 UNITS)

CLASS SIZE: MINIMUM: 20; MAXIMUM: 30

NHI Customer Service: (877) 558-6873 • nhicustomerservice@dot.gov

Course Number

FHWA-NHI-133099

Course Title

Managing Travel for Planned Special Events (2-Day)

The Rose Bowl, the Macy's Day Parade, and the Nation's numerous marathons, golf tournaments, and county fairs are just some of the planned special events that are held throughout the country every year. Managing travel to these and other events will allow event patrons to enjoy themselves from the moment they leave home. In addition, a well-designed transportation plan for these events accommodates the needs of the nearby residents and businesses.

This course provides practitioners with a working knowledge of the techniques and strategies they may wish to use for the successful planning and operation of a specific planned special event. Practitioners will gain an understanding of the collective tasks facing multidisciplinary and inter-jurisdictional stakeholder groups charged with developing and implementing solutions to acute and system-wide impacts on travel during a special event. Instructors will identify all potential tasks and stakeholder activities conducted within individual phases of managing planned special events.

This course will refer to FHWA's Managing Travel for Planned Special Events Handbook and guide participants on how to apply key concepts in the handbook. The handbook in CD format is provided with the course materials.

The 2-day version of the course will guide practitioners through all the phases of managing travel for planned special events for a specific event category, based upon an event scenario defined by the course participants. In addition, the goal of the 2-day course and group exercises is to meet the participant's needs in planning and managing a similar future event for a specific locale. Course participants will identify and apply pertinent planning steps, operations activities, and associated considerations in developing an action plan for the defined planned special event scenario.

This course is part of the Certificate of Accomplishment in Incident Management. To learn more about how you can achieve a certificate in Incident Management visit the NHI Web site at http://www.nhi.fhwa.dot.gov/training/cert_programs.aspx.

Outcomes

Upon completion of the course, participants will be able to:

- Name the main categories of planned special events
- State key phases of managing travel for planned special events
- Identify the goals of managing travel for planned special events
- Describe the benefits of proactively managing travel for planned special events
- Describe the purpose and value of an action plan for managing travel for a specific planned special event
- List key components of an action plan
- Identify key factors that influence the potential effect a planned special event may have on the performance of the surface transportation system
- List key components of a traffic management plan
- State opportunities or sources where resources could be obtained to initiate activities identified in a planned special event travel management action plan
- Name near-term or short-term actions that are priorities in a planned special event travel management action plan
- State potential activities involved with the implementation of a traffic management plan for a planned special event
- Name key activities performed by the traffic management team on the day of the event
- Explain how post-event activities may improve the management of travel for future planned special events

Target Audience

Transportation agencies that will be involved in developing the plans and implementing transportation management plans for upcoming events.This course and the corresponding workshop are designed for any individual engaged in or responsible for directing agency resources related to the following five key phases associated with managing travel for planned special events: (1) program planning, (2) event operations planning, (3) implementation activities, (4) day-of-event activities, and (5) post-event activities. This is an introductory course and workshop for individuals with limited or no experience with applying the recommended concepts and techniques in all of the phases involved with managing travel for planned special events. Participants could include traffic engineers and technicians, transportation planners, managers/supervisors, transit planners and operations supervisors, transportation management center staff, law enforcement personnel, public safety transportation coordinators (e.g., fire, emergency medical types of personnel, etc.), public information specialists, event operators (e.g., parking management, traffic control, etc.), emergency management personnel, consultants, and post-secondary students and faculty.

TRAINING LEVEL: Basic

FEE: 2016: $875 Per Person; 2017: N/A

LENGTH: 2 DAYS (CEU: 1.2 UNITS)

CLASS SIZE: MINIMUM: 20; MAXIMUM: 30

NHI Customer Service: (877) 558-6873 • nhicustomerservice@dot.gov

COURSE NUMBER

FHWA-NHI-133099A

COURSE TITLE

Managing Travel for Planned Special Events (1-Day)

The Rose Bowl, the Macy's Day Parade, and the Nation's numerous marathons, golf tournaments, and county fairs are just some of the planned special events that are held throughout the country every year. Managing travel to these and other events will allow event patrons to enjoy themselves from the moment they leave home. In addition, a well-designed transportation plan for these events accommodates the needs of the nearby residents and businesses.

This course provides practitioners with a working knowledge of the techniques and strategies they may wish to use for the successful planning and operation of a specific planned special event. Practitioners will gain an understanding of the collective tasks facing multidisciplinary and inter-jurisdictional stakeholder groups charged with developing and implementing solutions to acute and system-wide problems affecting travel during a special event. Instructors will identify all potential tasks and stakeholder activities conducted within individual phases of managing planned special events. The course will refer to FHWA's Managing Travel for Planned Special Events Handbook and guide participants on how to apply key concepts in the handbook. The handbook in CD format is provided with the course materials.

NOTE: See FHWA-NHI-133099 for the 2-day version of the course, which will provide scenario-based exercises and practices in a workshop format.

OUTCOMES

Upon completion of the course, participants will be able to:

- Name the main categories of planned special events
- State key phases of managing travel for planned special events
- Identify the goals of managing travel for planned special events
- Describe the benefits of proactively developing plans designed to manage travel for planned special events
- Describe the purpose and value of an action plan for managing travel for a specific planned special event
- List key components of an action plan
- Identify key factors that influence the potential effect a planned special event may have on the performance of the surface transportation system
- List key components of a traffic management plan

TARGET AUDIENCE

This course and the 2-day workshop are designed for any individual engaged in or responsible for directing agency resources related to the following five key phases associated with managing travel for planned special events: (1) program planning, (2) event operations planning, (3) implementation activities, (4) day-of-event activities, and (5) post-event activities. The 1-day introductory course is for individuals with limited or no experience with applying the recommended concepts and techniques in all of the phases involved with managing travel for a planned special event. Participants could include traffic engineers and technicians, transportation planners, managers/supervisors, transit planners and operations supervisors, transportation management center staff, law enforcement personnel, public safety transportation coordinators (e.g., fire, emergency medical personnel, etc.), public information specialists, event operators (e.g., parking management, traffic control, etc.), emergency management personnel, consultants, and post-secondary students and faculty.

TRAINING LEVEL: Basic

FEE: 2016: $425 Per Person; 2017: N/A

LENGTH: 1 DAYS (CEU: .6 UNITS)

CLASS SIZE: MINIMUM: 20; MAXIMUM: 30

NHI Customer Service: (877) 558-6873 • nhicustomerservice@dot.gov

COURSE NUMBER

FHWA-NHI-133107

COURSE TITLE

Principles of Evacuation Planning Tutorial (Web-Based)

Principles of Evacuation Planning Tutorial (133107) is a Web-based asynchronous/independent training that provides an introductory overview of evacuation planning topics and common considerations. It covers the roles and responsibilities of local, regional, and state agencies involved in the evacuation process, while highlighting the importance of collaboration.

This course also presents current and emerging evacuation planning tools, methodologies, and trends, and offers insight into special considerations that evacuation planning stakeholders should take into account when designing, reviewing, or contributing to evacuation planning efforts. Emphasis is placed on multi-agency/jurisdictional planning as part of identifying effective practices used in the U.S.

This training was developed at the request of the FHWA Transportation Pooled Fund Study Security and Emergency Management Update and Request. The pooled fund study states are California, Florida, Georgia, Kansas, Mississippi, Montana, New York, Texas, and Wisconsin. In addition, the TSA is a member of the pooled fund study.

OUTCOMES

Upon completion of the course, participants will be able to:

- Define evacuation planning from a transportation standpoint
- Describe how evacuation planning impacts local and state emergency management transportation operations
- Define the roles and responsibilities of local, regional, and state agencies
- List the benefits of working across agencies and localities to maximize the effectiveness of emergency planning efforts
- List evacuation planning considerations specific to Notice and No-Notice evacuations
- Describe other special considerations that evacuation planning stakeholders should take into account when executing evacuation plans
- Identify tools and methods for coordination and collaboration
- Identify current and emerging evacuation planning practices
- Describe effective emergency evacuation planning practices
- Explain the value of engaging other organizations and jurisdictions
- Identify resources available to emergency evacuation planning stakeholders and how to access them for further study

TARGET AUDIENCE

The Principles of Evacuation Planning Tutorial (133107) is designed for transportation and emergency planning stakeholders along with local leadership (e.g. local public and private emergency management stakeholders). This course also will be made available to a variety of other professionals with an interest in evacuation planning including Government jurisdictions below state level; transportation planners;metropolitan planning organizations; transportation planners (city/county); local emergency managers; transportation management center staff; state and local police planners; metro emergency planners; public works and public schools planners; and other contributing stakeholders.

TRAINING LEVEL: Basic

FEE: 2016: $50 Per Person; 2017: N/A

LENGTH: 6 HOURS (CEU: .5 UNITS)

CLASS SIZE: MINIMUM: 0; MAXIMUM: 0

NHI Customer Service: (877) 558-6873 • nhicustomerservice@dot.gov

COURSE NUMBER

FHWA-NHI-133109

COURSE TITLE

Strategies for Developing Work Zone Traffic Analyses

Strategies for Developing Work Zone Traffic Analyses is offered as a one-day instructor-led course utilizing lecture and small-group collaborative exercises to educate participants on how to develop effective transportation modeling strategies to support work zone-related decision-making. There is no hands-on computer based modeling work conducted in this course; rather it deals with developing analysis plans that combine people, data and tools to address work zone issues. The course is designed to cover:

Characterizing a work zone with respect to a prospective analysis

Classes of analytical tools and their capabilities within the context of work zones

Selecting an appropriate transportation modeling approach maximizing insight into potential impacts and mitigating technical risk

The course includes lecture, full-group interaction, and small group activities. The purpose of the course is three-fold. First, it will educate the participants regarding the constraints and opportunities of work zone analysis associated with available transportation modeling tools. Second, it will build familiarity for the participants with the various work zone factors influencing the development of a transportation analysis plan. Third, it will provide the participants with practical experience in developing analysis plans in a collaborative process considering issues ranging from work zone characteristics, performance measurement, technical risk assessment and resource constraints.

OUTCOMES

Upon completion of the course, participants will be able to:

- Define the need, scope, and role of work zone modeling and analysis
- Describe the work zone analysis decision-making engine and the interactions among scheduling, application, and transportation management plan decisions
- Explain how to characterize a work zone
- Identify the transportation modeling approaches available for work zone analysis
- Discuss how a transportation modeling approach can be used given a set of work zone characteristics
- Justify the selection of transportation modeling approach

TARGET AUDIENCE

A mix of experience with traffic analysis tools and work zone planning among participants is preferred. No prior experience with traffic analysis tools is required. The course is designed to promote interactions between participants. Therefore, the group is likely to benefit from a variety of viewpoints if participants have varied levels of analytical experience and diverse agency affiliations. The group may include:State Department of Transportation Staff (District Engineers, Corridor Planners, Project Eng., Traffic Eng., Work Zone Planners)FHWA staff (Division Staff, Transportation Engineers, Traffic Staff, Planners)Metropolitan Planning Organization Staff (Planners)Consultants

TRAINING LEVEL: Basic

FEE: 2016: $275 Per Person; 2017: N/A

LENGTH: 1 DAYS (CEU: .6 UNITS)

CLASS SIZE: MINIMUM: 12; MAXIMUM: 30

NHI Customer Service: (877) 558-6873 • nhicustomerservice@dot.gov

COURSE NUMBER

FHWA-NHI-133110

COURSE TITLE

Strategies for Developing Work Zone Traffic Analyses (Web-Based)

Strategies for Developing Work Zone Traffic Analyses is an interactive Web-based training course that provides an overview of how traffic analysis tools can be applied specifically to work zone analysis problems. Traffic analysis tools represent various transportation modeling approaches such as sketch planning, travel demand modeling, and traffic simulation (microscopic, mesoscopic, and simulation approaches).

The purpose of this course is three-fold. First, it will educate the participants regarding the constraints and opportunities of work zone analysis associated with available transportation modeling approaches. Second, it will build familiarity with the various work zone factors influencing the selection of a transportation modeling approach. Third, it will provide the participants with practical experience in developing a transportation modeling approach in a collaborative process that considers issues ranging from work zone characteristics, performance measurement, technical risk assessment, and resource constraints. In conclusion, participants will be able to characterize a work zone and select and justify a transportation modeling approach based upon the work zone characterization.

OUTCOMES

Upon completion of the course, participants will be able to:

- Define a role for work zone modeling
- Describe the work zone analysis decision-making engine
- Explain how to characterize a work zone
- Identify transportation modeling approaches
- Describe alternative transportation modeling approaches
- Justify selection of a transportation modeling approach

TARGET AUDIENCE

The Strategies for Developing Work Zone Traffic Analyses (WBT) is designed for professionals employed by State DOTs (district engineers, corridor planners, project engineers, traffic engineers, and work zone planners), FHWA Division Offices Staff, transportation engineers, traffic staff, planners, MPOs, and consultants.

TRAINING LEVEL: Basic

FEE: 2016: $50 Per Person; 2017: N/A

LENGTH: 6 HOURS (CEU: .4 UNITS)

CLASS SIZE: MINIMUM: 0; MAXIMUM: 0

NHI Customer Service: (877) 558-6873 • nhicustomerservice@dot.gov

COURSE NUMBER

FHWA-NHI-133112

COURSE TITLE

Design and Operation of Work Zone Traffic Control (1-Day)

This course provides participants with information on the safest and most efficient work zone traffic controls, including the application of effective design and installation concepts; and using signs and markings for detours, construction zones, and maintenance sites. The legal, administrative, and operational aspects also will be discussed. Classroom presentations include lectures, case histories, and workshops.

OUTCOMES

Upon completion of the course, participants will be able to:

- Describe each step involved in providing work zone traffic controls
- Identify and apply workable concepts and techniques for designing, installing, and maintaining controls in construction, maintenance, and utility operations
- Identify appropriate principles in the design of traffic control plans
- Apply traffic control plans to site conditions, monitor traffic controls, and make changes indicated by traffic accidents and incidents
- Discuss techniques and procedures used by different agencies
- Assess the legal consequences of action and inaction relative to work zone traffic control and identify risk management procedures

TARGET AUDIENCE

Design, construction, and maintenance personnel responsible for designing, installing, and monitoring work zone traffic control.

TRAINING LEVEL: Intermediate

FEE: 2016: $275 Per Person; 2017: N/A

LENGTH: 1 DAYS (CEU: .6 UNITS)

CLASS SIZE: MINIMUM: 20; MAXIMUM: 30

NHI Customer Service: (877) 558-6873 • nhicustomerservice@dot.gov

COURSE NUMBER

FHWA-NHI-133112A

COURSE TITLE

Design and Operation of Work Zone Traffic Control (3-Day)

This course provides participants with information on the safest and most efficient work zone traffic controls, including the application of effective design and installation concepts; and using signs and markings for detours, construction zones, and maintenance sites. The legal, administrative, and operational aspects also will be discussed. Classroom presentations include lectures, case histories, and workshops.

OUTCOMES

Upon completion of the course, participants will be able to:

- Describe each step involved in providing work zone traffic controls
- Identify and apply workable concepts and techniques for designing, installing, and maintaining controls in construction, maintenance, and utility operations
- Identify appropriate principles in the design of traffic control plans
- Apply traffic control plans to site conditions, monitor traffic controls, and make changes indicated by traffic accidents and incidents
- Discuss techniques and procedures used by different agencies
- Assess the legal consequences of action and inaction relative to work zone traffic control and identify risk management procedures

TARGET AUDIENCE

Design, construction, and maintenance personnel responsible for designing, installing, and monitoring work zone traffic control.

TRAINING LEVEL: Intermediate

FEE: 2016: $825 Per Person; 2017: N/A

LENGTH: 3 DAYS (CEU: 0 UNITS)

CLASS SIZE: MINIMUM: 20; MAXIMUM: 30

NHI Customer Service: (877) 558-6873 • nhicustomerservice@dot.gov

COURSE NUMBER

FHWA-NHI-133113

COURSE TITLE

Work Zone Traffic Control for Maintenance Operations

This course provides guidance and training for field personnel working in the planning, selection, application, and operation of short-term work zones. The course addresses typical short-term maintenance activities occurring on two-lane rural highways and multilane urban streets and highways. The course covers the applicable standards for work zone protection contained in the "Manual on Uniform Traffic Control Devices" (MUTCD), discussing the need for proper application of devices, while addressing liability issues of highway agencies and individuals. Classroom presentation includes practical exercises to plan, set up, operate, and remove work zone safety devices, including appropriate flagging procedures for these operations.

OUTCOMES

Upon completion of the course, participants will be able to:

- Apply traffic control through short-term and mobile work areas
- Use national work zone standards and requirements as contained in Part VI of the MUTCD
- Use standard traffic control devices in work zones
- Design and install traffic control schemes for short-term and mobile operations on rural two- and multilane streets and highways
- Apply proper flagging procedures

TARGET AUDIENCE

State, county, and utility personnel, such as maintenance crews, survey crews, and utility crews, who are responsible for establishing traffic controls through short-term, utility, and maintenance work areas.

TRAINING LEVEL: Accomplished

FEE: 2016: $275 Per Person; 2017: N/A

LENGTH: 1 DAYS (CEU: .6 UNITS)

CLASS SIZE: MINIMUM: 20; MAXIMUM: 30

NHI Customer Service: (877) 558-6873 • nhicustomerservice@dot.gov

COURSE NUMBER

FHWA-NHI-133114

COURSE TITLE

Construction Zone Safety Inspection (1-Day)

This course provides training in the management of traffic control plans and the inspection of construction zone safety devices. Participants receive instruction in traffic control plan review, inspection of traffic control procedures and safety devices, and the resolution of discrepancies from the traffic control plan, as well as on deficiencies in safety hardware maintenance. The following major topics are covered: Inspection of traffic control plan operation, maintenance of work zone signs and markings, inspection of construction safety hardware, and resolution of discrepancies from contract requirements.

OUTCOMES

Upon completion of the course, participants will be able to:

- Recognize the importance of construction zone safety devices
- Identify the contract requirements for selected devices
- Inspect the installation and operation of safety devices, including discrepancies and deficiencies in safety devices
- Resolve discrepancies from the contract requirements and ensure corrections in the deficient safety devices

TARGET AUDIENCE

FHWA safety engineers, FHWA highway engineers, and State and local personnel involved in the management of traffic control plans and the inspection of construction zone safety devices.

TRAINING LEVEL: Basic

FEE: 2016: $275 Per Person; 2017: N/A

LENGTH: 1 DAYS (CEU: .6 UNITS)

CLASS SIZE: MINIMUM: 20; MAXIMUM: 30

NHI Customer Service: (877) 558-6873 • nhicustomerservice@dot.gov

Course Number

FHWA-NHI-133114A

Course Title

Construction Zone Safety Inspection (1.5 Day)

This course provides training in the management of traffic control plans and the inspection of construction zone safety devices. Participants receive instruction in traffic control plan review, inspection of traffic control procedures and safety devices, and the resolution of discrepancies from the traffic control plan, as well as on deficiencies in safety hardware maintenance. The following major topics are covered: Inspection of traffic control plan operation, maintenance of work zone signs and markings, inspection of construction safety hardware, and resolution of discrepancies from contract requirements.

Outcomes

Upon completion of the course, participants will be able to:

- Recognize the importance of construction zone safety devices
- Identify the contract requirements for selected devices
- Inspect the installation and operation of safety devices, including discrepancies and deficiencies in safety devices
- Resolve discrepancies from the contract requirements and ensure corrections in the deficient safety devices

Target Audience

FHWA safety engineers, FHWA highway engineers, and State and local personnel involved in the management of traffic control plans and the inspection of construction zone safety devices.

Training Level: Basic

Fee: 2016: $300 Per Person; 2017: N/A

Length: 1.5 DAYS (CEU: .9 UNITS)

Class Size: MINIMUM: 20; MAXIMUM: 30

NHI Customer Service: (877) 558-6873 • nhicustomerservice@dot.gov

Course Number

FHWA-NHI-133115

Course Title

Advanced Work Zone Management and Design

This course provides participants with advanced levels of knowledge and competencies with technical and non-technical aspects of work zone traffic control practices including work zone planning, design, project management, and contract issues. The course is designed to provide maximum flexibility by including core, recommended, and optional lessons. Each participant receives a copy of the "Advanced Work Zone Management and Design" reference manual and a participant workbook that contains all lesson materials.

Outcomes

Upon completion of the course, participants will be able to:

- Apply the latest safety and mobility design concepts as it relates to temporary traffic control (TTC) plans for work zones
- Identify the latest MUTCD principles as it relates to TTC plans for planning, design, project management, and describe the various contracting issues that may need to be resolved
- Demonstrate knowledge of the latest concepts as related to Parts 1, 5 and 6 of the MUTCD
- Demonstrate knowledge of key concepts in the AASHTO Design Guide and other standards as related to such items as worker and flagger apparel (such as ANSI and similar standard guides)
- Evaluate work zone temporary traffic control designs for nighttime and daytime issues
- Analyze and evaluate operational, safety and mobility impacts of work zones, including scheduling, scope, phases and alternate routes
- Consider the application of ITS technologies and where applicable apply ITS technologies to work zone planning, design and execution
- Consider alternative innovations, best practices and recent research findings in work zone planning, design and execution
- Develop temporary transportation management plans for safety and mobility
- List elements necessary for successful contracts and identify strategies for resolving contract issues, including best practices in work zone contracting, also identify tools to resolve conflicts with contracting issues
- Identify and resolve community issues, including impacts of work zones on affected residential and business areas. Apply public participation, outreach, and work zone strategies to minimize or mitigate community impacts with respect to work zones
- Identify and analyze specific (key) issues and concerns that affect work zone design and demonstrate ability to explain safety and mobility issues, impacts and alternatives to peers, public and/or decision makers
- Summarize work zone safety and mobility impacts and alternatives

Target Audience

State, and local design engineers, traffic and safety engineers, senior work zone traffic engineers, transportation planners, employees of metropolitan planning organizations and board members, regional planners, regional construction engineers (with work zone experience), and senior engineering technicians.

Training Level: Accomplished

Fee: 2016: $825 Per Person; 2017: N/A

Length: 3 DAYS (CEU: 1.8 UNITS)

Class Size: MINIMUM: 20; MAXIMUM: 30

NHI Customer Service: (877) 558-6873 • nhicustomerservice@dot.gov

COURSE NUMBER

FHWA-NHI-133116

COURSE TITLE

Maintenance of Traffic for Technicians - WEB BASED

The Maintenance of Traffic for Technicians Web-based training presents information about the placement of, field maintenance required for, and inspection of traffic control devices. In addition, drafting work zone traffic control plans and flaggering are discussed.

We've broken this training into five modules:

1. General Terms and Procedures
2. Traffic Channelizing and Control Devices
3. Traffic Control Zones
4. Flagger Operations
5. Traffic Control Zone Operations

OUTCOMES

Upon completion of the course, participants will be able to:

- Identify the correct placement of work zone traffic control devices
- Perform field maintenance of work zone traffic control devices
- Inspect placement or operational functions of work zone traffic control devices
- Generate work zone traffic control plans
- Explain the basics of flagging

TARGET AUDIENCE

This training is designed for all persons with duties that include: Direct responsibility for placement of work zone traffic control devices; Direct responsibility for field maintenance of work zone traffic control devices; Inspection of the placement or operational function of work zone traffic control devices; and Drafting or electronic generation of work zone traffic control plans. The target audience could be geographically dispersed, in need of immediate training or information, or not have access to travel funds.

TRAINING LEVEL: Basic

FEE: 2016: $50 Per Person; 2017: N/A

LENGTH: 5 HOURS (CEU: 0 UNITS)

CLASS SIZE: MINIMUM: 1; MAXIMUM: 1

NHI Customer Service: (877) 558-6873 • nhicustomerservice@dot.gov

COURSE NUMBER

FHWA-NHI-133117

COURSE TITLE

Maintenance of Traffic for Supervisors - WEB BASED

The Maintenance of Traffic for Supervisors Web-based training presents information about the placement of, field maintenance required for, and inspection of traffic control devices. In addition, drafting work zone traffic control plans and flagging are discussed. This training focuses on the design of a traffic control plan, and how and why one needs to operate and implement traffic control in the work zone.

We've broken this training into five modules:

1. Fundamental Principles of Temporary Traffic Control Zones
2. Temporary Traffic Control Devices
3. Traffic Control Zones
4. Transportation Management Plans
5. Flagger Operations

OUTCOMES

Upon completion of the course, participants will be able to:

- Describe how to create clear, organized traffic control plans
- Identify acceptable temporary traffic control devices
- Determine good and bad flagging techniques

TARGET AUDIENCE

This training is designed for personnel with responsibility or authority to decide on the specific maintenance of traffic requirements to be implemented. These positions include engineers responsible for work zone traffic control development and work site traffic supervisors. The target audience could be geographically dispersed, in need of immediate training or information, or not have access to travel funds.

TRAINING LEVEL: Basic

FEE: 2016: $50 Per Person; 2017: N/A

LENGTH: 5 HOURS (CEU: 0 UNITS)

CLASS SIZE: MINIMUM: 1; MAXIMUM: 1

NHI Customer Service: (877) 558-6873 • nhicustomerservice@dot.gov

Course Number

FHWA-NHI-133118

Course Title

Flagger Training - WEB-BASED

Being a flagger is the most important job on the work site. Careless use of the sign or distraction from duty could cause serious injury to workers or the motoring public. Performing flagger duties diligently can prevent traffic incidents in the work area.

This is a basic training in the area of flagger training. It has been designed for someone learning the first steps in performing flagger duties. This training would be useful as a refresher course for all employees involved with work zone traffic control where flaggers are utilized.

This training does not go into individual state flagger training or certification requirements. For more information on flagger training requirements contact your State's safety office.

Outcomes

Upon completion of the course, participants will be able to:

- Identify the responsibilities of a flagger
- Describe the proper ways to place signs
- Describe the proper position for flagging
- Define the flagging procedures for stop, slow, and proceed
- Identify the correct procedures for various flagging situations
- Describe the proper conduct in flagging

Target Audience

This training is intended for individuals that will be performing or are engaging in flagger duties on construction/ maintenance projects. The course will assist them in better understanding the importance and duties involved with flagging on a project. It would be beneficial to the entry level employee as well as the experienced flagger.

Training Level: Basic

Fee: 2016: $25 Per Person; 2017: N/A

Length: 1 HOURS (CEU: 0 UNITS)

Class Size: MINIMUM: 1; MAXIMUM: 1

NHI Customer Service: (877) 558-6873 • nhicustomerservice@dot.gov

Course Number

FHWA-NHI-133119

Course Title

Safe and Effective Use of Law Enforcement Personnel in Work Zones - WEB-BASED

This training is provided to you at no cost by the FHWA Office of Operations.

NHI training 133119 Safe and Effective Use of Law Enforcement Personnel in Work Zones is an interactive Web-based training (WBT) course that provides law enforcement agencies with the practices and procedures to improve traffic safety in work zones. Work zone law enforcement is highly effective in reducing speeding, speed variability, and undesirable driving behaviors such as tailgating and unsafe lane changes, which improves both traffic and worker safety. The presence of work zone enforcement is also believed to raise driver awareness and overall alertness, further improving work zone safety.

The purpose of this course is to provide basic knowledge to help save lives, avoid work zone crashes, and improve safety when working in a work zone. This course will provide tips for safe practices for law enforcement officers (LEO's) in work zones as well as providing for a safer work zone environment. This Web-based training will educate participants on the standards and guidelines related to temporary traffic control in work zones; the role of LEO's in work zones; the components of a typical work zone; and the proper practices and procedures related to the use of law enforcement officers in work zones.

Outcomes

Upon completion of the course, participants will be able to:

- Describe the role of LEO's in work zones
- Explain proper practices and procedures related to the use of LEO's in work zones
- Explain safe operating practices of LEO's working in a Temporary Traffic Control (TTC) zone

Target Audience

133119 Safe and Effective User of Law Enforcement Personnel in Work Zones is a Web-based training course designed for LEO's. Specifically, this course targets state troopers, state, county, municipal officers, and highway patrol officers who will participate in work zone activities.

Training Level: Basic

Fee: 2016: $0 Per Person; 2017: N/A

Length: 2 HOURS (CEU: 0 UNITS)

Class Size: MINIMUM: 1; MAXIMUM: 1

NHI Customer Service: (877) 558-6873 • nhicustomerservice@dot.gov

Course Number

FHWA-NHI-133120

Course Title

Work Zone Traffic Analysis Applications and Decision Framework

Work Zone Traffic Analysis - Applications and Decision Framework is a two-day instructor-led course utilizing lecture and group collaborative exercises to provide guidance on work zone traffic analysis applications and decision framework. It will help work practitioners in understanding the analytical methods involved in conducting a work zone traffic analysis. This course is designed to cover establishing a work zone traffic analysis process; step-by-step guidance on determining the most suitable tools to perform a work zone analysis; key considerations when applying various modeling tools for work zone traffic analysis; a decision framework on how to select the best alternatives based on a set of performance measures; essential components of work zone traffic analysis report and a variety of case studies to demonstrate a diverse set of work zone traffic analysis applications.

The course provides an overview of the Federal Highway Administration's guidebook titled "Traffic Analysis Toolbox XII - Work Zone Traffic Analysis - Applications and Decision Framework. " Work Zone Traffic Analysis (WZTA) is the process of evaluating and determining the mobility and safety impacts within a transportation construction, maintenance, or rehabilitation project. The purpose of the course is to provide participants an understanding of the analytical methods involved in conducting and developing a WZTA as well as direction on where to go for more information.

Outcomes

Upon completion of the course, participants will be able to:

- Establish a work zone traffic analysis process
- Select the appropriate tool for work zone traffic analysis
- Identify and assess key considerations for modeling approach
- Apply modeling tools to work zone traffic analysis
- Apply road user costs
- Reconcile inconsistencies and conduct sensitivity analysis
- Establish a MOTAA decision framework
- Develop analysis report structure

Target Audience

Engineers, planners, modelers, and others responsible for framing a work zone traffic analysis, those who decide on and use work zone traffic analysis tools for which zone strategies to implement, and decision-makers considering work zone traffic analysis. These include State DOT staff, FHWA staff, Metropolitan Planning Organization staff, and consultants. This course is designed for those individuals seeking to supplement and expand their basic knowledge and understanding of work zone traffic analysis. This is a mid-level course and it focuses heavily on the analysis tools and methods for work zone traffic analysis and case study examples.

Training Level: Intermediate

Fee: 2016: $725 Per Person; 2017: N/A

Length: 2 DAYS (CEU: 1.2 UNITS)

Class Size: MINIMUM: 20; MAXIMUM: 30

NHI Customer Service: (877) 558-6873 • nhicustomerservice@dot.gov

COURSE NUMBER

FHWA-NHI-133121

COURSE TITLE

Traffic Signal Design and Operation

There is a need to understand that the congestion and delays that exist on our streets and roadways can be better managed with a thorough understanding of effective traffic signal timing and optimization. Well-developed, designed, implemented, maintained, and operated traffic signal control projects are essential to this process. Engineering tools are available to design, optimize, analyze, and simulate traffic flow. This course addresses the application of the "Manual of Uniform Traffic Control Devices" (MUTCD) to intersection displays, as well as signal timing, computerized traffic signal systems, control strategies, integrated systems, traffic control simulation, and optimization software. The course is divided into two primary parts: Traffic Signal Timing and Design, and Traffic Signal Systems.

Note: This course was previously numbered 133028.

OUTCOMES

Upon completion of the course, participants will be able to:

- List the steps required to plan, design, and implement a signalized intersection
- Devise an appropriate data collection plan for planning, designing, and operating a signalized intersection
- Perform a warrant analysis using the MUTCD warrants, including local policies
- Design basic phasing of the intersection - which movements will get a separate phase, and how they are numbered
- Calculate signal timing at the design stage for both actuated and coordinated operational strategies, including pedestrian clearance intervals
- Determine location of signal displays
- Select signal-related signs and pavement markings, including turning-movement signs and advance warning signs

TARGET AUDIENCE

Traffic engineering personnel from State, Federal, and local agencies involved in planning, design, operation or maintenance of traffic signals or traffic signal systems. The course will not assume any prior knowledge of computers and thus will describe the theory of operation and the manner in which it can be applied to traffic signal controls.

TRAINING LEVEL: Basic

FEE: 2016: $500 Per Person; 2017: N/A

LENGTH: 2 DAYS (CEU: 1.1 UNITS)

CLASS SIZE: MINIMUM: 20; MAXIMUM: 30

NHI Customer Service: (877) 558-6873 • nhicustomerservice@dot.gov

Course Number

FHWA-NHI-133122

Course Title

Traffic Signal Timing Concepts

Note: This course will offer special pricing to the following groups: Local Agencies - $75.00 per person; State DOT's - $200.00 per person. The reduced prices are being provided by the FHWA Office of Operations.

Traffic Signal Timing Concepts is a two-day course to assist in building technical expertise in signal timing by focusing on the relationship between network context and operational objectives to inform the design of signal timing parameters. The course will expand on the traditional signal timing process by incorporating an objectives and performance driven approach that leads to selection of appropriate computational methods for design and operation of traffic signal timing.

For many agencies, the design of signal timing parameters is an exercise in data collection and software driven optimization in response to citizen compliants. An ad-hoc complaint-driven processes with little documentation and infrequent attempts to quantify performance or improvements is not likely to lead to a well-managed, objective driven process for the timing and retiming of traffic signals, nor does it typically provide agencies with a good feel for the overall performance of their system. What is needed is an objective-driven, performance-oriented approach to traffic signal timing.

This course is very interactive and includes many exercises. Participants calculate various timing parameters by hand so they should bring a calculator to the course.

Completion of NHI #133121 is recommended but not required.

Outcomes

Upon completion of the course, participants will be able to:

- Discuss an objectives-based signal timing process
- Describe operations objectives in the context of network configuration and traffic conditions
- Review phasing and timing
- Design cycle lengths
- Design green time and fixed intervals
- Design phase sequence and offsets
- Develop operational mode parameters
- Evaluate signal timing outcomes

Target Audience

Traffic Signal Timing Concepts is a two-day course for practitioners involved in or responsible for design, operations, or management of traffic signals including State/MPOs/Local Government personnel and consultants and contractors.

Training Level: Intermediate

Fee: 2016: $500 Per Person; 2017: N/A

Length: 2 DAYS (CEU: 1.2 UNITS)

Class Size: MINIMUM: 20; MAXIMUM: 30

NHI Customer Service: (877) 558-6873 • nhicustomerservice@dot.gov

Course Number

FHWA-NHI-133123

Course Title

Systems Engineering for Signal Systems Including Adaptive Control

Note: This course will offer special pricing to the following groups: Local Agencies - $75.00 per person; State DOT's - $200.00 per person. The reduced prices are being provided by the FHWA Office of Operations.

Systems Engineering for Signal Systems Including Adaptive Control is a two-day course aimed to assist transportation professionals to identify the needs for improved traffic operations and utilize systems engineering principles for the implementation of traffic signal operational improvements. This course will provide traffic operations managers and personnel a comprehensive view of what is required before, during, and after the implementation of a new traffic control system. Adaptive signal control is used as the example throughout the course.

The overall goal of this course is to assist traffic operations staff in identifying traffic control system objectives and needs to facilitate planning, designing and implementing a new traffic control system. The FHWA document, Model Systems Engineering Documents for Adaptive Signal Control Technology (ASCT) Systems, (FHWA-HOP-11-027) is used for the exercises of this course.

Outcomes

Upon completion of the course, participants will be able to:

- Engage stakeholders
- Gather information needed for systems engineering process
- Evaluate and resolve constraints
- Assemble a concept of operations
- Extract requirements
- Document verification and validation process
- Develop a procurement strategy
- Assemble a systems engineering analysis
- Describe the systems engineering process

Target Audience

Professionals responsible for the planning, design, management or operation of traffic signal systems. This includes engineers, and technicians (advanced) of state/local agencies, consultants, and FHWA Operations staff.

Training Level: Basic

Fee: 2016: $500 Per Person; 2017: N/A

Length: 2 DAYS (CEU: 1.2 UNITS)

Class Size: MINIMUM: 20; MAXIMUM: 30

NHI Customer Service: (877) 558-6873 • nhicustomerservice@dot.gov

COURSE NUMBER

FHWA-NHI-133125

COURSE TITLE

Successful Traffic Signal Management: The Basic Service Approach

Note: This course will offer special pricing to the following groups: Local Agencies - $75.00 per person; State DOT's - $200.00 per person. The reduced prices are being provided by the FHWA Office of Operations.

Successful Traffic Signal Management: The Basic Service Approach is a two-day course aimed at helping agencies ensure that their limited resources are directed towards meeting the needs of the agencies most important stakeholders. A Traffic Signal Management Plan (TSMP) is a tool that documents and aligns an agencies traffic signal design, operation and maintenance strategies to achieve basic service objectives. The application of systematic business processes is integral to maintaining the resources and workforce capability that is necessary to sustain the operation and maintenance of traffic signal systems over long periods of time. Agencies that clearly articulate their operational objectives and meaningfully measure performance tend to operate and maintain traffic signal systems more effectively than agencies that fail to document this information.

The purpose of this course will be to describe and expand on the Basic Service Concept for use in developing an agency's Traffic Signal Management Plan. Emphasis will be placed on an agency developing a simply stated goal and then developing objectives, strategies and tactics enabling them to accomplish their stated goal. Each element of the traffic signal management plan will be thoroughly covered, resulting in a guideline that agencies can follow to develop their own TSMP.

OUTCOMES

Upon completion of the course, participants will be able to:

- Formulate clear objectives
- Select appropriate standards of performance
- Identify performance measures
- Relate organizational capabilities and resource allocation to objectives
- Assess infrastructure reliability
- Identify signal timing strategies
- Document communication policies
- Apply effective design strategies
- Develop a traffic signal management plan

TARGET AUDIENCE

Professionals involved in the design, management, operation or maintenance of traffic signal systems. This includes design engineers, operations engineers and technicians (advanced) of state/local agencies, consultants, and FHWA Operations staff.

TRAINING LEVEL: Basic

FEE: 2016: $500 Per Person; 2017: N/A

LENGTH: 2 DAYS (CEU: 1.2 UNITS)

CLASS SIZE: MINIMUM: 20; MAXIMUM: 30

NHI Customer Service: (877) 558-6873 • nhicustomerservice@dot.gov

Course Number

FHWA-NHI-133126

Course Title

National Traffic Incident Management Responder Training - Web-Based

This training was developed under the second Strategic Highway Research Program (SHRP2), and is being provided to you by the FHWA Office of Operations.

Three injury crashes occur every minute in the United States, putting nearly 39,000 incident responders potentially in harm's way every day. Congestion from these incidents often generates secondary crashes, further increasing traveler delay and frustration. The longer incident responders remain at the scene, the greater the risk they, and the traveling public, face. A cadre of well-trained responders helps improve traffic incident response. Better incident response improve the safety of responders and drivers, reduces crashes that occur because of incident-related congestion, decreases traffic delays caused by incidents, and can cut incident response time.

The National Traffic Incident Management Responder Training was created by responders for responders. This course provides first responders a shared understanding of the requirements for safe, quick clearance of traffic incident scenes; prompt, reliable and open communication; and motorist and responder safeguards. First responders learn how to operate more efficiently and collectively.

This training covers many TIM recommended procedures and techniques, including:

- TIM Fundamentals and Terminology
- Notification and Scene Size-Up
- Safe Vehicle Positioning
- Scene Safety
- Command Responsibilities
- Traffic Management
- Special Circumstances
- Clearance and Termination
- Telecommunicators

Prerequisite Note:

It is recommended that you take the following courses offered by FEMA:

IS 700 - National Management System (NIMS), An Introduction

ICS 100 - Introduction to Incident Command System (ICS)

ICS 200 - ICS for Single Resources and Initial Action Incidents

This training was developed through the second Strategic Highway Research Program (SHRP2).

Outcomes

Upon completion of the course, participants will be able to:

- Use a common set of practices and advance standards across all responder disciplines.
- The National Traffic Incident Management Training Program equips responders with a common set of core competencies and assists them in achieving the TIM National Unified Goal of strengthening TIM programs in the areas of: Responder safety; Safe, quick clearance; and Prompt, reliable, and interoperable communications.

Target Audience

The target audience for the training is individuals from all TIM responder disciplines, including: Law Enforcement, Fire/ Rescue, Emergency Medical Service, Towing and Recovery, Emergency Management, Communications, Highway/ Transportation and Dispatch within States, regions and localities.

TRAINING LEVEL: Basic

FEE: 2016: $0 Per Person; 2017: N/A

LENGTH: 4 HOURS (CEU: .4 UNITS)

CLASS SIZE: MINIMUM: 1; MAXIMUM: 1

NHI Customer Service: (877) 558-6873 • nhicustomerservice@dot.gov

Course Number

FHWA-NHI-134005

Course Title

Value Engineering Workshop (3-day)

Value Engineering (VE) is a systematic process of review and analysis of a project during the concept and design phases. VE is conducted by a multi-disciplined team of persons not involved in the project to provide recommendations such as: a) providing the needed functions safely, reliably, and at the lowest overall cost; b) improving the value and quality of the project; and c) reducing the time to complete the project.

This course begins with a Web-based training (WBT) component that is completed prior to the first day of the class (134005A). The 3-day workshop involves training participants to be valued contributors to the Value Engineering team, conducting a Value Engineering study in a team environment. It is preferable that the host agency provides actual project(s) to be used in this course, although The National Highway Institute (NHI) can provide projects upon request. Depending on the projects selected for use in the course, and based on the request of the host agency, the 3-day classroom session can be expanded to 4 or 5 days in length (NHI-134005B and NHI-134005C).

Upon successful course completion, participants will have acquired the training necessary to successfully participate in future Value Engineering studies for their agencies.

Outcomes

Upon completion of the course, participants will be able to:

- Explain how Value Engineering can improve project performance, reduce costs, and enhance value.
- Acquire the necessary behaviors and skills to be an effective Value Engineering team member with the ability to: Investigate the project and analyze project functions and costs; Creatively speculate on alternative ways to perform the various functions; Evaluate the most effective life-cycle alternatives; Develop viable alternatives into fully supported recommendations; Present the recommendations to stakeholders and agency management

Target Audience

The target audience for this course consists of FHWA and state highway agency personnel in management, administrative, and engineering disciplines who will participate as Value Engineering team members. Consultants or agency representatives of all technical disciplines associated with project design, development, construction, and maintenance can be included in order to provide the multiple perspectives needed to maximize the effectiveness of the team.

Training Level: Basic

Fee: 2016: $925 Per Person; 2017: $925 Per Person

Length: 3 DAYS (CEU: 1.8 UNITS)

Class Size: MINIMUM: 20; MAXIMUM: 30

NHI Customer Service: (877) 558-6873 • nhicustomerservice@dot.gov

Course Number

FHWA-NHI-134005A

Course Title

Introduction to Value Engineering - WEB-BASED

This training is a prerequisite of another NHI training and is offered at no cost.

Value Engineering (VE) is a systematic process of review and analysis of a project during the concept and design phases. VE is conducted by a multi-disciplined team of persons not involved in the project to provide recommendations such as: a) providing the needed functions safely, reliably, and at the lowest overall cost; b) improving the value and quality of the project; and c) reducing the time to complete the project.

This Web-based training is intended to provide an overview of the Value Engineering process, know as the Value Engineering study. Included in the training is a discussion of the benefits of utilizing VE, the keys to completing a successful VE study, and an overview of the objectives and tasks completed by the VE team at each phase.

Participants can complete this training independently. Those who plan on attending the 3-day Value Engineering classroom training must complete this online module prior to coming to class. Course certificates should be printed out and presented to the instructor on the first day to verify completion.

Outcomes

Upon completion of the course, participants will be able to:

- Identify the purpose of Value Engineering and its benefits to a highway transportation agency.
- Identify the critical skills required to participate successfully in the VE study.
- Describe each phase of creating a Value Engineering Job Plan in terms of the objective and tasks.

Target Audience

The target audience for this course consists of FHWA and state highway agency personnel in management, administrative, and engineering disciplines who will participate as Value Engineering team members or who are interested in learning more about the process. Consultants or agency representatives of all technical disciplines associated with project design, development, construction, and maintenance who will participate in a Value Engineering study should also attend.

Training Level: Basic

Fee: 2016: $0 Per Person; 2017: $0 Per Person

Length: .5 HOURS (CEU: 0 UNITS)

Class Size: MINIMUM: 20; MAXIMUM: 30

NHI Customer Service: (877) 558-6873 • nhicustomerservice@dot.gov

Course Number

FHWA-NHI-134005B

Course Title

Value Engineering Workshop (4-day)

Value Engineering (VE) is a systematic process of review and analysis of a project during the concept and design phases. VE is conducted by a multi-disciplined team of persons not involved in the project to provide recommendations such as: a) providing the needed functions safely, reliably, and at the lowest overall cost; b) improving the value and quality of the project; and c) reducing the time to complete the project.

This course begins with a Web-based training (WBT) component that is completed prior to the first day of the class. The 4-day workshop involves training participants to be valued contributors to the Value Engineering team, conducting a Value Engineering study in a team environment. It is preferable that the host agency provides actual project(s) to be used in this course, although The National Highway Institute (NHI) can provide projects upon request. Depending on the projects selected for use in the course, and based on the request of the host agency, the 3-day classroom session can be expanded to 3 or 5 days in length (NHI-134005 and NHI-134005C).

Upon successful course completion, participants will have acquired the training necessary to successfully participate in future Value Engineering studies for their agencies.

Outcomes

Upon completion of the course, participants will be able to:

- Explain how value engineering can improve project performance, reduce costs, and enhance value.
- Acquire the necessary behaviors and skills to be an effective Value Engineering Team member with the ability to: Investigate the project and analyze project functions and costs; Creatively speculate on alternative ways to perform the various functions; Evaluate the most effective life-cycle alternatives; Develop viable alternatives into fully supported recommendations; Present the recommendations to stakeholders and agency management

Target Audience

The target audience for this course consists of FHWA and state highway agency personnel in management, administrative, and engineering disciplines who will participate as Value Engineering team members. Consultants or agency representatives of all technical disciplines associated with project design, development, construction, and maintenance can be included in order to provide the multiple perspectives needed to maximize the effectiveness of the team.

Training Level: Basic

Fee: 2016: $1160 Per Person; 2017: $1160 Per Person

Length: 4 DAYS (CEU: 2.4 UNITS)

Class Size: MINIMUM: 20; MAXIMUM: 30

NHI Customer Service: (877) 558-6873 • nhicustomerservice@dot.gov

COURSE NUMBER

FHWA-NHI-134005C

COURSE TITLE

Value Engineering Workshop (5-day)

Value Engineering (VE) is a systematic process of review and analysis of a project during the concept and design phases. VE is conducted by a multi-disciplined team of persons not involved in the project to provide recommendations such as: a) providing the needed functions safely, reliably, and at the lowest overall cost; b) improving the value and quality of the project; and c) reducing the time to complete the project.

This course begins with a Web-based training (WBT) component that is completed prior to the first day of the class (134005A). The 3-day workshop involves training participants to be valued contributors to the Value Engineering team, conducting a Value Engineering study in a team environment. It is preferable that the host agency provides actual project(s) to be used in this course, although The National Highway Institute (NHI) can provide projects upon request. Depending on the projects selected for use in the course, and based on the request of the host agency, the 5-day classroom session can be shortened to 3 or 4 days in length (NHI-134005 and NHI-134005B).

Upon successful course completion, participants will have acquired the training necessary to successfully participate in future Value Engineering studies for their agencies.

OUTCOMES

Upon completion of the course, participants will be able to:

- Explain how value engineering can improve project performance, reduce costs, and enhance value.
- Acquire the necessary behaviors and skills to be an effective Value Engineering Team member with the ability to: Investigate the project and analyze project functions and costs; Creatively speculate on alternative ways to perform the various functions; Evaluate the most effective life-cycle alternatives; Develop viable alternatives into fully supported recommendations; Present the recommendations to stakeholders and agency management

TARGET AUDIENCE

The target audience for this course consists of FHWA and state highway agency personnel in management, administrative, and engineering disciplines who will participate as Value Engineering team members. Consultants or agency representatives of all technical disciplines associated with project design, development, construction, and maintenance can be included in order to provide the multiple perspectives needed to maximize the effectiveness of the team.

TRAINING LEVEL: Basic

FEE: 2016: $1395 Per Person; 2017: $1395 Per Person

LENGTH: 5 DAYS (CEU: 3 UNITS)

CLASS SIZE: MINIMUM: 20; MAXIMUM: 30

NHI Customer Service: (877) 558-6873 • nhicustomerservice@dot.gov

DESIGN AND TRAFFIC OPERATIONS

Course Number

FHWA-NHI-134109I

Course Title

Maintenance Training Series: Basics of Work Zone Traffic Control - WEB-BASED

Meeting the national requirements for work zone traffic control is a critically important responsibility of maintenance personnel. The national requirements, found in Part 6 of the Manual on Uniform Traffic Control Devices (MUTCD), promote driver and worker safety during roadway maintenance projects. This training, Basics of Work Zone Traffic Control, provides an introduction to the requirements outlined in Part 6 of the 2009 MUTCD. The course also offers an overview of the manual's structure and requirements regarding traffic control devices and their applications, flagging operations and procedures, and pedestrian and worker safety.

Through a series of work zone scenarios, this training uses the MUTCD Part 6 to review fundamental concepts of setting up work zones, including proper signage, taper lengths, and flagging procedures. Participants are encouraged to compare their State's standards, if available, to the guidance established in the MUTCD and determine what additional requirements may need to be met to establish safe, compliant work zones.

This training was developed as part of the Maintenance Training Series. To access all the courses in the series, enroll in the 134109 course.

Outcomes

Upon completion of the course, participants will be able to:

- Describe the content and use of The Manual on Uniform Traffic Control Devices (MUTCD) Part 6
- Use the MUTCD to correctly answer questions about the basics of work zone traffic control
- Differentiate among standard, guidance, and option conditions in the MUTCD
- Differentiate among standard, guidance, and option conditions in the MUTCD for work zone traffic control in rural and urban areas

Target Audience

This course is designed for State, regional, and county personnel who manage operations programs and deal with oversight and quality assurance across broad geographic areas. This target audience also is involved with handling materials, scheduling, budgeting, and planning.

Training Level: Basic

Fee: 2016: $25 Per Person; 2017: N/A

Length: 1 HOURS (CEU: 0 UNITS)

Class Size: MINIMUM: 0; MAXIMUM: 0

NHI Customer Service: (877) 558-6873 • nhicustomerservice@dot.gov

COURSE NUMBER

FHWA-NHI-380069

COURSE TITLE

Road Safety Audits/Assessments

Performing effective road safety audits/assessments, (RSAs), improves safety and demonstrates to the public an agency's dedication to crash reduction. An RSA is a formal safety performance examination of an existing or future road or intersection by an independent audit team. The RSA training provides practical information on how to conduct an RSA, select a location, and build an independent, multi-disciplinary team. The costs, time, benefits, and common myths and concerns surrounding RSAs will be discussed. Participants learn how to improve transportation safety by applying a new proactive approach. Emphasis is placed on using low cost safety improvements as well as understanding the interaction between the highway and all road users.

The training includes hands-on application of the training materials, which includes information on each stage of a road safety audit and easy-to-use-prompt lists. A copy of "FHWA Road Safety Audit Guidelines" is provided.

OUTCOMES

Upon completion of the course, participants will be able to:

- Express the road safety audit process terminology
- Perform a simple road safety audit, as a member of a team
- Assess the benefits of a road safety audit on a local or statewide basis

TARGET AUDIENCE

Personnel who are likely to serve on a road safety audit team including Federal, State, local transportation personnel, first responders and consultants who conduct highway safety studies should also attend.

TRAINING LEVEL: Accomplished

FEE: 2016: $400 Per Person; 2017: N/A

LENGTH: 2 DAYS (CEU: 1.2 UNITS)

CLASS SIZE: MINIMUM: 20; MAXIMUM: 30

NHI Customer Service: (877) 558-6873 • nhicustomerservice@dot.gov

Course Number

FHWA-NHI-380071

Course Title

Interactive Highway Safety Design Model

This course instructs highway design project managers, planners, designers, and traffic and safety reviewers in the application of the Interactive Highway Safety Design Model (IHSDM) software and provides guidance on interpretation of the output.

IHSDM is a suite of software tools to evaluate safety of two-lane rural highways. The software, developed for FHWA, was released in 2003 after several years of research and development to provide state-of-the-art techniques for safety analysis. IHSDM contains five tools that can be used to apply the most recent safety analysis techniques in a relatively straightforward and automated manner. For more information about IHSDM, go to http://www.tfhrc.gov/safety/ihsdm/ihsdm.htm.

Participants gain hands-on experience with the software. Therefore, the training facility must be equipped with computers. There should be no more than two participants per computer. Minimum system specifications for the computers are as follows: Operating System - Microsoft Vista, Windows XP or Windows 2000 Professional; HTML Browser - Microsoft Internet Explorer, Netscape Navigator, or Foxfire; Spreadsheet Program, Microsoft Excel or equivalent; Hardware - At least 450 MHz Pentium III (or equivalent) CPU, 256 MB RAM or greater desirable, 800x600 high colors (16 bit) display; and 300 MB free disk space

Outcomes

Upon completion of the course, participants will be able to:

- Describe key capabilities and limitations of IHSDM
- Evaluate a two-lane rural highway using IHSDM
- Recognize when and how IHSDM can be used in the project development process

Target Audience

Highway design project managers, planners, designers, and traffic and safety reviewers with at least one or two years of experience with highway design, preferably two-lane rural highway design.

Training Level: Accomplished

Fee: 2016: $400 Per Person; 2017: N/A

Length: 2 DAYS (CEU: 0 UNITS)

Class Size: MINIMUM: 20; MAXIMUM: 30

NHI Customer Service: (877) 558-6873 • nhicustomerservice@dot.gov

Course Number

FHWA-NHI-380077

Course Title

Intersection Safety Workshop

Beginning with an introduction to intersection and crash characteristics, this course provides information on ready-to-use, direct-application safety measures for rural unsignalized and signalized intersections. Participants are presented with a synthesis of countermeasures and their associated crash reduction factors as identified in the "AASHTO Strategic Highway Safety Plan - NCHRP 500 Guidebooks." The course focuses on the application of these countermeasures and design and safety operations best practices for substantive improvements to intersection safety. During the course, participants have the opportunity to present intersection safety situations that they are currently facing and discuss appropriate countermeasures and best practices to address those situations.

Outcomes

Upon completion of the course, participants will be able to:

- Apply models (equations) to predict the number of crashes for an intersection based upon traffic volumes
- Identify high crash intersections and recognize appropriate engineering countermeasures
- Identify crash reduction factors/crash modification factors associated with countermeasures
- Describe safety performance of intersection geometric design features and the models to quantify the safety effect
- List regulatory, warning, and guide signing and markings countermeasures and associated safety benefits
- List highway lighting countermeasures and associated safety benefits
- List traffic signal countermeasures and associated safety benefits

Target Audience

Federal, State, and local transportation traffic and safety engineers, and planners involved in reducing intersection crashes.

Training Level: Accomplished

Fee: 2016: $320 Per Person; 2017: N/A

Length: 1 DAYS (CEU: 0 UNITS)

Class Size: MINIMUM: 20; MAXIMUM: 30

NHI Customer Service: (877) 558-6873 • nhicustomerservice@dot.gov

Course Number

FHWA-NHI-380078

Course Title

Signalized Intersection Guidebook Workshop

This course provides an overview of the "Signalized Intersections: Informational Guide FHWA-HRT-04-091." The guide is a comprehensive document containing methods for evaluating the safety and operations of signalized intersections and tools to remedy deficiencies. It takes a holistic approach to signalized intersections and considers the safety and operational implications of a particular treatment on all system users, including motorists, pedestrians, bicyclists, and transit users. Using the guide, participants learn to make insightful intersection assessments, understand the tradeoffs of potential improvement measures, and apply guidebook measures and best practices to reduce the incidence of intersection crashes.

Outcomes

Upon completion of the course, participants will be able to:

- Recognize and apply fundamentals of signalized intersections in terms of user needs, geometric design, traffic design, and illumination
- Describe signalized intersection project process, safety analysis methods, and operational analysis methods
- Describe the more than 100 signalized intersection treatments and their advantages and disadvantages

Target Audience

Federal, State, and local transportation, traffic and safety engineers, and planners involved in planning, designing, operating, and remedying crash problems for signalized intersections.

Training Level: Intermediate

Fee: 2016: $330 Per Person; 2017: N/A

Length: 1 DAYS (CEU: .6 UNITS)

Class Size: MINIMUM: 20; MAXIMUM: 30

NHI Customer Service: (877) 558-6873 • nhicustomerservice@dot.gov

Course Number

FHWA-NHI-380095

Course Title

Geometric Design: Applying Flexibility and Risk Management

Highway designers often face complex trade-offs when developing projects. A "quality" design may be thought of as satisfying the needs of a wide variety of users while balancing the often competing interests of cost, safety, mobility, social and environmental impacts. Applying flexibility and risk management in highway design requires more than simply assembling geometric elements from the available tables, charts and equations of design criteria. This course provides participants with knowledge of the functional basis of critical design criteria to enable informed decisions when applying engineering judgment and flexibility. The course exercises and case studies provide practical applications of current knowledge from research and operational experience of human factors and safety effects for various design elements.

Outcomes

Upon completion of the course, participants will be able to:

- Define the relationship among design criteria, design guidelines and design standards
- Describe the concepts of design speed, target speed, posted speed and operating speed
- Describe the FHWA Policy for Design Standards and Design Exceptions
- List the 13 controlling geometric design criteria that require a formal written design exception from FHWA
- Evaluate the safety effects and qualitative risk of proposed design exceptions
- Evaluate the effectiveness and appropriateness of mitigation strategies for design exceptions
- Describe the relationship between safety and key geometric features of highway alignment and cross section
- Describe the applicability of a human-centered approach to geometric design considerations

Target Audience

This course is targeted toward engineers that are involved in applying engineering judgment in the selection of design criteria and in the assessment of design exceptions. It is most practical for practicing engineers and highway decision makers from state highway agencies, local agencies, design consultants, and FHWA field offices.

Training Level: Accomplished

Fee: 2016: $400 Per Person; 2017: N/A

Length: 2 DAYS (CEU: 1.2 UNITS)

Class Size: MINIMUM: 20; MAXIMUM: 30

NHI Customer Service: (877) 558-6873 • nhicustomerservice@dot.gov

COURSE NUMBER

FHWA-NHI-380100

COURSE TITLE

Using IHSDM

NHI delivers Web-conference Training to you!

The IHSDM course is a training that gives participants the opportunity to use the IHSDM software tools to evaluate and analyze highway designs.

The delivery format consists of 4 live Web Conference Trainings (WCT), which participants are required to attend. In between Web-conferences, participants must complete self-paced assignments.

The Interactive Highway Safety Design Model (IHSDM) is a suite of software analysis tools used to evaluate the safety and operational effects of geometric design decisions on highways.

IHSDM is a decision-support tool, which provides estimates of a highway design's expected safety and operational performance and checks existing or proposed highway designs against relevant design policy values. Results of the IHSDM support decisionmaking in the highway design process. Intended users include highway project managers, designers, and traffic and safety reviewers in State and local highway agencies and in engineering consulting firms.

The IHSDM, which supports the Data-Driven Safety Analysis initiative that is part of Federal Highway Administration's (FHWA's) Every Day Counts 3 efforts, includes six evaluation modules (Crash Prediction, Design Consistency, Intersection Review, Policy Review, Traffic Analysis, and Driver/Vehicle). This Web site summarizes the capabilities and applications of the IHSDM evaluation modules, and provides a library of the research reports documenting their development.

The IHSDM - HSM Predictive Method 2015 Release (version 11.0.1, October 2015) may be downloaded free of charge at http://www.ihsdm.org. The new version includes major enhancements to the Policy Review Module, which was expanded to include policy checks for rural multilane highways.

www.fhwa.dot.gov/research/tfhrc/projects/safety/comprehensive/ihsdm/index.cfm

OUTCOMES

Upon completion of the course, participants will be able to:

- Explain the scope and uses for the IHSDM tool.
- Input rural highway data to IHSDM.
- Explain the purpose of each of the six IHSDM modules.
- Demonstrate the workflow for each IHSDM module.
- Interpret and apply data from IHSDM reports and graphs to make rural highways safer.

TARGET AUDIENCE

The Using IHSDM Course is designed for personnel working on highway design projects who will be directly interacting with the IHSDM software tools or applying the data generated by them. The IHSDM course benefits highway design project managers, planners, designers, safety engineers, and other personnel responsible for reviewing operations and safety on rural highways. Participants should have general familiarity with highway design elements and terminology.

TRAINING LEVEL: Intermediate

FEE: 2016: $280 Per Person; 2017: N/A

LENGTH: 12 HOURS (CEU: 1.2 UNITS)

CLASS SIZE: MINIMUM: 10; MAXIMUM: 45

NHI Customer Service: (877) 558-6873 • nhicustomerservice@dot.gov

Course Number

FHWA-NHI-380118

Course Title

Integrating Geometric Design & Traffic Control for Improved Safety

This course provides an overview of the inter-relationship of geometric design and traffic control device applications. The primary focus of the course concerns interchange areas where lane elimination, lane configurations, and traffic control devices on freeways and expressways may present challenges for both designers and motorists. This course addresses lane balance effects, degree of control (markings) practices, arrows (signs and markings) usage, advance vehicle positioning, short auxiliary lanes, and geometric design influences on signing and marking. This course includes discussion and guidance for meeting driver expectations and the human factors associated with roadway geometry and the application, selection, and placement of traffic control devices. Participants engage in group exercises to strengthen and apply the principles covered in the workshop,

Outcomes

Upon completion of the course, participants will be able to:

- o Identify Human Factors concepts
- o Compare and evaluate lane configuration designs and methods of lane elimination
- o Explain the role of TCDs (signs and pavement markings)
- o Identify the basic signing and marking concepts, types and purposes from the MUTCD
- o Describe the flexibility and interdependence of geometric and traffic control design .

Target Audience

Engineers, engineering practitioners, technologists, and engineering assistants involved in freeway and expressway design, construction, and operations including Sections such as Roadway Design, Traffic Engineering, District personnel with responsible charge of plan review of TCDs (striping, signing, other markings), plan preparation, development/ revision of standards for the same and Consultant Management staff, as well as consultants performing work on such projects and/or related duties.

Training Level: Intermediate

Fee: 2016: $310 Per Person; 2017: N/A

Length: 1 DAYS (CEU: .6 UNITS)

Class Size: MINIMUM: 20; MAXIMUM: 30

NHI Customer Service: (877) 558-6873 • nhicustomerservice@dot.gov

COURSE NUMBER

FHWA-NHI-130088

COURSE TITLE

Bridge Construction Inspection

The Bridge Construction Inspection Course (BCIC) is one of the core curriculum initiatives cited by AASHTO, FHWA, and the five regional organizations. These core curriculum initiatives are being pursued in order to maximize regional, public, and industry resources in the development of core training and qualification-based certification programs, improve the quality of bridge construction, and promote uniformity in training content and qualification requirements.

Overall, the BCIC improves quality, ensures uniformity, and establishes minimum competencies for bridge construction inspection. The underlying themes of the course can be broken down into key segments. The BCIC will provide the construction inspector with:

1. The requisite knowledge of construction that will make him/her an effective inspector
2. An overall awareness of the problems and consequences that can arise during construction and how these factors will impact the safety and service life of the structure
3. A knowledge of the inspections that should be performed to confirm conformance to the contract documents, or document contract nonconformance

OUTCOMES

Upon completion of the course, participants will be able to:

- Explain the role of the construction inspector as part of the overall project team
- Interpret drawings and specifications
- Anticipate possible construction and materials problems
- Maintain bridge controls for location and elevation
- Describe construction sequence for various bridge systems (e.g. foundations, substructures, superstructures, and miscellaneous systems), bridge types and materials
- Conduct regular systematic inspections of materials and standards of construction, through the use of job aids, such as checklists
- Explain and perform basic inspection and testing of materials
- Perform accurate surveys and checking of dimensions
- Make and maintain sufficient records

TARGET AUDIENCE

Construction supervisors, transportation department field inspectors, field engineers, resident engineers, structural engineers, materials engineers, and other technical personnel involved in the construction inspection of bridges. The course is developed for participants without an in-depth engineering background. However, more knowledgeable persons can attend and will add to the overall effectiveness of the training through their active participation.

TRAINING LEVEL: Basic

FEE: 2016: $1100 Per Person; 2017: $1100 Per Person

LENGTH: 4.5 DAYS (CEU: 2.7 UNITS)

CLASS SIZE: MINIMUM: 20; MAXIMUM: 30

NHI Customer Service: (877) 558-6873 • nhicustomerservice@dot.gov

Course Number

FHWA-NHI-131050

Course Title

Asphalt Pavement In-Place Recycling Techniques

Transportation agencies focusing on the use of sustainable, cost effective, and environmentally conscious construction practices often consider in-place recycling techniques as a viable alternative to the more traditional rehabilitation techniques used on asphalt-surfaced pavements. NHI training 131050 Asphalt Pavement In-place Recycling Techniques is designed to help participants acquire necessary skills for selecting the appropriate in-place recycling technique for a given set of conditions, choosing the appropriate materials for the project, developing suitable specifications, and constructing those projects effectively.

The Asphalt Pavement In-place Recycling Techniques course includes two brief Web-based training (WBT) modules, and two days of instructor-led, classroom-based training (ILT). Through independent study, classroom interaction, and workshop activities, participants explore the current technologies available in the area of asphalt pavement in-place recycling. Two WBT lessons introduce pavement evaluation techniques and the three potential recycling techniques, along with the types of equipment commonly used for each. The classroom session focuses on project and technique selection and justification, materials considerations and mix design, construction specifications, and project control considerations during construction.

Outcomes

Upon completion of the course, participants will be able to:

- Describe the economic, environmental, and engineered performance benefits associated with using in-place asphalt recycling
- Identify the key factors that contribute to the selection of appropriate in-place asphalt recycling techniques under different traffic levels, pavement conditions, and environments
- Identify the key requirements in developing effective in-place asphalt recycling construction specifications, including method specification and end-result or performance specifications
- Demonstrate the ability to select the appropriate new materials and additives needed for each of three HMA pavement in-place recycling techniques
- List steps that can be taken to address a variety of issues that may impact the constructability of a project

Target Audience

This course is intended for State and local transportation agency engineers, such as pavement managers and maintenance engineers, and other agency personnel who are responsible for selecting, designing, or constructing the agency's asphalt pavement maintenance, resurfacing, rehabilitation, and reconstruction alternatives. The course particularly benefits those individuals responsible for selecting and designing asphalt in-place recycling projects, for writing effective specifications, or for inspecting asphalt in-place recycling projects during their construction. Contractors, consulting engineers, and industry representatives involved in asphalt pavement in-place recycling also will benefit from this course.

Training Level: Intermediate

Fee: 2016: $225 Per Person; 2017: $225 Per Person

Length: 2 DAYS (CEU: 1.3 UNITS)

Class Size: MINIMUM: 20; MAXIMUM: 30

NHI Customer Service: (877) 558-6873 • nhicustomerservice@dot.gov

Course Number

FHWA-NHI-131050A

Course Title

Asphalt Pavement In-Place Recycling Techniques--WEB-BASED

This training is a prerequisite of another NHI training and is offered at no cost.

Transportation agencies focusing on the use of sustainable, cost-effective, and environmentally conscious construction practices often consider in-place recycling techniques as a viable alternative to the more traditional rehabilitation techniques used on asphalt-surfaced pavements. NHI training 131050 Asphalt Pavement In-place Recycling Techniques is designed to help participants acquire necessary skills for selecting the appropriate in-place recycling technique for a given set of conditions, choosing the appropriate materials for the project, developing suitable specifications, and constructing those projects effectively.

The Asphalt Pavement In-place Recycling Techniques course includes two brief Web-based training (WBT) modules, and two days of instructor-led, classroom-based training (ILT). Through independent study, classroom interaction, and workshop activities, participants explore the current technologies available in the area of asphalt pavement in-place recycling. Two WBT lessons introduce pavement evaluation techniques and the three potential recycling techniques, along with the types of equipment commonly used for each. The classroom session focuses on project and technique selection and justification, materials considerations and mix design, construction specifications, and project control considerations during construction.

Outcomes

Upon completion of the course, participants will be able to:

- Describe the economic, environmental, and engineered performance benefits associated with using in-place asphalt recycling
- Identify the key factors that contribute to the selection of appropriate in-place asphalt recycling techniques under different traffic levels, pavement conditions, and environments
- Identify the key requirements in developing effective in-place asphalt recycling construction specifications, including method specification and end-result or performance specifications
- Demonstrate the ability to select the appropriate new materials and additives needed for each of three HMA pavement in-place recycling techniques
- List steps that can be taken to address a variety of issues that may impact the constructability of a project

Target Audience

This course is intended for State and local transportation agency engineers, such as pavement managers and maintenance engineers, and other agency personnel who are responsible for selecting, designing, or constructing the agency's asphalt pavement maintenance, resurfacing, rehabilitation, and reconstruction alternatives. The course particularly benefits those individuals responsible for selecting and designing asphalt in-place recycling projects, for writing effective specifications, or for inspecting asphalt in-place recycling projects during their construction. Contractors, consulting engineers, and industry representatives involved in asphalt pavement in-place recycling also will benefit from this course.

Training Level: Basic

Fee: 2016: $0 Per Person; 2017: $0 Per Person

Length: 2 HOURS (CEU: 0 UNITS)

Class Size: MINIMUM: 20; MAXIMUM: 30

NHI Customer Service: (877) 558-6873 • nhicustomerservice@dot.gov

COURSE NUMBER

FHWA-NHI-131100

COURSE TITLE

Pavement Smoothness: Use of Inertial Profiler Measurements for Construction Quality Control

Studies have shown that roughness is one of the biggest priorities of highway users. Additional studies have shown that pavements that are built smooth stay smoother longer and provide a longer pavement life. Most State highway agencies (SHAs) have some type of smoothness specification that is used to evaluate the smoothness of newly constructed or rehabilitated pavements during acceptance testing. Many agencies also have incentives or disincentives for new construction and rehabilitation, which are based on pavement smoothness.

Increasingly these agencies are turning to inertial profilers as the most reliable instrument for construction acceptance testing and verifying pavement smoothness. The intent of this course is to train inertial profiler operators in the basics of performing construction acceptance testing and to train those reviewing the data to comprehend how those data were obtained and what they represent in order to build smoother riding roadways.

The course has been developed to be delivered in a single day of instructor-led training. In order to keep the instructor-led portion of the training to a single day, the training includes two hours of independent study that should be completed prior to attending the instructor-led session.

OUTCOMES

Upon completion of the course, participants will be able to:

- Perform checks of the inertial profiler components to identify that the equipment is in proper working order.
- Determine the impact of current surface and environmental conditions on data collection.
- Collect profile data using appropriate operating techniques.
- Calculate a smoothness index using appropriate data processing techniques and computational procedures for use in construction quality control and specification compliance.
- Identify what features in a collected profile are manifested in a smoothness or roughness index.

TARGET AUDIENCE

The course was designed for an audience directly involved in the use of inertial profilers and the application of the data obtained from inertial profilers. This includes State and contractor road profiler operators who perform data collection, initial processing, and reporting of smoothness data. Paving superintendents, project engineers, pavement engineers, and inspectors who are performing data analysis, quality control, and acceptance will also benefit from this course. Ideally, each session of the course will include a mixture of State and contractor personnel, including those who collect data, those performing data processing, and those making decisions based upon data. ASSUMED TRAINING COMPETENCIESThe participants should have a basic understanding of how to operate a computer including turning it on and off, running programs, and saving data.

TRAINING LEVEL: Intermediate

FEE: 2016: $150 Per Person; 2017: $150 Per Person

LENGTH: 1 DAYS (CEU: .6 UNITS)

CLASS SIZE: MINIMUM: 20; MAXIMUM: 30

NHI Customer Service: (877) 558-6873 • nhicustomerservice@dot.gov

COURSE NUMBER

FHWA-NHI-131110

COURSE TITLE

Pavement Preservation Treatment Construction - WEB-BASED

FHWA, in partnership with Caltrans, the National Center for Pavement Preservation, and the Transportation Curriculum Coordination Council (TCCC) created the Pavement Preservation Treatment Construction Guide (PPTCG) as a resource for agency and industry pavement preservation practitioners. This course is designed to provide participants with an introduction to the PPTCG, so that they can better use it to familiarize themselves with general information on pavement preservation concepts and techniques. The guide covers basic pavement preservation concepts, as well as information on specific treatments to extend the life of asphalt pavements. The module topics are:

1. Introduction to Pavement Preservation (NHI-131110A)
2. Materials (NHI-131110B)
3. Crack Sealing, Crack Filling and Joint Sealing of Flexible and Rigid Pavements (NHI-131110C)
4. Patching and Edge Repairs (NHI-131110D)
5. Chip Seals (NHI-131110E)
6. Fog Seals (NHI-131110F)
7. Slurry Seals (NHI-131110G)
8. Micro-surfacing Projects (NHI-131110H)
9. Thin Functional and Maintenance Overlay Projects (NHI-131110I)
10. Ultra Thin, Hot-Mixed, Bonded Overlay Projects (NHI-131110J)
11. Selecting a Pavement Presentation Treatment (NHI-131110K)

Each of the modules is also offered as individual trainings and can be accessed by registering for the course number listed with each module.

OUTCOMES

Upon completion of the course, participants will be able to:

- Identify the components and value of a Pavement Preventive Maintenance (PPM) program
- Identify pavement conditions and other attributes that suggest whether preventive maintenance is appropriate
- Identify various pavement preservation strategies, techniques and materials
- State the performance characteristics of various pavement preservation strategies, techniques and materials
- Select the appropriate strategy(ies), technique(s) and material to extend the service life and retard the development of pavement distress

TARGET AUDIENCE

The primary audience for the Pavement Preservation Treatment Construction WBT course is Federal, State, and local highway construction and maintenance teams, specifically the highway workers and inspectors involved in the placement of pavement preservation treatments. Although not in the primary audience, design engineers will also benefit from the online guide and the associated training. The training course is primarily targeted at individuals unfamiliar with pavement preservation policy and technical information.

TRAINING LEVEL: Intermediate

FEE: 2016: $50 Per Person; 2017: N/A

LENGTH: 6.5 HOURS (CEU: 0 UNITS)

CLASS SIZE: MINIMUM: 1; MAXIMUM: 1

NHI Customer Service: (877) 558-6873 • nhicustomerservice@dot.gov

COURSE NUMBER

FHWA-NHI-131110A

COURSE TITLE

Pavement Preservation Treatment Series: Introduction to Pavement Preservation - WEB-BASED

This training is part of the "Pavement Preservation Treatment" series and is designed to provide participants with an introduction to the Pavement Preservation Treatment Construction Guide (PPTCG) and the basics of pavement preservation. Topics include: pavement structure, distresses, and differentiating pavement preservation from preventive maintenance.

As stated, this training draws on the PPTCG, which was created by FHWA, in partnership with Caltrans, the National Center for Pavement Preservation, and the Transportation Curriculum Coordination Council (TCCC) as a resource for agency and industry pavement preservation practitioners. It provides information on basic pavement preservation concepts and the different treatments available and how they should be applied, so agencies can make informed decisions when determining which treatments best fit their pavement preservation needs. The training is primarily targeted at individuals unfamiliar with pavement preservation policy and technical information.

To take the entire series of trainings for the PPTCG, access the NHI website and register for NHI-131110.

OUTCOMES

Upon completion of the course, participants will be able to:

- Identify common surface distresses in pavements.
- Distinguish between distresses caused by surface failure and those caused by subsurface layer failure.
- Recognize the difference between pavement preservation and pavement maintenance.

TARGET AUDIENCE

The primary audience for the Pavement Preservation Treatment Construction WBT course is Federal, State, and local highway construction and maintenance teams, specifically the highway workers and inspectors involved in the placement of pavement preservation treatments. Although not in the primary audience, design engineers will also benefit from the online guide and the associated training. The training course is primarily targeted at individuals unfamiliar with pavement preservation policy and technical information.

TRAINING LEVEL: Basic

FEE: 2016: $25 Per Person; 2017: N/A

LENGTH: .5 HOURS (CEU: 0 UNITS)

CLASS SIZE: MINIMUM: 1; MAXIMUM: 1

NHI Customer Service: (877) 558-6873 • nhicustomerservice@dot.gov

COURSE NUMBER

FHWA-NHI-131110B

COURSE TITLE

Pavement Preservation Treatment Series: Materials - WEB-BASED

This training is part of the "Pavement Preservation Treatment" series and is designed to provide participants with information on the materials used for preventive maintenance treatments. Topics include: materials comprising maintenance treatments, emulsions, and aggregates. This course is primarily intended for inspectors and technicians.

This training draws on the Pavement Preservation Treatment Construction Guide (PPTCG), which was created by FHWA, in partnership with Caltrans, the National Center for Pavement Preservation, and the Transportation Curriculum Coordination Council (TCCC) as a resource for agency and industry pavement preservation practitioners. It provides information on basic pavement preservation concepts and the different treatments available and how they should be applied, so agencies can make informed decisions when determining which treatments best fit their pavement preservation needs. The training is primarily targeted at individuals unfamiliar with pavement preservation policy and technical information.

To take the entire series of trainings for the PPTCG, access the NHI website and register for NHI-131110.

OUTCOMES

Upon completion of the course, participants will be able to:

- List the materials used in preventive maintenance treatments for flexible and rigid pavements.
- Recognize the differences between asphalt cement and emulsions and their use in pavement preservation treatments.
- List the six physical properties of aggregates that affect the performance of preservation treatments.

TARGET AUDIENCE

The primary audience for the Pavement Preservation Treatment Construction WBT course is Federal, State, and local highway construction and maintenance teams, specifically the highway workers and inspectors involved in the placement of pavement preservation treatments. Although not in the primary audience, design engineers will also benefit from the online guide and the associated training. The training course is primarily targeted at individuals unfamiliar with pavement preservation policy and technical information.

TRAINING LEVEL: Intermediate

FEE: 2016: $25 Per Person; 2017: $25 Per Person

LENGTH: 1 HOURS (CEU: 0 UNITS)

CLASS SIZE: MINIMUM: 1; MAXIMUM: 1

NHI Customer Service: (877) 558-6873 • nhicustomerservice@dot.gov

COURSE NUMBER

FHWA-NHI-131110C

COURSE TITLE

Pavement Preservation Treatment Series: Crack Sealing & Filling, and Joint Sealing - WEB-BASED

This training is part of the "Pavement Preservation Treatment" series and is designed to provide participants with information on crack sealing, crack filling, and joint sealing of flexible and rigid pavements. Topics include: working and non-working cracks, fatigue and longitudinal cracks, correct temperatures for crack sealant, crack repair sequence, hot sealant, and crack sealing or filling criteria. This course is primarily intended for inspectors and technicians.

This training draws on the Pavement Preservation Treatment Construction Guide (PPTCG), which was created by FHWA, in partnership with Caltrans, the National Center for Pavement Preservation, and the Transportation Curriculum Coordination Council (TCCC) as a resource for agency and industry pavement preservation practitioners. It provides information on basic pavement preservation concepts and the different treatments available and how they should be applied, so agencies can make informed decisions when determining which treatments best fit their pavement preservation needs. The training is primarily targeted at individuals unfamiliar with pavement preservation policy and technical information.

To take the entire series of trainings for the PPTCG, access the NHI website and register for NHI-131110.

OUTCOMES

Upon completion of the course, participants will be able to:

- Describe the difference between a working crack and a nonworking crack.
- List the types of distresses that crack sealing, crack filling, and joint sealing treatments will repair.
- Describe how proper storage and handling of sealants and fillers affect their constructability and performance.
- Describe the procedure of repairing surface cracks and rigid joints.
- Identify common problems associated with crack sealing, crack filling, and joint sealing treatments and recognize their solutions.
- List the capabilities and limitations of crack sealing, crack filling, and joint sealing treatments.

TARGET AUDIENCE

The primary audience for the Pavement Preservation Treatment Construction WBT course is Federal, State, and local highway construction and maintenance teams, specifically the highway workers and inspectors involved in the placement of pavement preservation treatments. Although not in the primary audience, design engineers will also benefit from the online guide and the associated training. The training course is primarily targeted at individuals unfamiliar with pavement preservation policy and technical information.

TRAINING LEVEL: Intermediate

FEE: 2016: $25 Per Person; 2017: $25 Per Person

LENGTH: 1 HOURS (CEU: 0 UNITS)

CLASS SIZE: MINIMUM: 1; MAXIMUM: 1

NHI Customer Service: (877) 558-6873 • nhicustomerservice@dot.gov

COURSE NUMBER

FHWA-NHI-131110D

COURSE TITLE

Pavement Preservation Treatment Series: Localized Pavement Repair - WEB-BASED

This training is part of the "Pavement Preservation Treatment" series and is designed to provide participants with information on localized pavement repair. Topics include: pothole formation and edge failure, seal or fill decisions, construction of, and problems with, pothole patching, dig outs, edge repairs, and skin patching, and capabilities and limitations of localized repairs. This course is primarily intended for inspectors and technicians.

This training draws on the Pavement Preservation Treatment Construction Guide (PPTCG), which was created by FHWA, in partnership with Caltrans, the National Center for Pavement Preservation, and the Transportation Curriculum Coordination Council (TCCC) as a resource for agency and industry pavement preservation practitioners. It provides information on basic pavement preservation concepts and the different treatments available and how they should be applied, so agencies can make informed decisions when determining which treatments best fit their pavement preservation needs. The training is primarily targeted at individuals unfamiliar with pavement preservation policy and technical information.

To take the entire series of trainings for the PPTCG, access the NHI website and register for NHI-131110.

OUTCOMES

Upon completion of the course, participants will be able to:

- Describe the mechanisms of pothole formation and edge failure.
- Select the type of localized pavement repair best suited to a given condition.
- Describe the process of pothole patching, dig outs, edge repairs, and skin patching.
- Identify common problems associated with pothole patching, dig outs, edge repairs, and skin patching and recognize their solutions.
- List the key capabilities and limitations of localized pavement repairs.

TARGET AUDIENCE

The primary audience for the Pavement Preservation Treatment Construction WBT course is Federal, State, and local highway construction and maintenance teams, specifically the highway workers and inspectors involved in the placement of pavement preservation treatments. Although not in the primary audience, design engineers will also benefit from the online guide and the associated training. The training course is primarily targeted at individuals unfamiliar with pavement preservation policy and technical information.

TRAINING LEVEL: Intermediate

FEE: 2016: $25 Per Person; 2017: $25 Per Person

LENGTH: 1 HOURS (CEU: 0 UNITS)

CLASS SIZE: MINIMUM: 1; MAXIMUM: 1

NHI Customer Service: (877) 558-6873 • nhicustomerservice@dot.gov

Course Number

FHWA-NHI-131110E

Course Title

Pavement Preservation Treatment Series: Chip Seals - WEB-BASED

This training is part of the "Pavement Preservation Treatment" series and is designed to provide participants with information on chip seals. Topics include: project selection, pavement and weather condition requirements, storage, traffic control, construction sequence, aggregate spreading distance, brooming, chip spreading process, distributor preparation, and troubleshooting.

This training draws on the Pavement Preservation Treatment Construction Guide (PPTCG), which was created by FHWA, in partnership with Caltrans, the National Center for Pavement Preservation, and the Transportation Curriculum Coordination Council (TCCC) as a resource for agency and industry pavement preservation practitioners. It provides information on basic pavement preservation concepts and the different treatments available and how they should be applied, so agencies can make informed decisions when determining which treatments best fit their pavement preservation needs. The training is primarily targeted at individuals unfamiliar with pavement preservation policy and technical information.

To take the entire series of trainings for the PPTCG, access the NHI website and register for NHI-131110.

Outcomes

Upon completion of the course, participants will be able to:

- Recognize pavement conditions best suited to the chip seal treatment.
- Identify how proper storage and handling of chip seal materials affect their constructability and performance.
- Describe the construction of chip seals.
- Identify common problems associated with chip seals and recognize their solutions.
- Recognize key capabilities and limitations of chip seals.

Target Audience

The primary audience for the Pavement Preservation Treatment Construction WBT course is Federal, State, and local highway construction and maintenance teams, specifically the highway workers and inspectors involved in the placement of pavement preservation treatments. Although not in the primary audience, design engineers will also benefit from the online guide and the associated training. The training course is primarily targeted at individuals unfamiliar with pavement preservation policy and technical information.

Training Level: Intermediate

Fee: 2016: $25 Per Person; 2017: $25 Per Person

Length: 1 HOURS (CEU: 0 UNITS)

Class Size: MINIMUM: 1; MAXIMUM: 1

NHI Customer Service: (877) 558-6873 • nhicustomerservice@dot.gov

COURSE NUMBER

FHWA-NHI-131110F

COURSE TITLE

Pavement Preservation Treatment Series: Fog Seals - WEB-BASED

This training is part of the "Pavement Preservation Treatment" series and is designed to provide participants with information on fog seals. Topics include: uses of fog seals, suitable pavement surfaces, storage and handling of materials, application process, and problems and causation. This course is primarily intended for inspectors and technicians.

This training draws on the Pavement Preservation Treatment Construction Guide (PPTCG), which was created by FHWA, in partnership with Caltrans, the National Center for Pavement Preservation, and the Transportation Curriculum Coordination Council (TCCC) as a resource for agency and industry pavement preservation practitioners. It provides information on basic pavement preservation concepts and the different treatments available and how they should be applied, so agencies can make informed decisions when determining which treatments best fit their pavement preservation needs. The training is primarily targeted at individuals unfamiliar with pavement preservation policy and technical information.

To take the entire series of trainings for the PPTCG, access the NHI website and register for NHI-131110.

OUTCOMES

Upon completion of the course, participants will be able to:

- Recognize pavement conditions most suitable for a fog seal.
- Describe how proper storage and handling of fog seal materials affect their constructability and performance.
- Describe the construction of a fog seal.
- Identify common problems associated with fog seals and recognize their solutions.
- List the key capabilities and limitations of fog seal treatments.

TARGET AUDIENCE

The primary audience for the Pavement Preservation Treatment Construction WBT course is Federal, State, and local highway construction and maintenance teams, specifically the highway workers and inspectors involved in the placement of pavement preservation treatments. Although not in the primary audience, design engineers will also benefit from the online guide and the associated training. The training course is primarily targeted at individuals unfamiliar with pavement preservation policy and technical information.

TRAINING LEVEL: Intermediate

FEE: 2016: $25 Per Person; 2017: $25 Per Person

LENGTH: 1 HOURS (CEU: 0 UNITS)

CLASS SIZE: MINIMUM: 1; MAXIMUM: 1

NHI Customer Service: (877) 558-6873 • nhicustomerservice@dot.gov

COURSE NUMBER

FHWA-NHI-131110G

COURSE TITLE

Pavement Preservation Treatment Series: Slurry Seals - WEB-BASED

This training is part of the "Pavement Preservation Treatment" series and is designed to provide participants with information on slurry seals. Topics include: reasons to use slurry seals, gradations of slurry seal aggregate, preparation and application process, and problems and solutions. This course is primarily intended for inspectors and technicians.

This training draws on the Pavement Preservation Treatment Construction Guide (PPTCG), which was created by FHWA, in partnership with Caltrans, the National Center for Pavement Preservation, and the Transportation Curriculum Coordination Council (TCCC) as a resource for agency and industry pavement preservation practitioners. It provides information on basic pavement preservation concepts and the different treatments available and how they should be applied, so agencies can make informed decisions when determining which treatments best fit their pavement preservation needs. The training is primarily targeted at individuals unfamiliar with pavement preservation policy and technical information.

To take the entire series of trainings for the PPTCG, access the NHI website and register for NHI-131110.

OUTCOMES

Upon completion of the course, participants will be able to:

- Identify the type of slurry seal appropriate to various traffic conditions.
- Describe the construction of slurry seals.
- Identify common problems associated with slurry seals and recognize their solutions.
- List the key capabilities and limitations of slurry seals.

TARGET AUDIENCE

The primary audience for the Pavement Preservation Treatment Construction WBT course is Federal, State, and local highway construction and maintenance teams, specifically the highway workers and inspectors involved in the placement of pavement preservation treatments. Although not in the primary audience, design engineers will also benefit from the online guide and the associated training. The training course is primarily targeted at individuals unfamiliar with pavement preservation policy and technical information.

TRAINING LEVEL: Intermediate

FEE: 2016: $25 Per Person; 2017: $25 Per Person

LENGTH: 1 HOURS (CEU: 0 UNITS)

CLASS SIZE: MINIMUM: 1; MAXIMUM: 1

NHI Customer Service: (877) 558-6873 • nhicustomerservice@dot.gov

COURSE NUMBER

FHWA-NHI-131110H

COURSE TITLE

Pavement Preservation Treatment Series: Micro-Surfacing - WEB-BASED

This training is part of the "Pavement Preservation Treatment" series and is designed to provide participants with information on micro-surfacing. Topics include: pavement and traffic condition considerations, construction, and troubleshooting.

This training draws on the Pavement Preservation Treatment Construction Guide (PPTCG), which was created by FHWA, in partnership with Caltrans, the National Center for Pavement Preservation, and the Transportation Curriculum Coordination Council (TCCC) as a resource for agency and industry pavement preservation practitioners. It provides information on basic pavement preservation concepts and the different treatments available and how they should be applied, so agencies can make informed decisions when determining which treatments best fit their pavement preservation needs. The training is primarily targeted at individuals unfamiliar with pavement preservation policy and technical information.

To take the entire series of trainings for the PPTCG, access the NHI website and register for NHI-131110.

OUTCOMES

Upon completion of the course, participants will be able to:

- Identify pavement conditions most suitable for a micro-surfacing treatment.
- Describe the construction of micro-surfacing.
- Identify common problems associated with micro-surfacing and recognize their solutions.
- List the key capabilities and limitations of micro-surfacing relative to various traffic conditions.

TARGET AUDIENCE

The primary audience for the Pavement Preservation Treatment Construction WBT course is Federal, State, and local highway construction and maintenance teams, specifically the highway workers and inspectors involved in the placement of pavement preservation treatments. Although not in the primary audience, design engineers will also benefit from the online guide and the associated training. The training course is primarily targeted at individuals unfamiliar with pavement preservation policy and technical information.

TRAINING LEVEL: Intermediate

FEE: 2016: $25 Per Person; 2017: $25 Per Person

LENGTH: 1 HOURS (CEU: 0 UNITS)

CLASS SIZE: MINIMUM: 1; MAXIMUM: 1

NHI Customer Service: (877) 558-6873 • nhicustomerservice@dot.gov

Course Number

FHWA-NHI-131110I

Course Title

Pavement Preservation Treatment Series: Thin Functional HMA Overlay - WEB-BASED

This training is part of the "Pavement Preservation Treatment" series and is designed to provide participants with information on thin functional hot-mix asphalt overlays. Topics include: proper usage, suitable pavement conditions, construction, and troubleshooting. This course is primarily intended for inspectors and technicians.

This training draws on the Pavement Preservation Treatment Construction Guide (PPTCG), which was created by FHWA, in partnership with Caltrans, the National Center for Pavement Preservation, and the Transportation Curriculum Coordination Council (TCCC) as a resource for agency and industry pavement preservation practitioners. It provides information on basic pavement preservation concepts and the different treatments available and how they should be applied, so agencies can make informed decisions when determining which treatments best fit their pavement preservation needs. The training is primarily targeted at individuals unfamiliar with pavement preservation policy and technical information.

To take the entire series of trainings for the PPTCG, access the NHI website and register for NHI-131110.

Outcomes

Upon completion of the course, participants will be able to:

- Identify pavement conditions best suited for a thin hot mix asphalt overlay.
- Describe the construction process for a thin hot mix asphalt overlay.
- Identify common problems associated with a thin hot mix asphalt overlay and recognize their solutions.
- List the key capabilities and benefits of a thin hot mix asphalt overlay relative to various traffic conditions.

Target Audience

The primary audience for the Pavement Preservation Treatment Construction WBT course is Federal, State, and local highway construction and maintenance teams, specifically the highway workers and inspectors involved in the placement of pavement preservation treatments. Although not in the primary audience, design engineers will also benefit from the online guide and the associated training. The training course is primarily targeted at individuals unfamiliar with pavement preservation policy and technical information.

Training Level: Intermediate

Fee: 2016: $25 Per Person; 2017: $25 Per Person

Length: 1 HOURS (CEU: 0 UNITS)

Class Size: MINIMUM: 1; MAXIMUM: 1

NHI Customer Service: (877) 558-6873 • nhicustomerservice@dot.gov

COURSE NUMBER

FHWA-NHI-131110J

COURSE TITLE

Pavement Preservation Treatment Series: Ultra Thin HMA Bonded Wearing Course - WEB-BASED

This training is part of the "Pavement Preservation Treatment" series and is designed to provide participants with information on ultra thin, hot-mixed asphalt bonded wearing course. Topics include: usage, distresses and application considerations, construction, and troubleshooting. This course is primarily intended for inspectors and technicians.

This training draws on the Pavement Preservation Treatment Construction Guide (PPTCG), which was created by FHWA, in partnership with Caltrans, the National Center for Pavement Preservation, and the Transportation Curriculum Coordination Council (TCCC) as a resource for agency and industry pavement preservation practitioners. It provides information on basic pavement preservation concepts and the different treatments available and how they should be applied, so agencies can make informed decisions when determining which treatments best fit their pavement preservation needs. The training is primarily targeted at individuals unfamiliar with pavement preservation policy and technical information.

To take the entire series of trainings for the PPTCG, access the NHI website and register for NHI-131110.

OUTCOMES

Upon completion of the course, participants will be able to:

- Identify pavement conditions best suited to ultra thin, hot-mixed asphalt bonded wearing course.
- Describe the construction of ultra thin, hot-mixed, asphalt bonded wearing course.
- Identify common problems associated with ultra thin, hot-mixed, asphalt bonded wearing course and recognize their solutions.
- List key capabilities and benefits of ultra thin, hot-mixed, asphalt bonded wearing course relative to various traffic conditions.

TARGET AUDIENCE

The primary audience for the Pavement Preservation Treatment Construction WBT course is Federal, State, and local highway construction and maintenance teams, specifically the highway workers and inspectors involved in the placement of pavement preservation treatments. Although not in the primary audience, design engineers will also benefit from the online guide and the associated training. The training course is primarily targeted at individuals unfamiliar with pavement preservation policy and technical information.

TRAINING LEVEL: Intermediate

FEE: 2016: $25 Per Person; 2017: $25 Per Person

LENGTH: 1 HOURS (CEU: 0 UNITS)

CLASS SIZE: MINIMUM: 1; MAXIMUM: 1

NHI Customer Service: (877) 558-6873 • nhicustomerservice@dot.gov

COURSE NUMBER

FHWA-NHI-131110K

COURSE TITLE

Pavement Preservation Treatment Series: Selecting the Right Treatment - WEB-BASED

This training is part of the "Pavement Preservation Treatment" series and is designed to provide participants with information on preservation treatment selection. This course is primarily intended for inspectors and technicians.

The training draws on the Pavement Preservation Treatment Construction Guide (PPTCG), which was created by FHWA, in partnership with Caltrans, the National Center for Pavement Preservation, and the Transportation Curriculum Coordination Council (TCCC) as a resource for agency and industry pavement preservation practitioners. It provides information on basic pavement preservation concepts and the different treatments available and how they should be applied, so agencies can make informed decisions when determining which treatments best fit their pavement preservation needs. The training is primarily targeted at individuals unfamiliar with pavement preservation policy and technical information.

To take the entire series of trainings for the PPTCG, access the NHI website and register for NHI-131110.

OUTCOMES

Upon completion of the course, participants will be able to:

- Select the appropriate pavement preservation treatment(s) after analyzing given pavement and traffic conditions.

TARGET AUDIENCE

The primary audience for the Pavement Preservation Treatment Construction WBT course is Federal, State, and local highway construction and maintenance teams, specifically the highway workers and inspectors involved in the placement of pavement preservation treatments. Although not in the primary audience, design engineers will also benefit from the online guide and the associated training. The training course is primarily targeted at individuals unfamiliar with pavement preservation policy and technical information.

TRAINING LEVEL: Intermediate

FEE: 2016: $25 Per Person; 2017: N/A

LENGTH: .5 HOURS (CEU: 0 UNITS)

CLASS SIZE: MINIMUM: 1; MAXIMUM: 1

NHI Customer Service: (877) 558-6873 • nhicustomerservice@dot.gov

Course Number

FHWA-NHI-131117

Course Title

Basic Materials for Highway and Structure Construction and Maintenance - WEB-BASED

This training is provided by the Transportation Curriculum Coordination Council (TCCC) in partnership with NHI to review basic materials for highway and structure construction and maintenance. The training was prepared by State DOT personnel for State DOT personnel. It contains good practices from various agencies. Each State agency/company has their own specifications, which the viewer needs to review and follow. This course is primarily intended for inspectors and technicians.

Although there are a number of materials used in the construction and maintenance process for both highways and structures, this course is focused on the three basic materials. They are Aggregate, Portland Cement Concrete (referred to as PCC), and Hot Mix Asphalt (referred to as HMA).

This training is directed toward the entry level technician, to give them a general view of the basic materials used in construction and maintenance. The course modules will address the procedures used in the production and sampling of aggregates.

Module 1 is called Basic Aggregates and includes quarry inspection, sand operation, stockpiling, and sampling. Module 2 covers Portland cement, including the production of Portland Cement, the hydration process, as well as other cementing materials used in concrete such as water, admixtures, and aggregates. Module 3 reviews Hot Mix Asphalt, including the asphalt binder and aggregates used in the production.

NHI is hosting this and other TCCC Web-based developments to serve a critical need for training. We need your feedback to determine whether we should continue posting other Web-based trainings like this one. Please take the time to complete the evaluation form provided at the end of the training, or email nhimarketing@dot.gov.

Outcomes

Upon completion of the course, participants will be able to:

- Identify aggregate production and sampling procedures
- Recognize the ingredients of PCC and the part each plays in concrete production
- Recognize the ingredients of HMA and the part each plays in hot mix asphalt production

Target Audience

This training is designed for Level I and Level II State/local public agency personnel and their industry counterparts involved in the construction, maintenance and testing process for highways and structures. Level I or Entry refers to employees/trainees with little to no experience in the subject area and perform his/her activities under direct supervision. Level II or Intermediate refers to employees that understand and demonstrate skills in one or more areas of the entry level and perform specific tasks under general supervision.

Training Level: Basic

Fee: 2016: $25 Per Person; 2017: N/A

Length: 3 HOURS (CEU: 0 UNITS)

Class Size: MINIMUM: 1; MAXIMUM: 1

NHI Customer Service: (877) 558-6873 • nhicustomerservice@dot.gov

Course Number

FHWA-NHI-131121

Course Title

Construction of Portland Cement Concrete Pavements - WEB-BASED

Improving and maintaining the quality of concrete is an important aspect of keeping pavements safe and long lasting. This training provides participants with an overview of the entire Portland cement concrete (PCC) paving and restoration process: setting forms, mixing, hauling, curing and applicable repair techniques. This training is presented in several modules:

1. Construction Quality
2. PCC Production Overview
3. Slipform Paving
4. Fixed Form Paving
5. Pavement Curing, Sawing, and Joint Sealing Operations
6. Concrete Pavement Restoration

This self-paced, Web-based training is designed for participants to progress at their own pace. The training focuses on the proper methods for construction of concrete paving and pavement restoration techniques with an emphasis on cause and effect.

Outcomes

Upon completion of the course, participants will be able to:

- Describe the differences between truck-mixed and ready-mixed concrete
- Identify factors in production and paving operations that contribute to achieving a smooth ride
- Describe the differences between slip-form and fixed-form paving
- Identify the factors that impact saw timing and crack control
- Recognize the importance and key factors in placing joint sealant materials
- Identify the components of concrete pavement restoration application and construction techniques
- Describe the purpose and appropriate use of full depth and partial depth repairs
- Indentify critical factors for curing and sawing operations that affect pavement performance
- Describe the purpose of grinding and dowel bar retrofit
- Identify applicable repair techniques for concrete pavement restoration
- Describe purpose of slab stabilization and joint and crack resealing

Target Audience

This training is designed for contractors, technicians, and inspectors who are involved in daily pavement operations for the placement and restoration of PCC pavements. Participants should have some working knowledge of concrete pavement construction.

Training Level: Intermediate

Fee: 2016: $50 Per Person; 2017: $50 Per Person

Length: 10 HOURS (CEU: 0 UNITS)

Class Size: MINIMUM: 1; MAXIMUM: 1

NHI Customer Service: (877) 558-6873 • nhicustomerservice@dot.gov

COURSE NUMBER

FHWA-NHI-131122

COURSE TITLE

Portland Cement Concrete Paving Inspection - WEB-BASED

This training is provided by the Transportation Curriculum Coordination Council (TCCC) in partnership with NHI to review inspection practices for Portland cement concrete paving projects. The training was originally developed by the Iowa Department of Transportation and more currently updated and reviewed by the TCCC and NHI. This course is recommended for the Transportation Curriculum Coordination Council levels I and II. This course is primarily intended for inspectors and technicians.

This training course has been prepared to provide guidance and instruction to inspectors involved in the construction of Portland cement concrete (PCC) pavements. The important tasks involved in this work are explained and proper procedures are described. The material is targeted for those who have not had experience in PCC paving construction.

OUTCOMES

Upon completion of the course, participants will be able to:

- Identify the materials in a PCC mixture and the concrete properties
- Comprehend Design Project Plans and recognize the joints types and saw cuts
- Identify the safety requirements and recognize safe Traffic Control practices
- Recognize and comprehend the use of the equipment in a PCC Paving project
- Recognize various sub grade treatments
- Inspect project tasks for compliance with pre-paving requirements, i.e., survey stakes, proof rolling, subgrade, and dowel baskets
- Inspect project tasks for compliance with PCC Paving requirements, i.e., string line, place and consolidate, finish, and texture
- Perform post-construction checks

TARGET AUDIENCE

This training is designed for FHWA, State, and local agencies and their industry counterparts involved in the process of placement and inspection of Portland cement concrete paving. It is applicable to anyone desiring a better understanding of activities and inspection procedures on Portland cement concrete paving projects.

TRAINING LEVEL: Intermediate

FEE: 2016: $50 Per Person; 2017: $50 Per Person

LENGTH: 5 HOURS (CEU: 0 UNITS)

CLASS SIZE: MINIMUM: 1; MAXIMUM: 1

NHI Customer Service: (877) 558-6873 • nhicustomerservice@dot.gov

CONSTRUCTION AND MAINTENANCE

Course Number

FHWA-NHI-131126

Course Title

Concrete Pavement Preservation Series (Includes NHI-131126A-K)

The Transportation Curriculum Coordination Council (TCCC) in partnership with NHI is pleased to offer this comprehensive training series (FHWA-NHI-131126) for concrete pavement preservation. The training was developed by the National Concrete Pavement Technology Center at Iowa State University in cooperation with FHWA.

The Concrete Pavement Preservation Series presents current guidelines and recommendations for the design, construction, and selection of cost-effective concrete pavement preservation strategies. It concentrates primarily on strategies and methods that are applicable at the project level, and not at the network level, where pavement management activities function and address such issues as prioritizing and budgeting.

Registration in NHI-131126 enrolls you in all 11 courses in the Concrete Pavement Preservation Series (NHI-131126A-K) plus gives you access to a downloadable version of the FHWA Concrete Pavement Preservation Guide! You can take some or all of these courses when it best suits your schedule.

NHI-131126 includes:

- Introduction module with downloadable version of the FHWA Concrete Pavement Preservation Guide
- NHI-131126A: Pavement Preservation Concepts
- NHI-131126B: Concrete Pavement Evaluation
- NHI-131126C: Slab Stabilization
- NHI-131126D: Partial-depth Repairs
- NHI-131126E: Full-depth Repairs
- NHI-131126F: Retrofitted Edge Drains
- NHI-131126G: Dowel Bar Retrofit
- NHI-131126H: Diamond Grinding and Grooving
- NHI-131126I: Joint Resealing and Crack Sealing
- NHI-131126J: Concrete Overlays
- NHI-131126K: Strategy Selection

Outcomes

Upon completion of the course, participants will be able to:

- Define pavement preservation
- List the major components of a pavement evaluation and the types of information gained from each
- Identify the purpose and suitable application of various concrete pavement preservation treatments
- Describe recommended materials and construction/installation practices for each treatment
- List factors to consider in the selection of concrete pavement preservation treatments

Target Audience

The Concrete Pavement Preservation Series meets the needs of a diverse audience to include design engineers, quality control personnel, contractors, suppliers, technicians, and trades people. While the course is aimed at those who have some familiarity with concrete pavements and pavement preservation, it should also be of value to those that are new to the field. This course is recommended for the Transportation Curriculum Coordination Council levels I - IV.

TRAINING LEVEL: Intermediate

FEE: 2016: $50 Per Person; 2017: N/A

LENGTH: 11 HOURS (CEU: 0 UNITS)

CLASS SIZE: MINIMUM: 1; MAXIMUM: 1

NHI Customer Service: (877) 558-6873 • nhicustomerservice@dot.gov

COURSE NUMBER

FHWA-NHI-131126A

COURSE TITLE

Concrete Pavement Preservation Series: Pavement Preservation Concepts

This training was prepared by the Transportation Curriculum Coordination Council (TCCC) in partnership with NHI to provide guidance on critical concrete pavement preservation issues. The training was developed by the National Concrete Pavement Technology Center at Iowa State University in cooperation with FHWA.

This module discusses how preventative maintenance impacts pavement preservation, good candidates for preservation, and the benefits to pavement preservation.

This module is part of the curriculum from the Concrete Pavement Preservation Series (FHWA-NHI-131126) which presents current guidelines and recommendations for the design, construction, and selection of cost-effective concrete pavement preservation strategies. The other Web-based training modules are:

- NHI-131126 Concrete Pavement Preservation Series with downloadable version of the FHWA Concrete Pavement Preservation Guide
- NHI-131126A: Pavement Preservation Concepts
- NHI-131126B: Concrete Pavement Evaluation
- NHI-131126C: Slab Stabilization
- NHI-131126D: Partial-depth Repairs
- NHI-131126E: Full-depth Repairs
- NHI-131126F: Retrofitted Edge Drains
- NHI-131126G: Dowel Bar Retrofit
- NHI-131126H: Diamond Grinding and Grooving
- NHI-131126I: Joint Resealing and Crack Sealing
- NHI-131126J: Concrete Overlays
- NHI-131126K: Strategy Selection

OUTCOMES

Upon completion of the course, participants will be able to:

- Define pavement preservation and preventive maintenance
- Describe characteristics of suitable pavements for preventive maintenance
- Describe the importance of selecting and placing the "right" treatment and placing it at the "right" time
- List the benefits of pavement preservation

TARGET AUDIENCE

The intended audience is quite diverse, and includes design engineers, quality control personnel, contractors, suppliers, technicians, and trades people. While the course is aimed at those who have some familiarity with concrete pavements and pavement preservation, it should also be of value to those that are new to the field. This course is recommended for the Transportation Curriculum Coordination Council levels I - IV.

TRAINING LEVEL: Intermediate

FEE: 2016: $25 Per Person; 2017: $25 Per Person

LENGTH: 1 HOURS (CEU: 0 UNITS)

CLASS SIZE: MINIMUM: 1; MAXIMUM: 1

NHI Customer Service: (877) 558-6873 • nhicustomerservice@dot.gov

Course Number

FHWA-NHI-131126B

Course Title

Concrete Pavement Preservation Series: Concrete Pavement Evaluation

This training was prepared by the Transportation Curriculum Coordination Council (TCCC) in partnership with NHI to provide guidance on critical concrete pavement preservation issues. The training was sponsored by the FHWA and developed by the National Concrete Pavement Technology Center at Iowa State University in cooperation with FHWA.

This module discusses how preventative maintenance impacts pavement preservation, good candidates for preservation, and the benefits to pavement preservation. This module also describes the common procedures associated with conducting thorough pavement evaluations.

This module is part of the curriculum from the Concrete Pavement Preservation Series (FHWA-NHI-131126) which presents current guidelines and recommendations for the design, construction, and selection of cost-effective concrete pavement preservation strategies. The other Web-based training modules are:

- NHI-131126 Concrete Pavement Preservation Series with downloadable version of the FHWA Concrete Pavement Preservation Guide
- NHI-131126A: Pavement Preservation Concepts
- NHI-131126B: Concrete Pavement Evaluation
- NHI-131126C: Slab Stabilization
- NHI-131126D: Partial-depth Repairs
- NHI-131126E: Full-depth Repairs
- NHI-131126F: Retrofitted Edge Drains
- NHI-131126G: Dowel Bar Retrofit
- NHI-131126H: Diamond Grinding and Grooving
- NHI-131126I: Joint Resealing and Crack Sealing
- NHI-131126J: Concrete Overlays
- NHI-131126K: Strategy Selection

Outcomes

Upon completion of the course, participants will be able to:

- Describe the need for a thorough pavement evaluation
- Name the common pavement evaluation components
- Describe what information is obtained from each pavement evaluation component

Target Audience

The intended audience is quite diverse, and includes design engineers, quality control personnel, contractors, suppliers, technicians, and trades people. While the course is aimed at those who have some familiarity with concrete pavements and pavement preservation, it should also be of value to those that are new to the field. This course is recommended for the Transportation Curriculum Coordination Council levels I - IV.

Training Level: Intermediate

Fee: 2016: $25 Per Person; 2017: $25 Per Person

Length: 2 HOURS (CEU: 0 UNITS)

Class Size: MINIMUM: 1; MAXIMUM: 1

NHI Customer Service: (877) 558-6873 • nhicustomerservice@dot.gov

Course Number

FHWA-NHI-131126C

Course Title

Concrete Pavement Preservation Series: Slab Stabilization

This training was prepared by the Transportation Curriculum Coordination Council (TCCC) in partnership with NHI to provide guidance on critical concrete pavement preservation issues. The training was developed by the National Concrete Pavement Technology Center at Iowa State University in cooperation with FHWA.

This module covers the use of slab stabilization (also known as undersealing) and slab jacking of concrete pavements. Slab stabilization restores support beneath slabs where voids have been detected, and slab jacking is used to raise depressed or settled slabs.

This module is part of the curriculum from the Concrete Pavement Preservation Series (FHWA-NHI-131126) which presents current guidelines and recommendations for the design, construction, and selection of cost-effective concrete pavement preservation strategies. The other Web-based training modules are:

- NHI-131126 Concrete Pavement Preservation Series with downloadable version of the FHWA Concrete Pavement Preservation Guide
- NHI-131126A: Pavement Preservation Concepts
- NHI-131126B: Concrete Pavement Evaluation
- NHI-131126C: Slab Stabilization
- NHI-131126D: Partial-depth Repairs
- NHI-131126E: Full-depth Repairs
- NHI-131126F: Retrofitted Edge Drains
- NHI-131126G: Dowel Bar Retrofit
- NHI-131126H: Diamond Grinding and Grooving
- NHI-131126I: Joint Resealing and Crack Sealing
- NHI-131126J: Concrete Overlays
- NHI-131126K: Strategy Selection

Outcomes

Upon completion of the course, participants will be able to:

- List benefits of slab stabilization and slab jacking
- Describe recommended materials and mixtures
- Describe recommended construction steps for both procedures
- Identify typical construction problems and remedies for slab stabilization

Target Audience

The intended audience is quite diverse, and includes design engineers, quality control personnel, contractors, suppliers, technicians, and trades people. While the course is aimed at those who have some familiarity with concrete pavements and pavement preservation, it should also be of value to those that are new to the field.This course is recommended for the Transportation Curriculum Coordination Council levels I - IV.

TRAINING LEVEL: Intermediate

FEE: 2016: $25 Per Person; 2017: $25 Per Person

LENGTH: 1 HOURS (CEU: 0 UNITS)

CLASS SIZE: MINIMUM: 1; MAXIMUM: 1

NHI Customer Service: (877) 558-6873 • nhicustomerservice@dot.gov

COURSE NUMBER

FHWA-NHI-131126D

COURSE TITLE

Concrete Pavement Preservation Series: Partial-depth Repairs

This training was prepared by the Transportation Curriculum Coordination Council (TCCC) in partnership with NHI to provide guidance on critical concrete pavement preservation issues. The training was developed by the National Concrete Pavement Technology Center at Iowa State University in cooperation with FHWA.

This module covers the procedures for partial-depth repairs (PDR) on PCC pavements. PDR is the removal and replacement of small, shallow areas of deteriorated PCC at spalled or distressed joints.

This module is part of the curriculum from the Concrete Pavement Preservation Series (FHWA-NHI-131126) which presents current guidelines and recommendations for the design, construction, and selection of cost-effective concrete pavement preservation strategies. The other Web-based training modules are:

- NHI-131126 Concrete Pavement Preservation Series with downloadable version of the FHWA Concrete Pavement Preservation Guide
- NHI-131126A: Pavement Preservation Concepts
- NHI-131126B: Concrete Pavement Evaluation
- NHI-131126C: Slab Stabilization
- NHI-131126D: Partial-depth Repairs
- NHI-131126E: Full-depth Repairs
- NHI-131126F: Retrofitted Edge Drains
- NHI-131126G: Dowel Bar Retrofit
- NHI-131126H: Diamond Grinding and Grooving
- NHI-131126I: Joint Resealing and Crack Sealing
- NHI-131126J: Concrete Overlays
- NHI-131126K: Strategy Selection

OUTCOMES

Upon completion of the course, participants will be able to:

- List benefits and appropriateness of partial-depth repairs
- List the advantages and disadvantages of different available repair materials
- Describe recommended construction procedures
- Identify typical construction problems and appropriate remedies

TARGET AUDIENCE

The intended audience is quite diverse, and includes design engineers, quality control personnel, contractors, suppliers, technicians, and trades people. While the course is aimed at those who have some familiarity with concrete pavements and pavement preservation, it should also be of value to those that are new to the field. This course is recommended for the Transportation Curriculum Coordination Council levels I - IV.

TRAINING LEVEL: Intermediate

FEE: 2016: $25 Per Person; 2017: $25 Per Person

LENGTH: 1 HOURS (CEU: 0 UNITS)

CLASS SIZE: MINIMUM: 1; MAXIMUM: 1

NHI Customer Service: (877) 558-6873 • nhicustomerservice@dot.gov

CONSTRUCTION AND MAINTENANCE

Course Number

FHWA-NHI-131126E

Course Title

Concrete Pavement Preservation Series: Full-depth Repairs

This training was prepared by the Transportation Curriculum Coordination Council (TCCC) in partnership with NHI to provide guidance on critical concrete pavement preservation issues. The training was developed by the National Concrete Pavement Technology Center at Iowa State University in cooperation with FHWA.

This module covers the procedures for cast-in-place Portland cement concrete (PCC) full-depth repair (FDR) of jointed concrete pavements (JCP) including jointed plain (JPCP) and jointed reinforced concrete pavements (JRCP). FDR techniques for continuously reinforced concrete pavements (CRCP) are discussed separately toward the end of the presentation. FDR is the cast-in-place concrete repairs that extend the full-depth of the existing slab.

This module is part of the curriculum from the Concrete Pavement Preservation Series (FHWA-NHI-131126) which presents current guidelines and recommendations for the design, construction, and selection of cost-effective concrete pavement preservation strategies. The other Web-based training modules are:

- NHI-131126 Concrete Pavement Preservation Series with downloadable version of the FHWA Concrete Pavement Preservation Guide
- NHI-131126A: Pavement Preservation Concepts
- NHI-131126B: Concrete Pavement Evaluation
- NHI-131126C: Slab Stabilization
- NHI-131126D: Partial-depth Repairs
- NHI-131126E: Full-depth Repairs
- NHI-131126F: Retrofitted Edge Drains
- NHI-131126G: Dowel Bar Retrofit
- NHI-131126H: Diamond Grinding and Grooving
- NHI-131126I: Joint Resealing and Crack Sealing
- NHI-131126J: Concrete Overlays
- NHI-131126K: Strategy Selection

Outcomes

Upon completion of the course, participants will be able to:

- List the benefits of full-depth repairs
- Describe primary design considerations in terms of dimensions, load transfer, and materials
- Describe recommended construction activities
- Identify typical construction problems and remedies

Target Audience

The intended audience is quite diverse, and includes design engineers, quality control personnel, contractors, suppliers, technicians, and trades people. While the course is aimed at those who have some familiarity with concrete pavements and pavement preservation, it should also be of value to those that are new to the field.This course is recommended for the Transportation Curriculum Coordination Council levels I - IV.

TRAINING LEVEL: Intermediate

FEE: 2016: $25 Per Person; 2017: $25 Per Person

LENGTH: 2 HOURS (CEU: 0 UNITS)

CLASS SIZE: MINIMUM: 1; MAXIMUM: 1

NHI Customer Service: (877) 558-6873 • nhicustomerservice@dot.gov

Course Number

FHWA-NHI-131126F

Course Title

Concrete Pavement Preservation Series: Retrofitted Edge Drains

This training was prepared by the Transportation Curriculum Coordination Council (TCCC) in partnership with NHI to provide guidance on critical concrete pavement preservation issues. The training was developed by the National Concrete Pavement Technology Center at Iowa State University in cooperation with FHWA.

This module presents design and construction information on retrofitted edge drains. This treatment is not as widely used as it once was, largely because it has limited applicability. Specifically, it must be targeted to those pavements that are 1) in good structural condition and 2) have bases with some degree of permeability that would allow water to be drained from beneath the pavement and to the edge drain.

This module is part of the curriculum from the Concrete Pavement Preservation Series (FHWA-NHI-131126) which presents current guidelines and recommendations for the design, construction, and selection of cost-effective concrete pavement preservation strategies. The other Web-based training modules are:

- NHI-131126 Concrete Pavement Preservation Series with downloadable version of the FHWA Concrete Pavement Preservation Guide
- NHI-131126A: Pavement Preservation Concepts
- NHI-131126B: Concrete Pavement Evaluation
- NHI-131126C: Slab Stabilization
- NHI-131126D: Partial-depth Repairs
- NHI-131126E: Full-depth Repairs
- NHI-131126F: Retrofitted Edge Drains
- NHI-131126G: Dowel Bar Retrofit
- NHI-131126H: Diamond Grinding and Grooving
- NHI-131126I: Joint Resealing and Crack Sealing
- NHI-131126J: Concrete Overlays
- NHI-131126K: Strategy Selection

Outcomes

Upon completion of the course, participants will be able to:

- List benefits of drainage
- List components of edge drain systems
- Describe recommended installation procedures
- Identify typical construction problems and remedies

Target Audience

The intended audience is quite diverse, and includes design engineers, quality control personnel, contractors, suppliers, technicians, and trades people. While the course is aimed at those who have some familiarity with concrete pavements and pavement preservation, it should also be of value to those that are new to the field.This course is recommended for the Transportation Curriculum Coordination Council levels I - IV.

TRAINING LEVEL: Intermediate

FEE: 2016: $25 Per Person; 2017: N/A

LENGTH: 1 DAYS (CEU: 0 UNITS)

CLASS SIZE: MINIMUM: 1; MAXIMUM: 1

NHI Customer Service: (877) 558-6873 • nhicustomerservice@dot.gov

COURSE NUMBER

FHWA-NHI-131126G

COURSE TITLE

Concrete Pavement Preservation Series: Dowel Bar Retrofit

This training was prepared by the Transportation Curriculum Coordination Council (TCCC) in partnership with NHI to provide guidance on critical concrete pavement preservation issues. The training was developed by the National Concrete Pavement Technology Center at Iowa State University in cooperation with FHWA.

This module presents design and construction information on load transfer restoration (LTR), sometimes referred to as retrofitted load transfer. In the introduction we will describe the difference between load transfer restoration (generic term) and dowel bar retrofitting (DBR) which is a specific means of achieving LTR. There are other methods available, but DBR is the most proven.

This module is part of the curriculum from the Concrete Pavement Preservation Series (FHWA-NHI-131126) which presents current guidelines and recommendations for the design, construction, and selection of cost-effective concrete pavement preservation strategies. The other Web-based training modules are:

- NHI-131126 Concrete Pavement Preservation Series with downloadable version of the FHWA Concrete Pavement Preservation Guide
- NHI-131126A: Pavement Preservation Concepts
- NHI-131126B: Concrete Pavement Evaluation
- NHI-131126C: Slab Stabilization
- NHI-131126D: Partial-depth Repairs
- NHI-131126E: Full-depth Repairs
- NHI-131126F: Retrofitted Edge Drains
- NHI-131126G: Dowel Bar Retrofit
- NHI-131126H: Diamond Grinding and Grooving
- NHI-131126I: Joint Resealing and Crack Sealing
- NHI-131126J: Concrete Overlays
- NHI-131126K: Strategy Selection

OUTCOMES

Upon completion of the course, participants will be able to:

- List benefits and applications of load transfer restoration
- Describe recommended materials and mixtures
- Describe recommended construction procedures
- Identify typical construction problems and remedies

TARGET AUDIENCE

The intended audience is quite diverse, and includes design engineers, quality control personnel, contractors, suppliers, technicians, and trades people. While the course is aimed at those who have some familiarity with concrete pavements and pavement preservation, it should also be of value to those that are new to the field.This course is recommended for the Transportation Curriculum Coordination Council levels I - IV.

TRAINING LEVEL: Intermediate

FEE: 2016: $25 Per Person; 2017: $25 Per Person

LENGTH: 1 HOURS (CEU: 0 UNITS)

CLASS SIZE: MINIMUM: 1; MAXIMUM: 1

NHI Customer Service: (877) 558-6873 • nhicustomerservice@dot.gov

Course Number

FHWA-NHI-131126H

Course Title

Concrete Pavement Preservation Series: Diamond Grinding and Grooving

This training was prepared by the Transportation Curriculum Coordination Council (TCCC) in partnership with NHI to provide guidance on critical concrete pavement preservation issues. The training was developed by the National Concrete Pavement Technology Center at Iowa State University in cooperation with FHWA.

This module describes recommended procedures for surface restoration of Portland cement concrete (PCC) pavements, specifically diamond grinding and diamond grooving operations.

This module is part of the curriculum from the Concrete Pavement Preservation Series (FHWA-NHI-131126) which presents current guidelines and recommendations for the design, construction, and selection of cost-effective concrete pavement preservation strategies. The other Web-based training modules are:

- NHI-131126 Concrete Pavement Preservation Series with downloadable version of the FHWA Concrete Pavement Preservation Guide
- NHI-131126A: Pavement Preservation Concepts
- NHI-131126B: Concrete Pavement Evaluation
- NHI-131126C: Slab Stabilization
- NHI-131126D: Partial-depth Repairs
- NHI-131126E: Full-depth Repairs
- NHI-131126F: Retrofitted Edge Drains
- NHI-131126G: Dowel Bar Retrofit
- NHI-131126H: Diamond Grinding and Grooving
- NHI-131126I: Joint Resealing and Crack Sealing
- NHI-131126J: Concrete Overlays
- NHI-131126K: Strategy Selection

Outcomes

Upon completion of the course, participants will be able to:

- Differentiate between diamond grinding and diamond grooving and list the benefits of each
- Identify appropriate blade spacing dimensions for grinding and grooving
- Describe recommended construction procedures
- Identify typical construction problems and remedies

Target Audience

The intended audience is quite diverse, and includes design engineers, quality control personnel, contractors, suppliers, technicians, and trades people. While the course is aimed at those who have some familiarity with concrete pavements and pavement preservation, it should also be of value to those that are new to the field. This course is recommended for the Transportation Curriculum Coordination Council levels I - IV.

TRAINING LEVEL: Intermediate

FEE: 2016: $25 Per Person; 2017: $25 Per Person

LENGTH: 1 HOURS (CEU: 0 UNITS)

CLASS SIZE: MINIMUM: 1; MAXIMUM: 1

NHI Customer Service: (877) 558-6873 • nhicustomerservice@dot.gov

Course Number

FHWA-NHI-131126I

Course Title

Concrete Pavement Preservation Series: Joint Sealing and Crack Resealing

This training was prepared by the Transportation Curriculum Coordination Council (TCCC) in partnership with NHI to provide guidance on critical concrete pavement preservation issues. The training was developed by the National Concrete Pavement Technology Center at Iowa State University in cooperation with FHWA.

This module covers joint resealing and crack sealing for concrete pavements. Joint resealing and crack sealing is defined as placement of an approved sealant material in an existing joint or crack to reduce moisture infiltration and prevent intrusion of incompressibles.

This module is part of the curriculum from the Concrete Pavement Preservation Series (FHWA-NHI-131126) which presents current guidelines and recommendations for the design, construction, and selection of cost-effective concrete pavement preservation strategies. The other Web-based training modules are:

- NHI-131126 Concrete Pavement Preservation Series with downloadable version of the FHWA Concrete Pavement Preservation Guide
- NHI-131126A: Pavement Preservation Concepts
- NHI-131126B: Concrete Pavement Evaluation
- NHI-131126C: Slab Stabilization
- NHI-131126D: Partial-depth Repairs
- NHI-131126E: Full-depth Repairs
- NHI-131126F: Retrofitted Edge Drains
- NHI-131126G: Dowel Bar Retrofit
- NHI-131126H: Diamond Grinding and Grooving
- NHI-131126I: Joint Resealing and Crack Sealing
- NHI-131126J: Concrete Overlays
- NHI-131126K: Strategy Selection

Outcomes

Upon completion of the course, participants will be able to:

- List the benefits of joint resealing
- Describe desirable sealant properties and characteristics
- Describe recommended installation procedures
- Identify typical construction problems and appropriate remedies

Target Audience

The intended audience is quite diverse, and includes design engineers, quality control personnel, contractors, suppliers, technicians, and trades people. While the course is aimed at those who have some familiarity with concrete pavements and pavement preservation, it should also be of value to those that are new to the field.This course is recommended for the Transportation Curriculum Coordination Council levels I - IV.

TRAINING LEVEL: Intermediate

FEE: 2016: $25 Per Person; 2017: $25 Per Person

LENGTH: 1 HOURS (CEU: 0 UNITS)

CLASS SIZE: MINIMUM: 1; MAXIMUM: 1

NHI Customer Service: (877) 558-6873 • nhicustomerservice@dot.gov

COURSE NUMBER

FHWA-NHI-131126J

COURSE TITLE

Concrete Pavement Preservation Series: Concrete Overlays

This training was prepared by the Transportation Curriculum Coordination Council (TCCC) in partnership with NHI to provide guidance on critical concrete pavement preservation issues. The training was developed by the National Concrete Pavement Technology Center at Iowa State University in cooperation with FHWA.

This module provides guidance on the selection of concrete pavement preservation strategies. Based on a collective review of a number of recent published documents, this module covers the seven step process that can be used to determine the most appropriate treatment (or combination of treatments) for a PCC pavement.

This module is part of the curriculum from the Concrete Pavement Preservation Series (FHWA-NHI-131126) which presents current guidelines and recommendations for the design, construction, and selection of cost-effective concrete pavement preservation strategies. The other Web-based training modules are:

- NHI-131126 Concrete Pavement Preservation Series with downloadable version of the FHWA Concrete Pavement Preservation Guide
- NHI-131126A: Pavement Preservation Concepts
- NHI-131126B: Concrete Pavement Evaluation
- NHI-131126C: Slab Stabilization
- NHI-131126D: Partial-depth Repairs
- NHI-131126E: Full-depth Repairs
- NHI-131126F: Retrofitted Edge Drains
- NHI-131126G: Dowel Bar Retrofit
- NHI-131126H: Diamond Grinding and Grooving
- NHI-131126I: Joint Resealing and Crack Sealing
- NHI-131126J: Concrete Overlays
- NHI-131126K: Strategy Selection

OUTCOMES

Upon completion of the course, participants will be able to:

- Describe the treatment selection process
- List the components of a life-cycle cost analysis
- List other factors that may enter the selection process

TARGET AUDIENCE

The intended audience is quite diverse, and includes design engineers, quality control personnel, contractors, suppliers, technicians, and trades people. While the course is aimed at those who have some familiarity with concrete pavements and pavement preservation, it should also be of value to those that are new to the field. This course is recommended for the Transportation Curriculum Coordination Council levels I - IV.

TRAINING LEVEL: Intermediate

FEE: 2016: $25 Per Person; 2017: $25 Per Person

LENGTH: 1 HOURS (CEU: 0 UNITS)

CLASS SIZE: MINIMUM: 1; MAXIMUM: 1

NHI Customer Service: (877) 558-6873 • nhicustomerservice@dot.gov

COURSE NUMBER

FHWA-NHI-131126K

COURSE TITLE

Concrete Pavement Preservation Series: Strategy Selection

This training was prepared by the Transportation Curriculum Coordination Council (TCCC) in partnership with NHI to provide guidance on critical concrete pavement preservation issues. The training was developed by the National Concrete Pavement Technology Center at Iowa State University in cooperation with FHWA.

This module provides guidance on the selection of concrete pavement preservation strategies. Based on a collective review of a number of recent published documents, this module covers the seven step process that can be used to determine the most appropriate treatment (or combination of treatments) for a PCC pavement.

This module is part of the curriculum from the Concrete Pavement Preservation Series (FHWA-NHI-131126) which presents current guidelines and recommendations for the design, construction, and selection of cost-effective concrete pavement preservation strategies. The other Web-based training modules are:

- NHI-131126 Concrete Pavement Preservation Series with downloadable version of the FHWA Concrete Pavement Preservation Guide
- NHI-131126A: Pavement Preservation Concepts
- NHI-131126B: Concrete Pavement Evaluation
- NHI-131126C: Slab Stabilization
- NHI-131126D: Partial-depth Repairs
- NHI-131126E: Full-depth Repairs
- NHI-131126F: Retrofitted Edge Drains
- NHI-131126G: Dowel Bar Retrofit
- NHI-131126H: Diamond Grinding and Grooving
- NHI-131126I: Joint Resealing and Crack Sealing
- NHI-131126J: Concrete Overlays
- NHI-131126K: Strategy Selection

OUTCOMES

Upon completion of the course, participants will be able to:

- Describe the treatment selection process
- List factors that might enter into the selection process
- Describe pavement deficiencies addressed by the different preservation treatments
- Describe how the benefits and costs of alternative treatment strategies are computed in a cost-effectiveness analysis
- Describe a process used to select the preferred treatment strategy

TARGET AUDIENCE

The intended audience is quite diverse, and includes design engineers, quality control personnel, contractors, suppliers, technicians, and trades people. While the course is aimed at those who have some familiarity with concrete pavements and pavement preservation, it should also be of value to those that are new to the field.This course is recommended for the Transportation Curriculum Coordination Council levels I - IV.

TRAINING LEVEL: Intermediate

FEE: 2016: $25 Per Person; 2017: N/A

LENGTH: .3 HOURS (CEU: 0 UNITS)

CLASS SIZE: MINIMUM: 1; MAXIMUM: 1

NHI Customer Service: (877) 558-6873 • nhicustomerservice@dot.gov

Course Number

FHWA-NHI-131127

Course Title

Concrete Series - WEB-BASED

The Transportation Curriculum Coordination Council (TCCC) in partnership with NHI is pleased to offer this comprehensive training series (FHWA-NHI-131127) for any engineer or supervisor working with Portland cement. The training was developed by the National Concrete Pavement Technology Center at Iowa State University. It is the first training of its kind offered by NHI, and we would like to give special recognition to the TCCC for their efforts. This course is recommended for the Transportation Curriculum Coordination Council levels II - IV.

The TCCC Concrete Series is part of a curriculum from the "Integrated Materials and Construction Practices for Concrete Pavement" manual developed through the National Concrete Pavement Technology Center at Iowa State University.

To streamline registration and enable you to take some or all of these courses when it best suits your schedule, we have created this new series option which automatically registers you for all 11 modules-it's that easy. They are as follows:

Module 1 - TCCC Design of Pavement (FHWA-NHI-134101)

Module 2 - TCCC Fundamentals of Materials Used for Concrete Pavements (FHWA-NHI-134084)

Module 3 - TCCC Mix Design Principles (FHWA-NHI-134087)

Module 4 - TCCC Fresh Concrete Properties (FHWA-NHI-134097)

Module 5 - TCCC Basics of Cement Hydration (FHWA-NHI-134096)

Module 6 - TCCC Incompatibility in Concrete Pavement Systems (FHWA-NHI-134085)

Module 7 - TCCC Early Age Cracking (FHWA-NHI-134095)

Module 8 - TCCC Hardened Concrete Properties- Durability (FHWA-NHI-134075)

Module 9 - TCCC Construction of Concrete Pavements (FHWA-NHI-134098)

Module 10 - TCCC QCQA for Concrete Pavements (FHWA-NHI-134100)

Module 11 - TCCC Troubleshooting for Concrete Pavements (FHWA-NHI-134102)

Outcomes

Upon completion of the course, participants will be able to:

- Explain concrete pavement construction as a complex, integrated system involving several discrete practices that interrelate and affect one another in various ways
- Recognize and implement technologies, tests, and best practices to identify materials, concrete properties, and construction practices that are known to optimize concrete performance
- Identify factors that lead to premature distress in concrete, and learn how to avoid or reduce those factors
- Apply appropriate how-to and troubleshooting information

Target Audience

This training is intended as both a training tool and a reference to help concrete paving engineers, quality control personnel, specifiers, contractors, suppliers, technicians, and tradespeople bridge the gap between recent research and practice regarding optimizing the performance of concrete for pavements.

TRAINING LEVEL: Intermediate

FEE: 2016: $50 Per Person; 2017: $50 Per Person

LENGTH: 12 HOURS (CEU: 0 UNITS)

CLASS SIZE: MINIMUM: 1; MAXIMUM: 1

NHI Customer Service: (877) 558-6873 • nhicustomerservice@dot.gov

CONSTRUCTION AND MAINTENANCE

COURSE NUMBER

FHWA-NHI-131129

COURSE TITLE

HMA Paving Field Inspection - WEB-BASED

This training was prepared by the Transportation Curriculum Coordination Council (TCCC) in partnership with NHI to provide guidance and instruction to inspectors involved in the construction of hot mix asphalt (HMA) pavements. The important tasks involved in this work are explained and proper procedures are described. This training is recommended for the Transportation Curriculum Coordination Council levels I, II, and III. This course is primarily intended for inspectors and technicians.

This training is arranged in a fashion to help the inspector first learn the various aspects of what is involved in a HMA paving operation and then become familiar with the duties that are a part of the HMA pavement grade inspection responsibilities. It also explains how to recognize the mix properties of a HMA mixture. The information included will assist the inspector in recognizing problems during a project and offering solutions to the problems. This training is not intended to cover every aspect of HMA paving.

OUTCOMES

Upon completion of the course, participants will be able to:

- Know various aspects of what is involved in a HMA paving operation
- Understand the duties of a HMA paving inspector
- Recognize the mix properties of a HMA mixture
- Recognize the problems that may occur on HMA paving projects
- Understand the product and project so solutions can be recommended

TARGET AUDIENCE

This training would be beneficial to anyone that is involved with an HMA paving project, but focuses on technicians/ inspectors that are involved with the production, placement, and inspection of HMA paving projects.

TRAINING LEVEL: Intermediate

FEE: 2016: $50 Per Person; 2017: $50 Per Person

LENGTH: 4.5 HOURS (CEU: 0 UNITS)

CLASS SIZE: MINIMUM: 1; MAXIMUM: 1

NHI Customer Service: (877) 558-6873 • nhicustomerservice@dot.gov

COURSE NUMBER

FHWA-NHI-131132

COURSE TITLE

Chip Seal Best Practices - WEB-BASED

The Chip Seal Best Practices course presents ways to assist in the development and implementation of pavement preservation programs by identifying the benefits of using chip seal as part of a preventive maintenance program.

This course has six modules. Module 1 is an introduction into chip seals, module 2 covers designing chip seal mixes, module 3 is selecting the proper materials for the chip seal mix, module 4 focuses on the use of the equipment, module 5 covers proper construction practices, and module 6 rounds out the course with performance measures of chip seals. The combination of all this information provides an excellent overview of successful chip seal practices worldwide.

OUTCOMES

Upon completion of the course, participants will be able to:

- Define chip seal
- Describe how chip seals are used as a preventive maintenance treatment for pavement
- Identify materials used in chip seals
- Describe the characteristics of chip seal design
- Identify types of chip seal
- Identify the important considerations of aggregate and binder selection
- Describe aggregate-binder compatibility
- Describe equipments used in chip seal practices
- Identify important variables in construction practice
- Define the measures of control implemented over the quality of materials and construction
- Identify construction best practices
- Describe the components of engineering-based performance measures
- Identify qualitative performance indicators for chip seal
- Define common visible chip seal distresses

TARGET AUDIENCE

This training is recommended for the Transportation Curriculum Coordination Council levels I, II and III. This training would benefit entry level construction inspectors, maintenance employees and contractor personnel as well as serve as refresher training for those already well versed in the selection and application of a chip seal as a preventive maintenance treatment.

TRAINING LEVEL: Basic

FEE: 2016: $25 Per Person; 2017: N/A

LENGTH: 3 HOURS (CEU: 0 UNITS)

CLASS SIZE: MINIMUM: 1; MAXIMUM: 1

NHI Customer Service: (877) 558-6873 • nhicustomerservice@dot.gov

Course Number

FHWA-NHI-131133

Course Title

Roller Compacted Concrete Pavements - WEB-BASED

The Roller Compacted Concrete (RCC) Pavements course provides detailed overviews of RCC properties and materials, mixture proportioning, structural design issues, and production and construction considerations, plus troubleshooting guidelines and an extensive reference list for more comprehensive information.

This course contains six modules. Module 1 is an introduction in RCC covering the characteristics, benefits, limitations, selection considerations, and typical uses. Module 2 discusses the property differences between RCC and conventional mixes, material requirements and testing. Module 3 covers mix proportioning of RCC, while Module 4 gets into structural design of RCC pavements. Module 5 acquaints the student with production and the proper handling and storage of materials, mixing and batching, and production planning. Module 6 covers the actual construction of a RCC pavement. All of the modules for this training were developed from the August 2010 "Guide for Roller-Compacted Concrete Pavements" which is available from the Portland Cement Association website www.cement.org/pavements.

Outcomes

Upon completion of the course, participants will be able to:

- Define RCC key elements and common uses
- Define RCC properties and materials
- Describe RCC mix proportioning
- Describe structural design of RCC pavement
- Identify RCC production
- Identify RCC pavement construction

Target Audience

This training provides agencies, contractors, materials suppliers, and others with a thorough introduction to and updated review of RCC and its many paving applications. This training is recommended for the Transportation Curriculum Coordination Council levels II through IV.

Training Level: Basic

Fee: 2016: $50 Per Person; 2017: N/A

Length: 6 HOURS (CEU: 0 UNITS)

Class Size: MINIMUM: 1; MAXIMUM: 1

NHI Customer Service: (877) 558-6873 • nhicustomerservice@dot.gov

Course Number

FHWA-NHI-131134

Course Title

Superpave for Construction - WEB-BASED

The Superpave for Construction Course contains information for field construction personnel on the Superpave mix design system and the control of field produced Hot Mix Asphalt.

There are two modules in this course. The first module introduces the Superpave Hot Mix Asphalt design testing and analysis. It will cover design testing procedures, design analysis methods, and will include calculations to analyze the volumetrics of paving samples. Module two includes relevant volumetric examples including the use of phase diagrams to calculate volumetric properties. Example problems are included. This course is an excellent learning tool to assist in understanding corrective actions for volumetric parameters.

Outcomes

Upon completion of the course, participants will be able to:

- Describe the benefits of Superpave over previous mix design methodologies
- Understand Superpave mix design procedures and testing
- Understand mix design analysis methods
- Perform the calculation necessary to analyze the volumetrics of paving samples for comparison
- Describe how to use phase diagrams to calculate volumetric properties
- Describe factors which can influence key mass-volume relationships and calculations
- Understand corrective action for volumetric parameters
- Calculate and evaluate volumetric properties through example problems

Target Audience

This training is targeted to intermediate and advanced technicians from both contractor and agency employment, which will be involved in construction of pavements using Superpave. This training is recommended for the Transportation Curriculum Coordination Council levels II and III.

Training Level: Basic

Fee: 2016: $25 Per Person; 2017: $25 Per Person

Length: 3.5 HOURS (CEU: 0 UNITS)

Class Size: MINIMUM: 1; MAXIMUM: 1

NHI Customer Service: (877) 558-6873 • nhicustomerservice@dot.gov

COURSE NUMBER

FHWA-NHI-134001

COURSE TITLE

Principles and Applications of Highway Construction Specifications

Well-written highway construction specifications are those that can be interpreted accurately to minimize confusion and reduce owner-contractor disputes. Across the country, current practices, standards, and requirements for writing specifications are changing. Agencies also are using effective specifications to manage risk and support alternative contracting methods.

NHI 134001 Principles of Writing Highway Construction Specifications is a highly engaging, two-day, instructor-led training session. It includes content that highlights the role of specifications as contract documents and tools for assigning risk. Course participants engage in lessons and practice sessions to identify types of specifications, select the most appropriate type for a given project, and generate an original, effective highway construction specification.

This is not a grammar course; however, adequate course content emphasizes the use of basic grammar and writing style so that the learners can generate specifications that are correct, consistent, clear, complete, and concise.

OUTCOMES

Upon completion of the course, participants will be able to:

- Explain the purposes of a specification.
- Explain how specifications are used to assign risk and influence the behavior of different parties, within a given a scenario.
- Compare the functions of Standard and Supplemental Specifications with the functions of Special Provisions.
- Explain how the "order of precedence" affects writing specifications and preparing plans.
- Describe the purpose of the General Provisions.
- Explain how a consistent writing style can affect the interpretation of specifications.
- Complete a checklist of the information needed before writing or revising a specification.
- Explain the potential benefits of writing in the active voice.
- Rewrite passive voice sentences into the active voice.
- Evaluate specifications to determine the need for imperative or indicative mood.
- State the five Cs used in specification writing. (Note: the five Cs include: correct; consistent; clear; complete; concise.)
- Explain each element of the AASHTO five-part format.
- Identify potential ambiguities in the wording, given a sample specification.
- Identify the potential benefits of each of the five Cs, given a sample specification.
- Apply the five Cs and the host agency's preferred format to revise the specification, given a sample specification.
- Write a new specification to a given set of criteria using the five Cs and the host agency's preferred format, given a sample specification.
- Compare method versus end-result specifications.
- Relate the type of specification to the allocation of risk.
- Write an end-result specification to replace a method specification, given an excerpt from a method specification.

TARGET AUDIENCE

This course is designed primarily for individuals who write, review, and implement an agency's contract specifications. Participants might represent Federal, State, and local transportation agencies; other public agencies; contractors; and consultant firms.Individuals who do not write specifications but may contribute to their development, as well as those who use specifications, could also benefit from this course and the interaction with their classmates. Such participants might include personnel from environmental, materials, or construction sections or units; legal departments; work zone and safety professionals; contractor personnel; and any others involved with the design and construction of transportation facilities.

TRAINING LEVEL: Intermediate

FEE: 2016: $475 Per Person; 2017: $475 Per Person

LENGTH: 2 DAYS (CEU: 1.2 UNITS)

CLASS SIZE: MINIMUM: 20; MAXIMUM: 30

NHI Customer Service: (877) 558-6873 • nhicustomerservice@dot.gov

Course Number

FHWA-NHI-134005

Course Title

Value Engineering Workshop (3-day)

Value Engineering (VE) is a systematic process of review and analysis of a project during the concept and design phases. VE is conducted by a multi-disciplined team of persons not involved in the project to provide recommendations such as: a) providing the needed functions safely, reliably, and at the lowest overall cost; b) improving the value and quality of the project; and c) reducing the time to complete the project.

This course begins with a Web-based training (WBT) component that is completed prior to the first day of the class (134005A). The 3-day workshop involves training participants to be valued contributors to the Value Engineering team, conducting a Value Engineering study in a team environment. It is preferable that the host agency provides actual project(s) to be used in this course, although The National Highway Institute (NHI) can provide projects upon request. Depending on the projects selected for use in the course, and based on the request of the host agency, the 3-day classroom session can be expanded to 4 or 5 days in length (NHI-134005B and NHI-134005C).

Upon successful course completion, participants will have acquired the training necessary to successfully participate in future Value Engineering studies for their agencies.

Outcomes

Upon completion of the course, participants will be able to:

- Explain how Value Engineering can improve project performance, reduce costs, and enhance value.
- Acquire the necessary behaviors and skills to be an effective Value Engineering team member with the ability to: Investigate the project and analyze project functions and costs; Creatively speculate on alternative ways to perform the various functions; Evaluate the most effective life-cycle alternatives; Develop viable alternatives into fully supported recommendations; Present the recommendations to stakeholders and agency management

Target Audience

The target audience for this course consists of FHWA and state highway agency personnel in management, administrative, and engineering disciplines who will participate as Value Engineering team members. Consultants or agency representatives of all technical disciplines associated with project design, development, construction, and maintenance can be included in order to provide the multiple perspectives needed to maximize the effectiveness of the team.

Training Level: Basic

Fee: 2016: $925 Per Person; 2017: $925 Per Person

Length: 3 DAYS (CEU: 1.8 UNITS)

Class Size: MINIMUM: 20; MAXIMUM: 30

NHI Customer Service: (877) 558-6873 • nhicustomerservice@dot.gov

COURSE NUMBER

FHWA-NHI-134005A

COURSE TITLE

Introduction to Value Engineering - WEB-BASED

This training is a prerequisite of another NHI training and is offered at no cost.

Value Engineering (VE) is a systematic process of review and analysis of a project during the concept and design phases. VE is conducted by a multi-disciplined team of persons not involved in the project to provide recommendations such as: a) providing the needed functions safely, reliably, and at the lowest overall cost; b) improving the value and quality of the project; and c) reducing the time to complete the project.

This Web-based training is intended to provide an overview of the Value Engineering process, know as the Value Engineering study. Included in the training is a discussion of the benefits of utilizing VE, the keys to completing a successful VE study, and an overview of the objectives and tasks completed by the VE team at each phase.

Participants can complete this training independently. Those who plan on attending the 3-day Value Engineering classroom training must complete this online module prior to coming to class. Course certificates should be printed out and presented to the instructor on the first day to verify completion.

OUTCOMES

Upon completion of the course, participants will be able to:

- Identify the purpose of Value Engineering and its benefits to a highway transportation agency.
- Identify the critical skills required to participate successfully in the VE study.
- Describe each phase of creating a Value Engineering Job Plan in terms of the objective and tasks.

TARGET AUDIENCE

The target audience for this course consists of FHWA and state highway agency personnel in management, administrative, and engineering disciplines who will participate as Value Engineering team members or who are interested in learning more about the process. Consultants or agency representatives of all technical disciplines associated with project design, development, construction, and maintenance who will participate in a Value Engineering study should also attend.

TRAINING LEVEL: Basic

FEE: 2016: $0 Per Person; 2017: $0 Per Person

LENGTH: .5 HOURS (CEU: 0 UNITS)

CLASS SIZE: MINIMUM: 20; MAXIMUM: 30

NHI Customer Service: (877) 558-6873 • nhicustomerservice@dot.gov

Course Number

FHWA-NHI-134005B

Course Title

Value Engineering Workshop (4-day)

Value Engineering (VE) is a systematic process of review and analysis of a project during the concept and design phases. VE is conducted by a multi-disciplined team of persons not involved in the project to provide recommendations such as: a) providing the needed functions safely, reliably, and at the lowest overall cost; b) improving the value and quality of the project; and c) reducing the time to complete the project.

This course begins with a Web-based training (WBT) component that is completed prior to the first day of the class. The 4-day workshop involves training participants to be valued contributors to the Value Engineering team, conducting a Value Engineering study in a team environment. It is preferable that the host agency provides actual project(s) to be used in this course, although The National Highway Institute (NHI) can provide projects upon request. Depending on the projects selected for use in the course, and based on the request of the host agency, the 3-day classroom session can be expanded to 3 or 5 days in length (NHI-134005 and NHI-134005C).

Upon successful course completion, participants will have acquired the training necessary to successfully participate in future Value Engineering studies for their agencies.

Outcomes

Upon completion of the course, participants will be able to:

- Explain how value engineering can improve project performance, reduce costs, and enhance value.
- Acquire the necessary behaviors and skills to be an effective Value Engineering Team member with the ability to: Investigate the project and analyze project functions and costs; Creatively speculate on alternative ways to perform the various functions; Evaluate the most effective life-cycle alternatives; Develop viable alternatives into fully supported recommendations; Present the recommendations to stakeholders and agency management

Target Audience

The target audience for this course consists of FHWA and state highway agency personnel in management, administrative, and engineering disciplines who will participate as Value Engineering team members. Consultants or agency representatives of all technical disciplines associated with project design, development, construction, and maintenance can be included in order to provide the multiple perspectives needed to maximize the effectiveness of the team.

Training Level: Basic

Fee: 2016: $1160 Per Person; 2017: $1160 Per Person

Length: 4 DAYS (CEU: 2.4 UNITS)

Class Size: MINIMUM: 20; MAXIMUM: 30

NHI Customer Service: (877) 558-6873 • nhicustomerservice@dot.gov

COURSE NUMBER

FHWA-NHI-134005C

COURSE TITLE

Value Engineering Workshop (5-day)

Value Engineering (VE) is a systematic process of review and analysis of a project during the concept and design phases. VE is conducted by a multi-disciplined team of persons not involved in the project to provide recommendations such as: a) providing the needed functions safely, reliably, and at the lowest overall cost; b) improving the value and quality of the project; and c) reducing the time to complete the project.

This course begins with a Web-based training (WBT) component that is completed prior to the first day of the class (134005A). The 3-day workshop involves training participants to be valued contributors to the Value Engineering team, conducting a Value Engineering study in a team environment. It is preferable that the host agency provides actual project(s) to be used in this course, although The National Highway Institute (NHI) can provide projects upon request. Depending on the projects selected for use in the course, and based on the request of the host agency, the 5-day classroom session can be shortened to 3 or 4 days in length (NHI-134005 and NHI-134005B).

Upon successful course completion, participants will have acquired the training necessary to successfully participate in future Value Engineering studies for their agencies.

OUTCOMES

Upon completion of the course, participants will be able to:

- Explain how value engineering can improve project performance, reduce costs, and enhance value.
- Acquire the necessary behaviors and skills to be an effective Value Engineering Team member with the ability to: Investigate the project and analyze project functions and costs; Creatively speculate on alternative ways to perform the various functions; Evaluate the most effective life-cycle alternatives; Develop viable alternatives into fully supported recommendations; Present the recommendations to stakeholders and agency management

TARGET AUDIENCE

The target audience for this course consists of FHWA and state highway agency personnel in management, administrative, and engineering disciplines who will participate as Value Engineering team members. Consultants or agency representatives of all technical disciplines associated with project design, development, construction, and maintenance can be included in order to provide the multiple perspectives needed to maximize the effectiveness of the team.

TRAINING LEVEL: Basic

FEE: 2016: $1395 Per Person; 2017: $1395 Per Person

LENGTH: 5 DAYS (CEU: 3 UNITS)

CLASS SIZE: MINIMUM: 20; MAXIMUM: 30

NHI Customer Service: (877) 558-6873 • nhicustomerservice@dot.gov

Course Number

FHWA-NHI-134006

Course Title

Utility Coordination for Highway Projects

This is a blended course, with both Web-based and instructor-led components. The Web-based training component (NHI 134006A) must be completed before attending the instructor-led training session.

Every State highway agency participates in construction projects that include accommodation and relocation of utilities along public rights-of-way. 134006 Utility Coordination for Highway Projects considers how communication, cooperation, and coordination between transportation agencies and utility companies can mitigate or avoid common challenges. Participants in this blended course (combination of Web-based and instructor-led formats) learn how, when, and where in the project development process to identify and conduct effective utility coordination.

Participants first take a self-paced, Web-based training to learn about regulatory requirements for both public and private utilities, subsurface utility engineering (SUE), and their own State's Utility Accommodation Policy. During the 2-day classroom event, participants learn to identify risks and potential issues associated with utilities, and then work together to evaluate ways to avoid or mitigate those risks and issues. (Please note: An optional lesson on utility challenges in projects using design-build delivery and other alternative contracting methods is available to be taught at the discretion of the State.) By putting these lessons into practice, utility-related complications in many cases can be predicted and mitigated at the most appropriate stage of project development, which can reduce potential negative impacts to timeline and budget.

Outcomes

Upon completion of the course, participants will be able to:

- Explain the importance of early and effective cooperation, communication, and coordination of utility-related activities throughout a project's lifecycle.
- Identify successful techniques that could be used to avoid or mitigate utility challenges throughout the project development and delivery process.
- Explain the major impacts of identified conflicts or issues on the schedule or budget of a project.
- Explain the basic skills necessary to identify utility conflicts and develop a utility conflict matrix.
- Generate a personal resource toolkit for each of six major areas of project development (planning, design, environmental considerations, right-of-way, construction, and maintenance).

Target Audience

The course targets Federal, State, and local personnel who are responsible for planning, designing, constructing, operating, and maintaining transportation facilities that involve the accommodation or relocation of utilities. It is most effectively delivered with participation from representatives of public and private utility companies, DOT contractors, risk managers, right-of-way staff, mid-to-senior level managers, and engineering consultants. The participation of utility company representatives in particular will be integral to the success of the course. Therefore, course organizers need to make every effort to include utility company personnel, as well as the Federal, State, and local transportation practitioners who comprise the more traditional audience for NHI training. Contractors and risk managers are appropriate and vital attendees as well. A minimum of 10% of every class should come from outside Federal, State, and local transportation agencies.

Training Level: Basic

Fee: 2016: $425 Per Person; 2017: $425 Per Person

Length: 2 DAYS (CEU: 1.2 UNITS)

Class Size: MINIMUM: 20; MAXIMUM: 30

NHI Customer Service: (877) 558-6873 • nhicustomerservice@dot.gov

Course Number

FHWA-NHI-134006A

Course Title

Introduction to Utility Coordination for Highway Projects--WEB-BASED

NHI 134006 is a blended course, with both Web-based and instructor-led components. The Web-based training component (NHI 134006A) must be completed before attending the instructor-led training session.

This training is a prerequisite of another NHI training and is offered at no cost.

Every State highway agency participates in construction projects that include accommodation and relocation of utilities along public rights-of-way. 134006 Utility Coordination for Highway Projects considers how communication, cooperation, and coordination between transportation agencies and utility companies can mitigate or avoid common challenges.

In the Web-based training, participants learn about regulatory requirements for both public and private utilities, subsurface utility engineering (SUE), and their own State's Utility Accommodation Policy. By putting these lessons into practice, utility-related complications in many cases can be predicted and mitigated at the most appropriate stage of project development, which can reduce potential negative impacts to timeline and budget.

Outcomes

Upon completion of the course, participants will be able to:

- Explain the importance of early and effective cooperation, communication, and coordination of utility-related activities throughout a project's lifecycle.
- Identify successful techniques that could be used to avoid or mitigate utility challenges throughout the project development and delivery process.
- Explain the major impacts of identified conflicts or issues on the schedule or budget of a project.

Target Audience

The course targets Federal, State, and local personnel who are responsible for planning, designing, constructing, operating, and maintaining transportation facilities that involve the accommodation or relocation of utilities. It is most effectively delivered with additional participation from representatives of public and private utility companies, DOT contractors, risk managers, right-of-way staff, mid-to senior-level managers, and engineering consultants.

Training Level: Basic

Fee: 2016: $0 Per Person; 2017: $0 Per Person

Length: 4 HOURS (CEU: 0 UNITS)

Class Size: MINIMUM: 0; MAXIMUM: 0

NHI Customer Service: (877) 558-6873 • nhicustomerservice@dot.gov

Course Number

FHWA-NHI-134037A

Course Title

Managing Highway Contract Claims: Analysis and Avoidance

Construction contract claims are the result of the owner and the contractor being unable to come to agreement regarding an alleged change. Reducing or eliminating claims requires (1) a reduction in the number of potential changes, and (2) the implementation of practices that increase the likelihood of an owner and contractor resolving a dispute. This course provides the basic tools to address both elements of reducing or eliminating contract claims.

In this course, participants first walk step-by-step through the evaluation of a contract claim, looking at each component. Separate course modules are devoted to these three components of a claim: entitlement, impact, and cost. The "Entitlement" module focuses on the contract and the proper interpretation of common contract clauses. The "Impacts" module focuses on delay and inefficiency--the two most difficult impacts to measure and, consequently, most difficult to resolve. The "Cost" module explores costs that can prove difficult for the project team to resolve.

Next, the participants identify and review best practices associated with successful dispute resolution. In addition, there is a module devoted solely to claims avoidance techniques and dispute resolution processes.

By completing this course, participants will have the opportunity to master techniques that can help them manage and avoid claims.

Outcomes

Upon completion of the course, participants will be able to:

- Define "claim"
- List the three parts of a claim
- Describe the difference between a directed and constructive change
- List examples of directed and constructive changes
- List basic contract principles and rules of contract interpretation
- List the contract clauses most relevant to the evaluation of claims
- Define essential scheduling terms
- Explain the differences among the six types of delays
- List five methods for analyzing delays
- Explain how to perform a contemporaneous schedule analysis
- List five methods for measuring productivity/inefficiency
- Explain how to perform a measured mile analysis
- Describe how to avoid constructive acceleration
- List five methods for calculating costs
- List the four assumptions upon which a total cost calculation is based
- Identify project costs that are affected by delays
- Calculate extended home office overhead costs by the Eichleay and Canadian methods
- Identify acceleration costs
- Identify inefficiency costs
- Identify common miscellaneous costs
- Explain the key steps necessary to evaluate claims
- Describe the False Claims Act
- Demonstrate an ability to evaluate a contractor's claim
- Describe FHWA policy regarding participation in paying damages for contractor claims
- Explain the importance of a claims avoidance system

- Describe a claims avoidance and dispute resolution system
- Explain the strengths and weaknesses of dispute review board

TARGET AUDIENCE

This an intermediate level course.It is designed specifically for State DOTs, but is also appropriate for LPOs and MPOs. It is a valuable course for contractors, design consultants, project managers, and attorneys involved in the evaluation, management, and resolution of disputes on highway construction projects.

TRAINING LEVEL: Intermediate

FEE: 2016: $475 Per Person; 2017: $475 Per Person

LENGTH: 2.5 DAYS (CEU: 1.5 UNITS)

CLASS SIZE: MINIMUM: 20; MAXIMUM: 30

NHI Customer Service: (877) 558-6873 • nhicustomerservice@dot.gov

COURSE NUMBER

FHWA-NHI-134061

COURSE TITLE

Construction Program Management and Inspection

The Federal Highway Administration's (FHWA) responsibilities for construction project and program oversight has changed considerably throughout the years. Today, the FHWA field engineers are typically involved in a diverse array of issues that were not common in construction projects of decades past. Changes in legislation, declines in staffing resources and expertise, and increased complexity of the Federal-aid construction program have all had an impact on how the FHWA conducts construction program management and oversight. Today's FHWA field engineers must have a more focused and programmatic approach in fulfilling construction stewardship and oversight responsibilities.

This 2-day training workshop highlights the FHWA roles and resources to assist the State in delivering a quality construction program. The training will assist the FHWA field engineers in maintaining and improving technical competence and in selecting a balanced program of construction management techniques.

The workshop uses the "Construction Program Management and Inspection Guide" as instructional material. While the workshop is focused primarily at FHWA's staff and FHWA oversight activities, participation by State partners and other relevant entities is highly encouraged to further educate and train Federal Aide partners to "act on FHWA's behalf in line with the Divisions/State DOT Stewardship Agreement.

OUTCOMES

Upon completion of the course, participants will be able to:

- Manage and oversee Federal-aid construction programs.

TARGET AUDIENCE

This training is targeted at FHWA Division field engineers and State agencies, and will provide staff with the background and knowledge they need for managing and overseeing their Federal-aid construction programs. The training is geared towards the new FHWA generalist employee but is also intended as a refresher for the veteran FHWA engineer.

TRAINING LEVEL: Basic

FEE: 2016: $335 Per Person; 2017: N/A

LENGTH: 2 DAYS (CEU: 0 UNITS)

CLASS SIZE: MINIMUM: 15; MAXIMUM: 40

NHI Customer Service: (877) 558-6873 • nhicustomerservice@dot.gov

COURSE NUMBER

FHWA-NHI-134062

COURSE TITLE

Bridge Evaluation for Rehabilitation Design Considerations 4.5 Day

The ultimate goal of this effort is the development of a nationally accepted program that will serve to improve quality, ensure uniformity, and establish a minimum standard for bridge rehabilitation. The course will present innovative and state-of-the-art bridge rehabilitation technologies and procedures for a broad array of structural elements including bridge decks, girders, piers, and abutments.

Core curriculum for the course is 4.5 days and covers the outcomes listed below.

OUTCOMES

Upon completion of the course, participants will be able to:

- Describe conditions that suggest the need for rehabilitation
- Identify the need for, and capacity of, destructive and/or non destructive testing (NDT) for assessment of existing conditions
- Prescribe analysis and load testing to determine the effect of existing conditions on the structure
- Distinguish root causes of distress and deterioration
- Formulate appropriate rehabilitation strategies
- Select procedures and materials for rehabilitation
- Develop effective rehabilitation construction documents
- Prepare and implement quality assurance for construction
- Monitor and resolve construction and material problems

TARGET AUDIENCE

The target audience includes design engineers, field engineers, resident engineers, structural engineers, materials engineers, and other technical personnel involved in the construction and rehabilitation design of bridges. Participants with an engineering background are expected to constitute the target audience. People knowledgeable in new bridge design, but not necessarily bridge rehabilitation, should attend.

TRAINING LEVEL: Intermediate

FEE: 2016: $1100 Per Person; 2017: $1100 Per Person

LENGTH: 4.5 DAYS (CEU: 2.7 UNITS)

CLASS SIZE: MINIMUM: 20; MAXIMUM: 30

NHI Customer Service: (877) 558-6873 • nhicustomerservice@dot.gov

Course Number

FHWA-NHI-134062A

Course Title

Bridge Evaluation for Rehabilitation Design Considerations 5-Day

The ultimate goal of this effort is the development of a nationally accepted program that will serve to improve quality, ensure uniformity, and establish a minimum standard for bridge rehabilitation. The course will present innovative and state-of-the-art bridge rehabilitation technologies and procedures for a broad array of structural elements including bridge decks, girders, piers, and abutments.

The 5-day version of this course includes two additional modules on the rehabilitation of timber and masonry structures.

Outcomes

Upon completion of the course, participants will be able to:

- Describe conditions that suggest the need for rehabilitation
- Identify the need for, and capacity of, destructive and/or non destructive testing (NDT) for assessment of existing conditions
- Prescribe analysis and load testing to determine the effect of existing conditions on the structure
- Distinguish root causes of distress and deterioration
- Formulate appropriate rehabilitation strategies
- Select procedures and materials for rehabilitation
- Develop effective rehabilitation construction documents
- Prepare and implement quality assurance for construction
- Monitor and resolve construction and material problems

Target Audience

The target audience includes design engineers, field engineers, resident engineers, structural engineers, materials engineers, and other technical personnel involved in the construction and rehabilitation design of bridges. Participants with an engineering background are expected to constitute the target audience. People knowledgeable in new bridge design, but not necessarily bridge rehabilitation should attend.

Training Level: Intermediate

Fee: 2016: $1175 Per Person; 2017: $1175 Per Person

Length: 5 DAYS (CEU: 3 UNITS)

Class Size: MINIMUM: 20; MAXIMUM: 30

NHI Customer Service: (877) 558-6873 • nhicustomerservice@dot.gov

Course Number

FHWA-NHI-134063

Course Title

Maintenance Leadership Academy

The Maintenance Leadership Academy provides an intensive training program to individuals who hold positions as State, district, and county maintenance supervisors. The Academy can help decrease the time it takes to acclimate new managers and provide an opportunity for career development.

Participants acquire an understanding of the various processes, methods, and materials that are applied to maintain their organization's bridge and highway systems. Participants develop a knowledge base of planning, scheduling, quality control, customer focus, program presentation, asset management, contract management and performance improvement. See sample outcomes below for each of the six modules that comprise the Academy.

The Academy curriculum consists of self-paced lessons accessed via the Web, as well as instructor-led classroom sessions. Self-paced lessons are completed prior to attending each of the two classroom sessions. Upon enrolling for the Maintenance Leadership Academy, participants attend a 1-hour Orientation Web-conference that provides an overview of the Academy's schedule and information on how to access the self-paced lessons.

An example of the structure of the Academy, which ideally allows 12 weeks from the date of the Orientation Web-conference through final day of instructor-led training:

- Enroll and attend a 1-hour Orientation Web-conference
- Complete 22 hours of independent study materials (paper-based and Web-based)
- Attend 8 days of instructor-led, classroom training
- Complete 10.5 hours of independent study material and attend a 1-hour homework review Web-conference
- Attend the final 4 days of instructor-led, classroom training

NOTE: Interested hosts should submit their course requests at least four months in advance of the desired start date. Contact NHI with any questions about the Maintenance Leadership Academy course structure.

Outcomes

Upon completion of the course, participants will be able to:

- Describe the use of maintenance administration in achieving highway agency goals. (Module A)
- Describe how various treatments fit into an overall system preservation program and when to implement them. (Module B)
- Identify appropriate drainage maintenance and roadside management techniques. (Module C)
- Describe the maintenance manager's roles and responsibilities for developing, implementing, and managing a comprehensive plan for dealing with weather-related events. (Module D)
- Explain the maintenance and use of traffic control devices (including work zone plans, work zone traffic control devices, signs, striping, guardrails, and median barriers) in maintenance operations. (Module E)
- Describe how environmental protection issues, regulations and control measures affect highway maintenance activities. (Module F)

Target Audience

This course was designed for State, regional, or county personnel who manage operations programs and deal with oversight and quality assurance over broader geographic areas. They are involved with handling materials, scheduling, budgeting and planning. Participants have an advanced skill in maintenance activities. Participants enrolling in the Academy will need to have taken NHI-134064 "Transportation Construction Quality Assurance" and NHI-131110 "Pavement Preservation Treatment Construction" or had equivalent training or experience in these content areas.

TRAINING LEVEL: Accomplished

FEE: 2016: $3100 Per Person; 2017: $3100 Per Person

LENGTH: 12 DAYS (CEU: 10.5 UNITS)

CLASS SIZE: MINIMUM: 20; MAXIMUM: 30

NHI Customer Service: (877) 558-6873 • nhicustomerservice@dot.gov

COURSE NUMBER

FHWA-NHI-134064

COURSE TITLE

Transportation Construction Quality Assurance (1.5-Day)

The Federal Highway Administration (FHWA) identified the need for transportation construction and materials personnel to increase their knowledge of the fundamentals of effective transportation construction Quality Assurance (QA). This course was developed to ensure that agency, contractor, producer, and consultant personnel responsible for interpreting and applying quality assurance specifications in transportation construction are properly qualified. The course will utilize a Quality Assurance Reference Manual, adapted from the current NETTCP manual.

This one and a half-day version of the course covers Chapters 1 through 6 of the course materials and will be available to, and appropriate for, all audiences including management level personnel. The content covered in this first day includes how quality assurance is featured in a transportation construction quality assurance program, quality assurance program elements, the evolution of quality assurance specifications, measuring quality, and the roles and responsibilities of both contractor and agency personnel.

OUTCOMES

Upon completion of the course, participants will be able to:

- Consistently apply fundamental Quality Assurance concepts, terminology, and definitions
- Differentiate QA specifications from other specifications
- Explain each of the six core elements of a QA program and how each is essential to successful implementation of Quality Assurance
- Describe the respective roles and responsibilities of the project decision makers (Contractor QC and Agency Acceptance personnel) and how their interaction contributes to construction quality

TARGET AUDIENCE

This is an intermediate-level course for personnel who are implementing QA specifications on construction projects. Necessary background knowledge for participants is 3-5 years minimum in transportation construction specifications inspections. The suggested list of personnel that may consider attending, if they have the requisite background knowledge are Contractor/Consultant Personnel (QC managers/QC Plan Administrators, Senior Production Facility QC Technician/Inspectors, Senior QC Laboratory Personnel, and Senior Field QC Technicians/Inspectors) and Agency Personnel (Project Managers/Resident Engineers, Senior Production Facility Acceptance Technicians/Inspectors, Senior Acceptance Laboratory Personnel, and Senior Field Acceptance Technicians/Inspectors.

TRAINING LEVEL: Intermediate

FEE: 2016: $250 Per Person; 2017: $250 Per Person

LENGTH: 1.5 DAYS (CEU: 1.1 UNITS)

CLASS SIZE: MINIMUM: 20; MAXIMUM: 30

NHI Customer Service: (877) 558-6873 • nhicustomerservice@dot.gov

COURSE NUMBER

FHWA-NHI-134064A

COURSE TITLE

Transportation Construction Quality Assurance (3-Day)

The Federal Highway Administration (FHWA) identified the need for transportation construction and materials personnel to increase their knowledge of the fundamentals of effective transportation construction Quality Assurance (QA). This course was developed to ensure that agency, contractor, producer, and consultant personnel responsible for interpreting and applying quality assurance specifications in transportation construction are properly qualified. The course will utilize a Quality Assurance Reference Manual, adapted from the current NETTCP manual.

This three-day version of the course covers Chapters 1 through 10 of the course materials and will be available to, and appropriate for, production, laboratory, and field QC and Acceptance technicians and inspectors. This version contains mathematical terms and principles used in QA sampling, testing, and decision-making. The content also includes how quality assurance is featured in a transportation construction quality assurance program, quality assurance program elements, the evolution of quality assurance specifications, measuring quality, and the roles and responsibilities of both contractor and agency personnel.

OUTCOMES

Upon completion of the course, participants will be able to:

- Consistently apply fundamental Quality Assurance concepts, terminology, and definitions
- Differentiate QA specifications from other specifications
- Explain each of the six core elements of a QA program and how each is essential to successful implementation of Quality Assurance
- Describe the respective roles and responsibilities of the project decision makers (Contractor QC and Agency Acceptance personnel) and how their interaction contributes to construction quality
- Apply the mathematical concepts of variability, statistical distribution, and sampling protocols to measure construction quality
- Describe the primary components of inspection, properly document the results of inspection, and utilize inspection data to quantify quality of workmanship

TARGET AUDIENCE

This is an intermediate-level course for personnel who are implementing QA specifications on construction projects. Necessary background knowledge for participants: 3-5 years minimum in transportation construction specifications inspections, basic statistical knowledge/training, some usage of tools necessary to the Quality Assurance process (contractor test results). The suggested list of personnel that may consider attending, if they have the requisite background knowledge are Contractor/Consultant Personnel (QC managers/QC Plan Administrators, Senior Production Facility QC Technician/Inspectors, Senior QC Laboratory Personnel, and Senior Field QC Technicians/Inspectors) and Agency Personnel (Project Managers/Resident Engineers, Senior Production Facility Acceptance Technicians/Inspectors, Senior Acceptance Laboratory Personnel, andSenior Field Acceptance Technicians/Inspectors).

TRAINING LEVEL: Intermediate

FEE: 2016: $350 Per Person; 2017: $350 Per Person

LENGTH: 3 DAYS (CEU: 1.8 UNITS)

CLASS SIZE: MINIMUM: 20; MAXIMUM: 30

NHI Customer Service: (877) 558-6873 • nhicustomerservice@dot.gov

Course Number

FHWA-NHI-134067

Course Title

Construction Inspection of Bridge Rehabilitation Projects

This 4-day course has been designed to improve quality, ensure uniformity, and establish a minimum standard for bridge rehabilitation.

The keys to successfully ensuring quality on rehab jobs are: knowing what should happen on a given job; identifying problems when they do happen; and correctly using available resources to solve the problem. This course presents innovative and best practice inspection techniques for each structural element of a bridge.

This course will introduce participants to distress and deterioration they may encounter when working with concrete or steel that requires repair. It is essential to identify the issues that harm these materials because it is often poor construction techniques that lead to reduced structural condition or shortened service life. The focus then turns to construction and inspection practices pertaining to concrete decks, steel superstructures, concrete superstructures and substructures, joints, and bearings.

The course is activity-rich, using discussions of best practices, small and large group activities for identifying critical inspection moments, and a wide array of case studies from real projects to emphasize the importance of applying these techniques in the field.

Outcomes

Upon completion of the course, participants will be able to:

- Relate observable deterioration of bridge structural elements to distress mechanisms
- Associate potential construction and materials problems
- Explain the role of the construction inspector as part of the overall project team
- Interpret drawings and specifications
- Describe rehabilitation sequences for various bridge systems, bridge types, and materials
- Explain basic inspection and testing of materials
- Make and maintain sufficient records

Target Audience

This course will be appropriate for inspectors with 1-5 years of experience who are seeking a better foundation in bridge rehabilitation techniques. They will likely have a basic grasp of construction and inspection methods, bridge terminology, and causes of distress and deterioration, although this information will be reviewed at the beginning of the course.The course will be appropriate for experienced bridge inspectors who are seeking to learn about innovative methods in bridge rehabilitation and obtain a refresher on familiar inspection methods. Construction supervisors, transportation department field inspectors, construction inspectors, field engineers, resident engineers, structural engineers, materials engineers, and other technical personnel involved in the inspection of bridge rehabilitation projects will benefit from this course. The course is designed for participants without an in-depth engineering background. However, those with engineering backgrounds are welcome to attend and can provide valuable perspective in the context of group activities and discussions.

Training Level: Basic

Fee: 2016: $1050 Per Person; 2017: $1050 Per Person

Length: 4 DAYS (CEU: 2.4 UNITS)

Class Size: MINIMUM: 20; MAXIMUM: 30

NHI Customer Service: (877) 558-6873 • nhicustomerservice@dot.gov

Course Number

FHWA-NHI-134069

Course Title

Ethics Awareness for the Transportation Industry - WEB-BASED

This training is provided by the Transportation Curriculum Coordination Council (TCCC) in partnership with NHI to provide good practices for ethical behavior of transportation employees. The training was prepared by State DOT personnel for State DOT personnel. This course is primarily intended for inspectors and technicians.

The training contains good practices from various agencies. The topics of discussion in this training are: conflict of interest, safety, fraud, falsification of documentation, reporting ethical concerns, gifts and favors, fairness, personal use of agency property, and consequences.

Not all State agencies' codes of conduct are the same but they all demand similar ethical behavior of their employees. Be sure to access to your agency's codes or check with your supervisor for more information specific to your organization. Each State agency/company has their own work rules, which the viewer needs to review and follow.

NHI is hosting this and other TCCC Web-based developments to serve a critical need for training. We need your feedback to determine whether we should continue posting other Web-based trainings like this one. Please take the time to complete the evaluation form provided at the end of the training, or email NHIMarketing@dot.gov with your feedback.

Outcomes

Upon completion of the course, participants will be able to:

- Describe agency expectations on ethics
- Give an example of a current code of conduct policy
- Recognize and practice good ethics as an employee in the transportation industry
- Explain the consequences when rules and regulations are not followed

Target Audience

This training is designed for Level I and Level II State and local public agency personnel and their industry counterparts involved in the construction, maintenance and testing process for highways and structures. Level I or Entry refers to employees/ trainees with little to no experience in the subject area and perform his/her activities under direct supervision. Level II or Intermediate refers to employees that understand and demonstrate skills in one or more areas of the entry level and perform specific tasks under general supervision.

Training Level: Basic

Fee: 2016: $25 Per Person; 2017: $25 Per Person

Length: 1 HOURS (CEU: 0 UNITS)

Class Size: MINIMUM: 1; MAXIMUM: 1

NHI Customer Service: (877) 558-6873 • nhicustomerservice@dot.gov

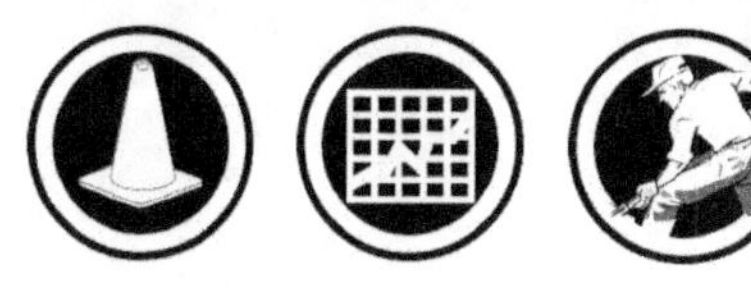

COURSE NUMBER

FHWA-NHI-134070

COURSE TITLE

SpecRisk Quality Assurance Specification Development and Validation Course - WEB-BASED

This course will provide an introduction to statistical analysis and the development of statistically valid quality assurance specifications, introducing general guidelines established and put forth by the Federal Government and FHWA policy. The course also provides participants with an introduction to SpecRisk, the resource that is necessary to successfully develop statistically valid specifications. The course is designed and delivered to motivate members of the target audience to use SpecRisk software to develop their specifications. Although the course demonstrates basic functions of the software, it is not intended to be an in-depth training on how to use SpecRisk.

This course requires a solid foundation in basic statistics. Completion of FHWA-NHI 134042, or equivalent training, is also recommended. NHI 134042 trains participants to identify the importance of organizing data and how to plot frequency histograms. It explains how a sample relates to the population, the relationship between single and multiple samples, and the use of random stratified sampling tables. This knowledge provides an excellent foundation for this course.

OUTCOMES

Upon completion of the course, participants will be able to:

- Recognize key concepts to develop an effective, statistically valid Quality Assurance (QA) specification.
- Make an informed selection among available options when developing an acceptance plan.
- Develop QA specifications in alignment with best practices, Federal regulations, and FHWA policy.
- Apply SpecRisk software to understand risks and develop statistically valid specifications.

TARGET AUDIENCE

Personnel involved in specification development: Federal, State, and local highway agency engineers and technicians in materials, construction, and research. The training is also appropriate for industry personnel that are involved in reviewing and providing input to the specification development process.

TRAINING LEVEL: Basic

FEE: 2016: $50 Per Person; 2017: $50 Per Person

LENGTH: 8 HOURS (CEU: 0 UNITS)

CLASS SIZE: MINIMUM: 1; MAXIMUM: 1

NHI Customer Service: (877) 558-6873 • nhicustomerservice@dot.gov

COURSE NUMBER

FHWA-NHI-134071

COURSE TITLE

Basic Construction and Maintenance Documentation - Improving the Daily Diary - WEB-BASED

This training is provided by the Transportation Curriculum Coordination Council (TCCC) in partnership with NHI to help improve documentation on construction and maintenance projects. The training was prepared by State DOT personnel for State DOT personnel. This course is primarily intended for inspectors and technicians.

It contains good practices from various agencies. This training is intended to assist you with proper documentation on a construction or maintenance project. It is important that the information in the daily diary kept for projects are accurate, correct, and factual to insure proper payment and to avoid lawsuits.

Please note that the terminology may differ slightly from DOT to DOT; for example, the document may also be referred to as a Daily Work Report. Each State agency/company has their own requirements, which the viewer needs to review and follow.

NHI is hosting this and other TCCC Web-based developments to serve a critical need for training. We need your feedback to determine whether we should continue posting other Web-based trainings like this one. Please take the time to complete the evaluation form provided at the end of the training, or email NHIMarketing@dot.gov with your feedback.

OUTCOMES

Upon completion of the course, participants will be able to:

- Compose a complete and correct daily diary
- Recognize the importance of daily diary entries

TARGET AUDIENCE

This training is designed for Level I and Level II State and local public agency personnel and their industry counterparts involved in the construction, maintenance and testing process for highways and structures. Level I or Entry refers to employees/ trainees with little to no experience in the subject area and perform his/her activities under direct supervision. Level II or Intermediate refers to employees that understand and demonstrate skills in one or more areas of the entry level and perform specific tasks under general supervision.

TRAINING LEVEL: Basic

FEE: 2016: $25 Per Person; 2017: $25 Per Person

LENGTH: 1 HOURS (CEU: 0 UNITS)

CLASS SIZE: MINIMUM: 1; MAXIMUM: 1

NHI Customer Service: (877) 558-6873 • nhicustomerservice@dot.gov

COURSE NUMBER

FHWA-NHI-134072

COURSE TITLE

Math Module - WEB-BASED

This training is provided by the Transportation Curriculum Coordination Council (TCCC) in partnership with NHI to review math basics such as, math functions, algebra, and geometry. This course includes instruction that would be applicable to variety of end users. The basic math functions would be appropriate to the entry level technician or as a review. This course is primarily intended for inspectors and technicians.

The more complex areas of algebra and geometry would be appropriate for the more advanced technician. Problems covered in this course can be applied to further an employee's education or use the principals to solve everyday work problems. This course can be used as both a learning tool and/or as an excellent refresher.

OUTCOMES

Upon completion of the course, participants will be able to:

- Perform basic and intermediate calculations using mathematics, algebra, and geometry
- Understand the impact of mathematics, algebra, and geometry in their job functions
- Build upon a foundation for applying operations and engineering concepts on the job
- Understand the impact that their actions may have on the safe and reliable operation of DOT components and systems

TARGET AUDIENCE

This course is designed for FHWA, State, and Local Agencies and their industry counterparts that are involved in construction and maintenance practices. It is applicable to anyone that will be performing everyday calculations for inspection, testing, and a variety of other job functions.

TRAINING LEVEL: Basic

FEE: 2016: $50 Per Person; 2017: $50 Per Person

LENGTH: 6 HOURS (CEU: 0 UNITS)

CLASS SIZE: MINIMUM: 1; MAXIMUM: 1

NHI Customer Service: (877) 558-6873 • nhicustomerservice@dot.gov

COURSE NUMBER

FHWA-NHI-134074

COURSE TITLE

Bolted Connections - WEB-BASED

This training was prepared by the Transportation Curriculum Coordination Council (TCCC) in partnership with NHI to introduce the basics of structural bolted connections. The information presented is useful for non-structural applications as well. Bolting is a common method of making connections, care should be exercised both in their design, installation and maintenance. For the purposes of this course, we are focusing on the installation of bolts. This training is recommended for the Transportation Curriculum Coordination Council levels II through IV. This course is primarily intended for inspectors and technicians.

This module consists of three lessons:

Bolted joints reviews the basic connection types, types of holes, faying/contact surfaces, use of washers, tightening patterns, and fastener documentation.

Installation procedures explain how important it is to protect the fastener assemblies and surfaces during construction. We will review fastener assemblies' pre-installation verification and explain the rotational-capacity testing.

Installation methods discuss basic guidelines to achieve quality fastener installations. There are several accepted methods for installing structural bolts. The methods covered in this training are turn-of-nut, calibrated wrench, direct tension indicator, twist-off bolt, and lock pin and collar.

OUTCOMES

Upon completion of the course, participants will be able to:

- Identify various fastener connection types
- Describe installation procedures
- Identify and describe various accepted installation methods

TARGET AUDIENCE

This training is designed for FHWA, State, and local agencies and their industry counterparts involved in the installation and inspection of bolts and bolted connections on construction projects.

TRAINING LEVEL: Intermediate

FEE: 2016: $25 Per Person; 2017: $25 Per Person

LENGTH: 2 HOURS (CEU: 0 UNITS)

CLASS SIZE: MINIMUM: 1; MAXIMUM: 1

NHI Customer Service: (877) 558-6873 • nhicustomerservice@dot.gov

COURSE NUMBER

FHWA-NHI-134075

COURSE TITLE

Hardened Concrete Properties - Durability - WEB-BASED

This training is provided by the Transportation Curriculum Coordination Council (TCCC) in partnership with NHI to review integrated materials and construction practices for concrete pavement. The training was developed by the National Concrete Pavement Technology Center at Iowa State University. This training is recommended for the Transportation Curriculum Coordination Council levels II - IV. This course is primarily intended for inspectors and technicians.

Durability as a property of hardened concrete is essential for long-lasting pavements. This workshop discusses factors that contribute to durable concrete and covers permeability, frost resistance, sulfate resistance, alkali silica attack, and a brief look at abrasion resistance.

This module is part of a curriculum from the "Integrated Materials and Construction Practices for Concrete Pavement" manual developed through the National Concrete Pavement Technology Center at Iowa State University. The other Web-based training modules include:

FHWA-NHI-134084 TCCC Fundamentals of Materials Used for Concrete Pavements

FHWA-NHI-134085 TCCC Incompatibility in Concrete Pavement Systems

FHWA-NHI-134087 TCCC Mix Design Principles

FHWA-NHI-134095 TCCC Early Age Cracking

FHWA-NHI-134096 TCCC Basics of Cement Hydration

FHWA-NHI-134097 TCCC Fresh Concrete Properties

FHWA-NHI-134098 TCCC Construction of Concrete Pavements

FHWA-NHI-134100 TCCC QCQA for Concrete Pavements

FHWA-NHI-134101 TCCC Design of Pavement

FHWA-NHI-134102 TCCC Troubleshooting for Concrete Pavements

OUTCOMES

Upon completion of the course, participants will be able to:

- Recognize factors contribute to durable concrete
- Explain the importance of permeability, alkali-silica reaction, abrasion resistance and, in certain regions in the country, frost resistance and sulfate resistance of hardened concrete
- Identify tests that can be performed to determine the variables affecting the durability of hardened concrete

TARGET AUDIENCE

This training is designed for FHWA, State, and local agencies and their industry counterparts involved in the process to assure that concrete meets all the requirements for durability. It is applicable to anyone desiring a better understanding of the factors of durability.

TRAINING LEVEL: Basic

FEE: 2016: $25 Per Person; 2017: $25 Per Person

LENGTH: 1 HOURS (CEU: 0 UNITS)

CLASS SIZE: MINIMUM: 1; MAXIMUM: 1

NHI Customer Service: (877) 558-6873 • nhicustomerservice@dot.gov

COURSE NUMBER

FHWA-NHI-134077

COURSE TITLE

Contract Administration Core Curriculum

More than 10,000 Federal-aid construction contracts are authorized by FHWA each year. Those contracts are subsequently administered by State departments of transportation and local public agencies that may not be familiar with FHWA's complex requirements for construction contracts. Recent FHWA program reviews of projects administered by local public agencies indicated that contract administration is a continuing high-risk area that needs additional focus.

Therefore, a newly revised, 2-day instructor-led training course was developed in concert with updates and revisions to the Contract Administration Core Curriculum (CACC) Manual (revised October 2014). The training was developed to explain basic Federal-aid requirements; promote awareness of FHWA policy; facilitate familiarity with the newly reorganized, revised, and expanded CACC manual; and allow supervised practice activities using the manual to find information. By engaging in a variety of in-class exercises and case studies, participants become quite familiar with the CACC Manual and learn how to best use it as a daily resource.

A basic understanding of the background and structure of the Federal-aid Highway Program (FAHP) is required for participants attending this course. Participants who are new to administering Federal-aid contracts should take one of NHI's introductory courses to the Federal-aid Highway Program (NHI 310110, 310109, or the Web-based training 310115) prior to attending this course. Anyone needing a refresher on the FAHP is encouraged to take NHI 310115 (Federal-aid 101) before attending the CACC course.

Prior to attending class, all participants are expected to watch the Federal-aid Essentials video Stewardship and Oversight. This video is approximately 8.5-minutes long and can be accessed at www.fhwa.dot.gov/federal-aidessentials/.

OUTCOMES

Upon completion of the course, participants will be able to:

- Use the Contract Administration Core Curriculum Manual (CACC) and other FHWA resources in order to answer questions regarding program-level and project-level requirements on Federal Aid (FA) projects
- Describe the impact program-level contract requirements have on individual FA projects
- Identify the contract requirements associated with administering FA projects for Federal and State entities at the pre-award, advertising and award, and post-award and constructions stages

TARGET AUDIENCE

This course is designed for Federal Highway Administration (FHWA) Division Office personnel who must read, interpret, and apply Federal regulations and guidance that affects administration of Federal-aid contracts, as well as any State and local government agency personnel who must interpret and apply Federal regulations and guidance that affects administration of Federal-aid contracts.

TRAINING LEVEL: Basic

FEE: 2016: $235 Per Person; 2017: $235 Per Person

LENGTH: 2 DAYS (CEU: 1.3 UNITS)

CLASS SIZE: MINIMUM: 20; MAXIMUM: 30

NHI Customer Service: (877) 558-6873 • nhicustomerservice@dot.gov

Course Number

FHWA-NHI-134080

Course Title

Environmental Factors in Construction and Maintenance

NOTE: This course is intended for highway construction inspectors, maintenance supervisors, and other inspection and field personnel.

This is a blended course that comprises approximately 6 hours of independent study work and a 1.5-day instructor-led session. Participants must complete the independent study materials before attending the instructor-led session.

Mandated environmental considerations are an important part of all highway agencies' roadway construction and maintenance activities. NHI 134080 Environmental Factors in Construction and Maintenance focuses on balancing the need to fulfill environmental protections and the need to complete project activities in a safe, timely, and financially responsible manner.

This course emphasizes common environmental agency regulations, adherence to plans, early and frequent communication regarding construction and maintenance commitments, and the potential for encountering unexpected issues. Course activities help participants understand how to build environmental considerations into their standard practice. Learning to relate environmental commitments to construction and maintenance processes and practices can help transportation personnel ensure compliance with numerous and increasingly complex Federal, State, and local environmental regulations.

Course content is delivered via approximately 6 hours of independent study workbook materials and a 1.5-day classroom-based, instructor-led session. Two FHWA instructors relate their construction experience and environmental knowledge to help ensure that participants in this course will be able to apply the training content immediately to their projects and duties.

Outcomes

Upon completion of the course, participants will be able to:

- Relate design-phase environmental commitments to construction documents
- Explain your role in early and continuous communication to support commitments that occurred during design phase
- Recognize the importance of environmental protection during construction and maintenance operations
- Describe quality control measures and documentation that can be implemented through the construction sequence to provide environmental mitigation measures
- Recognize the role of the project inspectors (and environmental inspectors, when used) in addressing environmental issues
- Describe a variety of environmental compliance and commitment tracking tools
- Identify resources for consultation on environmental issues

Target Audience

This course is intended primarily for Federal, State, and local highway construction inspectors, maintenance supervisors, and other inspection and field personnel who must ensure that identified environmental impacts are mitigated during construction and maintenance operations. This may include FHWA employees, as well as State employees and local agencies and consultants that oversee such activities.

Training Level: Basic

Fee: 2016: $235 Per Person; 2017: $235 Per Person

Length: 1.5 DAYS (CEU: 1.5 UNITS)

Class Size: MINIMUM: 20; MAXIMUM: 30

NHI Customer Service: (877) 558-6873 • nhicustomerservice@dot.gov

Course Number

FHWA-NHI-134084

Course Title

Fundamentals of Materials Used for Concrete Pavements - WEB-BASED

This training is provided by the Transportation Curriculum Coordination Council (TCCC) in partnership with NHI to review integrated materials and construction practices for concrete pavement. It is the first training of its kind offered by NHI, and we would like to give special recognition to the TCCC for their efforts. This training is recommended for the Transportation Curriculum Coordination Council levels II - IV. This course is primarily intended for inspectors and technicians.

The materials used in Portland cement concrete play an extremely valuable role in the performance of the concrete. This training covers both the non-reactive and reactive materials used in Portland cement concrete. This would include the aggregates, curing compound, reinforcement, and the materials that are chemically reactive.

This module is part of a curriculum from the "Integrated Materials and Construction Practices for Concrete Pavement" manual developed through the National Concrete Pavement Technology Center at Iowa State University. The other Web-based training modules include:

FHWA-NHI-134075 TCCC Hardened Concrete Properties - Durability

FHWA-NHI-134085 TCCC Incompatibility in Concrete Pavement Systems

FHWA-NHI-134087 TCCC Mix Design Principles

FHWA-NHI-134095 TCCC Early Age Cracking

FHWA-NHI-134096 TCCC Basics of Cement Hydration

FHWA-NHI-134097 TCCC Fresh Concrete Properties

FHWA-NHI-134098 TCCC Construction of Concrete Pavements

FHWA-NHI-134100 TCCC QCQA for Concrete Pavements

FHWA-NHI-134101 TCCC Design of Pavement

FHWA-NHI-134102 TCCC Troubleshooting for Concrete Pavements

Outcomes

Upon completion of the course, participants will be able to:

- Identify materials used in Portland cement concrete
- Describe the importance of each material and the role it plays in the performance of the concrete
- Describe how each material reacts with the other materials to obtain strength, permeability, workability, etc.

Target Audience

This training is designed for FHWA, State, and local agencies and their industry counterparts involved in the process to assure that the materials used in Portland cement concrete meet specification requirements and are compatible to provide good, durable concrete. It is applicable to anyone desiring a better understanding of the materials used in Portland cement concrete.

Training Level: Intermediate

Fee: 2016: $25 Per Person; 2017: $25 Per Person

Length: 2 HOURS (CEU: 0 UNITS)

Class Size: MINIMUM: 1; MAXIMUM: 1

NHI Customer Service: (877) 558-6873 • nhicustomerservice@dot.gov

Course Number

FHWA-NHI-134085

Course Title

Incompatibility in Concrete Pavement Systems - WEB-BASED

This training is provided by the Transportation Curriculum Coordination Council (TCCC) in partnership with NHI to review integrated materials and construction practices for concrete pavement. It is the first training of its kind offered by NHI, and we would like to give special recognition to the TCCC for their efforts. This training is recommended for the Transportation Curriculum Coordination Council levels II - IV.

The materials used in Portland cement concrete play an extremely valuable role in the performance of the concrete. This training covers the incompatibilities of materials used in Portland cement concrete. Although certain materials may be perfectly acceptable on their own, when they are combined they are not compatible with each other. This can cause early stiffening, retardation, cracking, and the lack of a quality of air void system.

This module is part of a curriculum from the "Integrated Materials and Construction Practices for Concrete Pavement" manual developed through the National Concrete Pavement Technology Center at Iowa State University. The other Web-based training modules include:

FHWA-NHI-134075 TCCC Hardened Concrete Properties - Durability

FHWA-NHI-134084 TCCC Fundamentals of Materials Used for Concrete Pavements

FHWA-NHI-134087 TCCC Mix Design Principles

FHWA-NHI-134095 TCCC Early Age Cracking

FHWA-NHI-134096 TCCC Basics of Cement Hydration

FHWA-NHI-134097 TCCC Fresh Concrete Properties

FHWA-NHI-134098 TCCC Construction of Concrete Pavements

FHWA-NHI-134100 TCCC QCQA for Concrete Pavements

FHWA-NHI-134101 TCCC Design of Pavement

FHWA-NHI-134102 TCCC Troubleshooting for Concrete Pavements

Outcomes

Upon completion of the course, participants will be able to:

- Identify the causes of incompatible conditions leading to early stiffening or setting and occasional early age cracking
- Recognize the importance to use the correct air void system
- Describe test methods used to identify incompatibilities

Target Audience

This training is designed for FHWA, State, and local agencies and their industry counterparts involved in the process to assure that the materials used in Portland cement concrete meet specification requirements and are compatible to provide good, durable concrete. It is applicable to anyone desiring a better understanding of the materials used in Portland cement concrete.

Training Level: Intermediate

Fee: 2016: $25 Per Person; 2017: N/A

Length: 1 HOURS (CEU: 0 UNITS)

Class Size: MINIMUM: 1; MAXIMUM: 1

NHI Customer Service: (877) 558-6873 • nhicustomerservice@dot.gov

Course Number

FHWA-NHI-134087

Course Title

Mix Design Principles - WEB-BASED

This training is provided by the Transportation Curriculum Coordination Council (TCCC) in partnership with NHI to review integrated materials and construction practices for concrete pavement. This training is recommended for the Transportation Curriculum Coordination Council levels II - IV.

This module discusses mix design and mix proportioning. Mix design is the process of choosing the characteristics we are looking for in the concrete mixture. Mix proportioning, on the other hand, involves taking the information provided by the mix design process and using that information to determine the actual proportions of ingredients in the mixture. This course discusses theoretical, laboratory, and field testing to determine the Portland cement concrete mix that will achieve the best possible durability, strength, constructability, economy, and uniformity.

This module is part of a curriculum from the "Integrated Materials and Construction Practices for Concrete Pavement" manual developed through the National Concrete Pavement Technology Center at Iowa State University. The other Web-based training modules include:

FHWA-NHI-134075 TCCC Hardened Concrete Properties - Durability

FHWA-NHI-134084 TCCC Fundamentals of Materials Used for Concrete Pavements

FHWA-NHI-134085 TCCC Incompatibility in Concrete Pavement Systems

FHWA-NHI-134095 TCCC Early Age Cracking

FHWA-NHI-134096 TCCC Basics of Cement Hydration

FHWA-NHI-134097 TCCC Fresh Concrete Properties

FHWA-NHI-134098 TCCC Construction of Concrete Pavements

FHWA-NHI-134100 TCCC QCQA for Concrete Pavements

FHWA-NHI-134101 TCCC Design of Pavement

FHWA-NHI-134102 TCCC Troubleshooting for Concrete Pavements

Outcomes

Upon completion of the course, participants will be able to:

- Describe the overall goal of mix design
- Define the difference between mix design and mix proportioning
- Recognize field and laboratory testing plans
- Describe test methods used to identify incompatibilities

Target Audience

This training is designed for FHWA, State, and local agencies and their industry counterparts involved in the process to assure that the mix design and proportioning of Portland cement concrete materials meet specification requirements and provide good, durable concrete. It is applicable to anyone desiring a better understanding of the mix design of Portland cement concrete.

Training Level: Intermediate

Fee: 2016: $25 Per Person; 2017: $25 Per Person

Length: 1 HOURS (CEU: 0 UNITS)

Class Size: MINIMUM: 1; MAXIMUM: 1

NHI Customer Service: (877) 558-6873 • nhicustomerservice@dot.gov

Course Number

FHWA-NHI-134095

Course Title

Early Age Cracking - WEB-BASED

This training is provided by the Transportation Curriculum Coordination Council (TCCC) in partnership with NHI to review integrated materials and construction practices for concrete pavement. The training was developed by the National Concrete Pavement Technology Center at Iowa State University. It is the first training of its kind offered by NHI, and we would like to give special recognition to the TCCC for their efforts. This training is recommended for the Transportation Curriculum Coordination Council levels II - IV. This course is primarily intended for inspectors and technicians.

Cracks are not a problem as long as they are controlled through jointing; ideally the concrete will crack below the saw joint to relieve the stress. Uncontrolled random cracks are not aesthetically acceptable and can reduce ride quality, durability, and particularly load transfer. Early cracking in this module is defined as those cracks that occur before the concrete is open to public traffic. In this module, we will be talking about early age cracking. Primarily, why does it occur and how can it be eliminated or at least controlled?

This module is part of a curriculum from the "Integrated Materials and Construction Practices for Concrete Pavement" manual developed through the National Concrete Pavement Technology Center at Iowa State University. The other Web-based training modules include:

FHWA-NHI-134075 TCCC Hardened Concrete Properties - Durability

FHWA-NHI-134084 TCCC Fundamentals of Materials Used for Concrete Pavements

FHWA-NHI-134085 TCCC Incompatibility in Concrete Pavement Systems

FHWA-NHI-134087 TCCC Mix Design Principles

FHWA-NHI-134096 TCCC Basics of Cement Hydration

FHWA-NHI-134097 TCCC Fresh Concrete Properties

FHWA-NHI-134098 TCCC Construction of Concrete Pavements

FHWA-NHI-134100 TCCC QCQA for Concrete Pavements

FHWA-NHI-134101 TCCC Design of Pavement

FHWA-NHI-134102 TCCC Troubleshooting for Concrete Pavements

Outcomes

Upon completion of the course, participants will be able to:

- Describe the various mechanisms that can lead to early age cracking
- Define and understand why curling and warping occur
- Recognize how curling and warping affect early age cracking
- Recognize the proper use of the materials and maintaining good construction practices can control early age cracking
- Describe how certain material properties and construction methods can affect early age cracking and can help prevent the cracking from occurring

Target Audience

This training is designed for FHWA, State, and local agencies and their industry counterparts involved in the process to assure that concrete meets all the requirements to prevent early age cracking. It is applicable to anyone desiring a better understanding of the causes and prevention of early age cracking.

TRAINING LEVEL: Intermediate

FEE: 2016: $25 Per Person; 2017: $25 Per Person

LENGTH: 1 HOURS (CEU: 0 UNITS)

CLASS SIZE: MINIMUM: 1; MAXIMUM: 1

NHI Customer Service: (877) 558-6873 • nhicustomerservice@dot.gov

COURSE NUMBER

FHWA-NHI-134096

COURSE TITLE

Basics of Cement Hydration - WEB-BASED

This training is provided by the Transportation Curriculum Coordination Council (TCCC) in partnership with NHI to review integrated materials and construction practices for concrete pavement. The training was developed by the National Concrete Pavement Technology Center at Iowa State University. This training is recommended for the Transportation Curriculum Coordination Council levels III and IV. This course is primarily intended for inspectors and technicians.

This module covers how a concrete mixture changes from a plastic state to become a solid concrete slab in a relatively short period of time. Central to this transformation is a complex process called hydration, an irreversible series of chemical reactions between water and cement.

This module is part of a curriculum from the "Integrated Materials and Construction Practices for Concrete Pavement" manual developed through the National Concrete Pavement Technology Center at Iowa State University. The other Web-based training modules include:

FHWA-NHI-134075 TCCC Hardened Concrete Properties - Durability

FHWA-NHI-134084 TCCC Fundamentals of Materials Used for Concrete Pavements

FHWA-NHI-134085 TCCC Incompatibility in Concrete Pavement Systems

FHWA-NHI-134087 TCCC Mix Design Principles

FHWA-NHI-134095 TCCC Early Age Cracking

FHWA-NHI-134097 TCCC Fresh Concrete Properties

FHWA-NHI-134098 TCCC Construction of Concrete Pavements

FHWA-NHI-134100 TCCC QCQA for Concrete Pavements

FHWA-NHI-134101 TCCC Design of Pavement

FHWA-NHI-134102 TCCC Troubleshooting for Concrete Pavements

OUTCOMES

Upon completion of the course, participants will be able to:

- Knowledge of physical and chemical occurrences during cement hydration
- Identify various factors that can adversely affect these occurrences
- Recognize the different temperature changes during particular stages of hydration

TARGET AUDIENCE

This training is designed for FHWA, State, and local agencies and their industry counterparts involved in the process to assure that the mix design and proportioning of Portland cement concrete materials meet specification requirements and provide good, durable concrete. It is applicable to anyone desiring a better understanding of the mix design of Portland cement concrete.

TRAINING LEVEL: Intermediate

FEE: 2016: $25 Per Person; 2017: N/A

LENGTH: 1 HOURS (CEU: 0 UNITS)

CLASS SIZE: MINIMUM: 1; MAXIMUM: 1

NHI Customer Service: (877) 558-6873 • nhicustomerservice@dot.gov

Course Number

FHWA-NHI-134097

Course Title

Fresh Concrete Properties - WEB-BASED

This training is provided by the Transportation Curriculum Coordination Council (TCCC) in partnership with NHI to review integrated materials and construction practices for concrete pavement. The training was developed by the National Concrete Pavement Technology Center at Iowa State University. This training is recommended for the Transportation Curriculum Coordination Council levels III and IV. This course is primarily intended for inspectors and technicians.

This module covers the properties of fresh concrete needed to produce high-quality, long lasting pavements and how to monitor these properties.

This module is part of a curriculum from the "Integrated Materials and Construction Practices for Concrete Pavement" manual developed through the National Concrete Pavement Technology Center at Iowa State University. The other Web-based training modules include:

FHWA-NHI-134075 TCCC Hardened Concrete Properties - Durability

FHWA-NHI-134084 TCCC Fundamentals of Materials Used for Concrete Pavements

FHWA-NHI-134085 TCCC Incompatibility in Concrete Pavement Systems

FHWA-NHI-134087 TCCC Mix Design Principles

FHWA-NHI-134095 TCCC Early Age Cracking

FHWA-NHI-134096 TCCC Basics of Cement Hydration

FHWA-NHI-134098 TCCC Construction of Concrete Pavements

FHWA-NHI-134100 TCCC QCQA for Concrete Pavements

FHWA-NHI-134101 TCCC Design of Pavement

FHWA-NHI-134102 TCCC Troubleshooting for Concrete Pavements

Outcomes

Upon completion of the course, participants will be able to:

- List the main properties of fresh concrete
- Describe what affects each property
- Recognize how to monitor these properties through concrete testing

Target Audience

This training is designed for FHWA, State, and local agencies and their industry counterparts involved in the process to assure that the properties of a concrete mixture provide ease in placement, ease of consolidation, and long lasting pavement. It is applicable to anyone desiring a better understanding of the properties of Portland cement concrete.

Training Level: Intermediate

Fee: 2016: $25 Per Person; 2017: N/A

Length: 1 HOURS (CEU: 0 UNITS)

Class Size: MINIMUM: 1; MAXIMUM: 1

NHI Customer Service: (877) 558-6873 • nhicustomerservice@dot.gov

WBT

Course Number

FHWA-NHI-134101

Course Title

Design of Pavement - WEB-BASED

This training is provided by the Transportation Curriculum Coordination Council (TCCC) in partnership with NHI to review integrated materials and construction practices for concrete pavement. The training was developed by the National Concrete Pavement Technology Center at Iowa State University. This training is recommended for the Transportation Curriculum Coordination Council levels III and IV. This course is primarily for inspectors and technicians.

This module covers pavement design and subgrade concepts as they relate to materials and construction. It does not provide sufficient detail to actually design or evaluate a design. It covers the primary goal of pavement design, which is to provide a pavement with the following characteristics: safe, long lasting, cost effective, low maintenance, and constructible.

This module is part of a curriculum from the "Integrated Materials and Construction Practices for Concrete Pavement" manual developed through the National Concrete Pavement Technology Center at Iowa State University. The other Web-based training modules include:

FHWA-NHI-134075 TCCC Hardened Concrete Properties - Durability

FHWA-NHI-134084 TCCC Fundamentals of Materials Used for Concrete Pavements

FHWA-NHI-134085 TCCC Incompatibility in Concrete Pavement Systems

FHWA-NHI-134087 TCCC Mix Design Principles

FHWA-NHI-134095 TCCC Early Age Cracking

FHWA-NHI-134096 TCCC Basics of Cement Hydration

FHWA-NHI-134097 TCCC Fresh Concrete Properties

FHWA-NHI-134098 TCCC Construction of Concrete Pavements

FHWA-NHI-134100 TCCC QCQA for Concrete Pavements

FHWA-NHI-134102 TCCC Troubleshooting for Concrete Pavements

Outcomes

Upon completion of the course, participants will be able to:

- Identify pavement types and design features
- Recognize what design variables are controlled by field operations
- Discuss the two primary types of pavement distresses (performance measures)
- Recognize how subgrades and bases effect construction operations and long-term pavement performance

Target Audience

This training is designed for FHWA, State, and local agencies and their industry counterparts involved in designing, constructing, and inspecting Portland cement concrete pavements.

Training Level: Intermediate

Fee: 2016: $25 Per Person; 2017: $25 Per Person

Length: 1 HOURS (CEU: 0 UNITS)

Class Size: MINIMUM: 1; MAXIMUM: 1

NHI Customer Service: (877) 558-6873 • nhicustomerservice@dot.gov

Course Number

FHWA-NHI-134105

Course Title

Pipe Installation, Inspection, and Quality - WEB-BASED

This training was developed by the Transportation Curriculum Coordination Council (TCCC) in partnership with State DOT personnel. It helps transportation professionals involved in the installation, inspection, and quality of pipe on highway construction projects improve their understanding of the factors that contribute to high-quality installations. The training was prepared by State DOT personnel for State DOT personnel. It contains good practices from various agencies. Each State agency/company has their own specifications, which the viewer needs to review and follow for the specified pipe. This course is primarily intended for inspectors and technicians.

This course is focused on the three basic pipe materials. They are Concrete, Metal, and Plastic. This course contains important instructional material, procedures and guidance that has been developed to maintain uniformity among pipe inspectors. This course will cover what you need to know, do, and look for during the inspection of pipe installation.

This training is directed toward the intermediate level technician, to give them an in-depth view of the basic materials used in pipe construction. The course modules will address the different types of pipe as well as the foundation work, bedding selection, placement, joint sealants, backfilling and documentation for concrete, metal and plastic pipe.

NHI is hosting this and other TCCC Web-based developments to serve a critical need for training. We need your feedback to determine whether we should continue posting other Web-based trainings like this one. Please take the time to complete the evaluation form provided at the end of the training, or email nhimarketing@dot.gov.

Outcomes

Upon completion of the course, participants will be able to:

- Identify basic material pipe types
- Recognize proper foundation and bedding requirements for pipe
- Link different types of pipe with its required specifications for installation
- Identify common errors to avoid when dealing with placement, joints and backfilling of pipe
- Recognize the importance of accurate records and reporting

Target Audience

This course targets field personnel involved in all aspects of highway construction from engineers to technicians. The ideal audience will have a mix of experience and responsibility levels so that agency-specific practices can be shared by more experienced participants with those who are newer to the field. The course materials also are appropriate for project manager/resident engineer involvement.

Training Level: Intermediate

Fee: 2016: $50 Per Person; 2017: N/A

Length: 7 HOURS (CEU: 0 UNITS)

Class Size: MINIMUM: 1; MAXIMUM: 1

NHI Customer Service: (877) 558-6873 • nhicustomerservice@dot.gov

COURSE NUMBER

FHWA-NHI-134106

COURSE TITLE

Basic Construction Surveying - WEB-BASED

This training was prepared by the Transportation Curriculum Coordination Council (TCCC) in partnership with NHI to review the basics of construction surveying. This training has been prepared to provide guidance and instruction to those involved in construction surveying. The important surveying tasks involved in this work and the surveying procedures to be followed are also described in this training. This course is primarily intended for inspectors and technicians.

This training is targeted for those who are new to the construction surveying experience or for anyone needing a refresher. This training is recommended for the Transportation Curriculum Coordination Council levels I and II.

We've broken this training into three modules:

1. Basic Surveying Concepts
2. Measurement and Construction Surveying
3. Survey Mathematics

OUTCOMES

Upon completion of the course, participants will be able to:

- Describe basic surveying concepts
- Understand measurement and construction surveying
- List the instruments and techniques used in measurement
- Perform stationing and staking operations
- Perform basic survey mathematics

TARGET AUDIENCE

This training is designed for FHWA, State, and local agencies and their industry counterparts involved in construction survey. This training is targeted for those who have not had construction surveying experience or anyone needing a review over the key concepts of surveying.

TRAINING LEVEL: Basic

FEE: 2016: $25 Per Person; 2017: $25 Per Person

LENGTH: 3 HOURS (CEU: 0 UNITS)

CLASS SIZE: MINIMUM: 1; MAXIMUM: 1

NHI Customer Service: (877) 558-6873 • nhicustomerservice@dot.gov

Course Number

FHWA-NHI-134107

Course Title

Recognizing Roadside Weeds (Southeastern States) - WEB-BASED

This training was prepared by the Transportation Curriculum Coordination Council (TCCC) in partnership with NHI and has been designed for someone learning the first steps in the vegetation management. However, it does not go into the education of weed prevention. This training is recommended for the Transportation Curriculum Coordination Council levels I, and II. This course is primarily intended for inspectors and technicians.

The first step in determining an appropriate weed control strategy is to identify the weed plant. There are numerous different plants growing along many roadsides that can be considered weeds. This is a basic course in the area of weed identification. Most weeds are territorial to different climates and regions, therefore, making it difficult to identify nationally weeds that are dealt with by different State DOT's. This training does focus on southeastern states and is organized in alphabetical order of the weeds that will be covered.

For more information on how stop the migration of weeds contact your State Vegetation Management Program.

Outcomes

Upon completion of the course, participants will be able to:

- Understand the definition of a weed
- Describe the reasons for weed control
- Identify several of the most common weeds

Target Audience

This course is designed for entry level individuals working in vegetation management.

Training Level: Basic

Fee: 2016: $25 Per Person; 2017: $25 Per Person

Length: 1 HOURS (CEU: 0 UNITS)

Class Size: MINIMUM: 1; MAXIMUM: 1

NHI Customer Service: (877) 558-6873 • nhicustomerservice@dot.gov

Course Number

FHWA-NHI-134108

Course Title

Plan Reading Series - WEB-BASED

The Transportation Curriculum Coordination Council (TCCC) in partnership with NHI is pleased to offer this comprehensive training series (FHWA-NHI-134108) for highway plan reading. This training is recommended for the Transportation Curriculum Coordination Council levels II - IV. This course is primarily intended for inspectors and technicians.

The ability to read plans is essential for anyone involved in highway and/or bridge construction. This training contains modules covering both basic plan reading instructions, as well as, providing a more in-depth level of instruction for anyone seeking more information and/or a review of plan reading.

To streamline registration and enable you to take some or all of these trainings when it best suits your schedule, we have created this new series option which automatically registers you for all 8 modules-it's that easy. They are as follows:

Module 1: Highway Plan Reading Basics (134108A) - This module describes the foundational information needed to begin reading and understanding highway plans. This includes an overview of the title page and its components, station numbers, townships, and quantity estimates.

Module 2: Grading Plans (134108B) - This module reviews the information found in the Grading Plans (sheets that begin with "B") section of a highway plan.

Module 3: Traffic Control Plans (134108C) - This module reviews the information found in the Traffic Control Plans (sheets that begin with "C") section of a highway plan.

Module 4: Erosion and Sediment Control Plans (134108D) - This module reviews the information found in the Erosion and Sediment Control Plans (sheets that begin with "D") section of a highway plan.

Module 5: Right of Way Plans (134108E) - This module reviews the information found in Right-of-Way Plans for a highway project.

Module 6: County Plans (134108F) - This module reviews the information found in a county plan.

Module 7: Bridge Plans (134108G) - This module reviews the information found in a bridge plan.

Module 8: Culvert Plans (134108H) - This module reviews the information found in a culvert plan.

Outcomes

Upon completion of the course, participants will be able to:

- Recognize plan sheets for highway, county, bridge, culvert construction
- Recognize station locations and calculate; cross section, profile, and plan views; centerline location; point of intersection; and a variety of plan details
- Recognize plan sheet for all parts of both a bridge substructure and superstructure
- Comprehend the terminology and symbols used when reading plans

Target Audience

This training is designed for FHWA, State, and local agencies and their industry counterparts involved in the construction process of highways, county, bridges, or culverts. It is applicable to anyone desiring a better understanding of plan reading.

TRAINING LEVEL: Basic

FEE: 2016: $50 Per Person; 2017: $50 Per Person

LENGTH: 8 HOURS (CEU: 0 UNITS)

CLASS SIZE: MINIMUM: 1; MAXIMUM: 1

NHI Customer Service: (877) 558-6873 • nhicustomerservice@dot.gov

COURSE NUMBER

FHWA-NHI-134108A

COURSE TITLE

Plan Reading: Highway Plan Reading Basics - WEB-BASED

This training is provided by the Transportation Curriculum Coordination Council (TCCC) in partnership with NHI to review the basics for highway plan reading. This course is recommended for the Transportation Curriculum Coordination Council levels II - IV. This course is primarily intended for inspectors and technicians.

The ability to read plans is essential for anyone involved in highway and/or bridge construction. This training describes the foundational information needed to begin reading and understanding highway plans. This includes an overview of the title page and its components, station numbers, townships, and quantity estimates.

This training is part of the curriculum from the Plan Reading Series (FHWA-NHI-134108) which covers both basic plan reading instructions, as well as, providing a more in-depth level of instruction for anyone seeking more information and/ or a review of plan reading. The other Web-based training modules include:

FHWA-NHI-134108B Grading Plans

FHWA-NHI-134108C Traffic Control Plans

FHWA-NHI-134108D Erosion and Sediment Control Plans

FHWA-NHI-134108E Right-of-Way Plans

FHWA-NHI-134108F County Plans

FHWA-NHI-134108G Bridge Plans

FHWA-NHI-134108H Culvert Plans

OUTCOMES

Upon completion of the course, participants will be able to:

- Describe the components of a plan's title sheet
- Calculate the distance between two station numbers
- Explain how a township is designated in a plan
- Identify quantity estimates for given supplies and materials

TARGET AUDIENCE

This training is designed for FHWA, State, and local agencies and their industry counterparts involved in the construction process of highways and/or bridges. It is applicable to anyone desiring a better understanding of plan reading.

TRAINING LEVEL: Basic

FEE: 2016: $25 Per Person; 2017: $25 Per Person

LENGTH: 1 HOURS (CEU: 0 UNITS)

CLASS SIZE: MINIMUM: 1; MAXIMUM: 1

NHI Customer Service: (877) 558-6873 • nhicustomerservice@dot.gov

Course Number

FHWA-NHI-134108B

Course Title

Plan Reading: Grading Plans - WEB-BASED

This training is provided by the Transportation Curriculum Coordination Council (TCCC) in partnership with NHI to review the basics for highway plan reading. This course is recommended for the Transportation Curriculum Coordination Council levels II - IV. This course is primarily intended for inspectors and technicians.

The ability to read plans is essential for anyone involved in highway and/or bridge construction. This training reviews the information found in the Grading Plans (sheets that begin with "B") section of a highway plan.

This training is part of the curriculum from the Plan Reading Series (FHWA-NHI-134108) which covers both basic plan reading instructions, as well as, providing a more in-depth level of instruction for anyone seeking more information and/or a review of plan reading. The other Web-based training modules include:

FHWA-NHI-134108A Highway Plan Reading Basics

FHWA-NHI-134108C Traffic Control Plans

FHWA-NHI-134108D Erosion and Sediment Control Plans

FHWA-NHI-134108E Right-of-Way Plans

FHWA-NHI-134108F County Plans

FHWA-NHI-134108G Bridge Plans

FHWA-NHI-134108H Culvert Plans

Outcomes

Upon completion of the course, participants will be able to:

- Describe the information provided in the grading plans
- Identify grade characteristics provided in the typical grading sections sheets
- Explain the importance of plan and profile sheets
- Describe the different elements that can be depicted in plan and profile sheets

Target Audience

This training is designed for FHWA, State, and local agencies and their industry counterparts involved in the construction process of highways and/or bridges. It is applicable to anyone desiring a better understanding of plan reading.

Training Level: Basic

Fee: 2016: $25 Per Person; 2017: N/A

Length: 1.5 HOURS (CEU: 0 UNITS)

Class Size: MINIMUM: 1; MAXIMUM: 1

NHI Customer Service: (877) 558-6873 • nhicustomerservice@dot.gov

COURSE NUMBER

FHWA-NHI-134108C

COURSE TITLE

Plan Reading: Traffic Control Plans - WEB-BASED

This training is provided by the Transportation Curriculum Coordination Council (TCCC) in partnership with NHI to review the basics for highway plan reading. This course is recommended for the Transportation Curriculum Coordination Council levels II - IV. This course is primarily intended for inspectors and technicians.

The ability to read plans is essential for anyone involved in highway and/or bridge construction. This training reviews the information found in the Traffic Control Plans (sheets that begin with "C") section of a highway plan.

This training is part of the curriculum from the Plan Reading Series (FHWA-NHI-134108) which covers both basic plan reading instructions, as well as, providing a more in-depth level of instruction for anyone seeking more information and/or a review of plan reading. The other Web-based training modules include:

FHWA-NHI-134108A Highway Plan Reading Basics

FHWA-NHI-134108B Grading Plans

FHWA-NHI-134108D Erosion and Sediment Control Plans

FHWA-NHI-134108E Right-of-Way Plans

FHWA-NHI-134108F County Plans

FHWA-NHI-134108G Bridge Plans

FHWA-NHI-134108H Culvert Plans

OUTCOMES

Upon completion of the course, participants will be able to:

- Describe the information provided in the traffic control plans
- Identify signs to be used in the project
- Identify sign locations

TARGET AUDIENCE

This training is designed for FHWA, State, and local agencies and their industry counterparts involved in the construction process of highways and/or bridges. It is applicable to anyone desiring a better understanding of plan reading.

TRAINING LEVEL: Basic

FEE: 2016: $25 Per Person; 2017: N/A

LENGTH: .5 HOURS (CEU: 0 UNITS)

CLASS SIZE: MINIMUM: 1; MAXIMUM: 1

NHI Customer Service: (877) 558-6873 • nhicustomerservice@dot.gov

COURSE NUMBER

FHWA-NHI-134108D

COURSE TITLE

Plan Reading: Erosion and Sediment Control Plans - WEB-BASED

This training is provided by the Transportation Curriculum Coordination Council (TCCC) in partnership with NHI to review the basics for highway plan reading. This course is recommended for the Transportation Curriculum Coordination Council levels II - IV. This course is primarily intended for inspectors and technicians.

The ability to read plans is essential for anyone involved in highway and/or bridge construction. This training reviews the information found in the Erosion and Sediment Control Plans (sheets that begin with "D") section of a highway plan.

This training is part of the curriculum from the Plan Reading Series (FHWA-NHI-134108) which covers both basic plan reading instructions, as well as, providing a more in-depth level of instruction for anyone seeking more information and/ or a review of plan reading. The other Web-based training modules include:

FHWA-NHI-134108A Highway Plan Reading Basics

FHWA-NHI-134108B Grading Plans

FHWA-NHI-134108C Traffic Control Plans

FHWA-NHI-134108E Right-of-Way Plans

FHWA-NHI-134108F County Plans

FHWA-NHI-134108G Bridge Plans

FHWA-NHI-134108H Culvert Plans

OUTCOMES

Upon completion of the course, participants will be able to:

- Describe the information provided in the erosion and sediment control plans
- Explain the erosion and sediment control items used in the plan

TARGET AUDIENCE

This training is designed for FHWA, State, and local agencies and their industry counterparts involved in the construction process of highways and/or bridges. It is applicable to anyone desiring a better understanding of plan reading.

TRAINING LEVEL: Basic

FEE: 2016: $25 Per Person; 2017: N/A

LENGTH: .5 HOURS (CEU: 0 UNITS)

CLASS SIZE: MINIMUM: 1; MAXIMUM: 1

NHI Customer Service: (877) 558-6873 • nhicustomerservice@dot.gov

Course Number

FHWA-NHI-134108E

Course Title

Plan Reading: Right-of-Way Plans - WEB-BASED

This training is provided by the Transportation Curriculum Coordination Council (TCCC) in partnership with NHI to review the basics for highway plan reading. This course is recommended for the Transportation Curriculum Coordination Council levels II - IV. This course is primarily intended for inspectors and technicians.

The ability to read plans is essential for anyone involved in highway and/or bridge construction. This training reviews the information found in right-of-way plans for a highway project.

This training is part of the curriculum from the Plan Reading Series (FHWA-NHI-134108) which covers both basic plan reading instructions, as well as, providing a more in-depth level of instruction for anyone seeking more information and/or a review of plan reading. The other Web-based training modules include:

FHWA-NHI-134108A Highway Plan Reading Basics

FHWA-NHI-134108B Grading Plans

FHWA-NHI-134108C Traffic Control Plans

FHWA-NHI-134108D Erosion and Sediment Control Plans

FHWA-NHI-134108F County Plans

FHWA-NHI-134108G Bridge Plans

FHWA-NHI-134108H Culvert Plans

Outcomes

Upon completion of the course, participants will be able to:

- Explain the purpose of right-of-way plans
- Explain when right-of-way is needed
- Describe the information provided in right-of-way plans
- Describe when land is acquired for easements
- Explain how parcels are used in right-of-way plans
- Describe how utilities will be handled for the project

Target Audience

This training is designed for FHWA, State, and local agencies and their industry counterparts involved in the construction process of highways and/or bridges. It is applicable to anyone desiring a better understanding of plan reading.

Training Level: Basic

Fee: 2016: $25 Per Person; 2017: N/A

Length: 1 HOURS (CEU: 0 UNITS)

Class Size: MINIMUM: 1; MAXIMUM: 1

NHI Customer Service: (877) 558-6873 • nhicustomerservice@dot.gov

COURSE NUMBER

FHWA-NHI-134108F

COURSE TITLE

Plan Reading: County Plans - WEB-BASED

This training is provided by the Transportation Curriculum Coordination Council (TCCC) in partnership with NHI to review the basics for highway plan reading. This course is recommended for the Transportation Curriculum Coordination Council levels II - IV. This course is primarily intended for inspectors and technicians.

The ability to read plans is essential for anyone involved in highway and/or bridge construction. This training reviews the information found in a county plan.

This training is part of the curriculum from the Plan Reading Series (FHWA-NHI-134108) which covers both basic plan reading instructions, as well as, providing a more in-depth level of instruction for anyone seeking more information and/or a review of plan reading. The other Web-based training modules include:

FHWA-NHI-134108A Highway Plan Reading Basics

FHWA-NHI-134108B Grading Plans

FHWA-NHI-134108C Traffic Control Plans

FHWA-NHI-134108D Erosion and Sediment Control Plans

FHWA-NHI-134108E Right-of-Way Plans

FHWA-NHI-134108G Bridge Plans

FHWA-NHI-134108H Culvert Plans

OUTCOMES

Upon completion of the course, participants will be able to:

- Describe the information provided in a county plan
- Given a county plan, explain the details of the project

TARGET AUDIENCE

This training is designed for FHWA, State, and local agencies and their industry counterparts involved in the construction process of highways and/or bridges. It is applicable to anyone desiring a better understanding of plan reading.

TRAINING LEVEL: Basic

FEE: 2016: $25 Per Person; 2017: N/A

LENGTH: 1 HOURS (CEU: 0 UNITS)

CLASS SIZE: MINIMUM: 1; MAXIMUM: 1

NHI Customer Service: (877) 558-6873 • nhicustomerservice@dot.gov

Course Number

FHWA-NHI-134108G

Course Title

Plan Reading: Bridge Plans - WEB-BASED

This training is provided by the Transportation Curriculum Coordination Council (TCCC) in partnership with NHI to review the basics for highway plan reading. This course is recommended for the Transportation Curriculum Coordination Council levels II - IV. This course is primarily intended for inspectors and technicians.

The ability to read plans is essential for anyone involved in highway and/or bridge construction. This training reviews the information found in a bridge plan.

This training is part of the curriculum from the Plan Reading Series (FHWA-NHI-134108) which covers both basic plan reading instructions, as well as, providing a more in-depth level of instruction for anyone seeking more information and/or a review of plan reading. The other Web-based training modules include:

FHWA-NHI-134108A Highway Plan Reading Basics

FHWA-NHI-134108B Grading Plans

FHWA-NHI-134108C Traffic Control Plans

FHWA-NHI-134108D Erosion and Sediment Control Plans

FHWA-NHI-134108E Right-of-Way Plans

FHWA-NHI-134108F County Plans

FHWA-NHI-134108H Culvert Plans

Outcomes

Upon completion of the course, participants will be able to:

- Identify the major components of a bridge structure
- Describe the information provided in a bridge plan
- Using a bridge plan, explain details of the project

Target Audience

This training is designed for FHWA, State, and local agencies and their industry counterparts involved in the construction process of highways and/or bridges. It is applicable to anyone desiring a better understanding of plan reading.

Training Level: Basic

Fee: 2016: $25 Per Person; 2017: N/A

Length: 1.5 HOURS (CEU: 0 UNITS)

Class Size: MINIMUM: 1; MAXIMUM: 1

NHI Customer Service: (877) 558-6873 • nhicustomerservice@dot.gov

COURSE NUMBER

FHWA-NHI-134108H

COURSE TITLE

Plan Reading: Culvert Plans - WEB-BASED

This training is provided by the Transportation Curriculum Coordination Council (TCCC) in partnership with NHI to review the basics for highway plan reading. This course is recommended for the Transportation Curriculum Coordination Council levels II - IV. This course is primarily intended for inspectors and technicians.

The ability to read plans is essential for anyone involved in highway and/or bridge construction. This training reviews the information found in a culvert plan.

This training is part of the curriculum from the Plan Reading Series (FHWA-NHI-134108) which covers both basic plan reading instructions, as well as, providing a more in-depth level of instruction for anyone seeking more information and/or a review of plan reading. The other Web-based training modules include:

FHWA-NHI-134108A Highway Plan Reading Basics

FHWA-NHI-134108B Grading Plans

FHWA-NHI-134108C Traffic Control Plans

FHWA-NHI-134108D Erosion and Sediment Control Plans

FHWA-NHI-134108E Right-of-Way Plans

FHWA-NHI-134108F County Plans

FHWA-NHI-134108G Bridge Plans

OUTCOMES

Upon completion of the course, participants will be able to:

- Identify the major components of a culvert
- Describe the information provided in a culvert plan
- Using a culvert plan, explain details of the project

TARGET AUDIENCE

This training is designed for FHWA, State, and local agencies and their industry counterparts involved in the construction process of highways and/or bridges. It is applicable to anyone desiring a better understanding of plan reading.

TRAINING LEVEL: Basic

FEE: 2016: $25 Per Person; 2017: N/A

LENGTH: 1.5 HOURS (CEU: 0 UNITS)

CLASS SIZE: MINIMUM: 1; MAXIMUM: 1

NHI Customer Service: (877) 558-6873 • nhicustomerservice@dot.gov

COURSE NUMBER

FHWA-NHI-134109

COURSE TITLE

Maintenance Training Series - WEB-BASED

The Maintenance Training Series was created to train individuals responsible for the maintenance of our Nation's roadways. The series consists of 11 self-paced, Web-based trainings (WBTs) on various maintenance operations topics, ranging from the conceptual (pavement preservation) to the practical (management of underground storage tanks). The trainings included in the series are listed below and each will take approximately 1 hour to complete.

Participants who wish to complete all 11 trainings in the Maintenance Training Series should enroll in course 134109. Those who are interested in specific topics may enroll in each training individually.

- Pavement Preservation Program (134109A)
- Shaping and Shoulders (134109B)
- Thin HMA Overlays and Leveling (134109C)
- Base and Subbase Stabilization and Repair (134109D)
- Drainage (134109E)
- Outdoor Advertising and Litter Control (134109F)
- Roadside Vegetation Management (134109G)
- Weather-related Operations (134109H)
- Basics of Work Zone Traffic Control (134109I)
- Underground Storage Tanks (134109J)
- Cultural and Historic Preservation (134109K)

OUTCOMES

Upon completion of the course, participants will be able to:

- Learning outcomes have been established at the module level. Please see the individual modules for the specific learning outcomes.

TARGET AUDIENCE

This course was designed for State, regional, and county personnel who manage operations programs and deal with oversight and quality assurance over broader geographic areas. The target audience is also involved with handling materials, scheduling, budgeting and planning.

TRAINING LEVEL: Basic

FEE: 2016: $50 Per Person; 2017: N/A

LENGTH: 11 HOURS (CEU: 0 UNITS)

CLASS SIZE: MINIMUM: 1; MAXIMUM: 1

NHI Customer Service: (877) 558-6873 • nhicustomerservice@dot.gov

COURSE NUMBER

FHWA-NHI-134109A

COURSE TITLE

Maintenance Training Series: Pavement Preservation Program - WEB-BASED

Pavement preservation represents a major paradigm shift in the way many transportation agencies view and operate their highway networks. The Pavement Preservation Programs course provides basic information on what comprises a pavement preservation program and how it is implemented. It places particular emphasis on changes in practice and assignment of dedicated funding.

Additionally, the training covers the benefits and challenges to a preservation program; Federal and State resources available to support a preservation program; and approaches for communicating the advantages of pavement preservation to stakeholders.

This training was developed as part of the Maintenance Training Series. To access all the courses in the series, enroll in the 134109 course.

OUTCOMES

Upon completion of the course, participants will be able to:

- Identify the benefits and challenges of implementing a pavement preservation program
- Determine ways to develop support for a pavement preservation program

TARGET AUDIENCE

This course is designed for State, regional, and county personnel who manage operations programs and deal with oversight and quality assurance across broad geographic areas. This target audience also is involved with handling materials, scheduling, budgeting, and planning.

TRAINING LEVEL: Basic

FEE: 2016: $25 Per Person; 2017: N/A

LENGTH: 1 HOURS (CEU: 0 UNITS)

CLASS SIZE: MINIMUM: 0; MAXIMUM: 0

NHI Customer Service: (877) 558-6873 • nhicustomerservice@dot.gov

COURSE NUMBER

FHWA-NHI-134109B

COURSE TITLE

Maintenance Training Series: Shaping and Shoulders - WEB-BASED

Shoulders play an important role in both pavement performance and roadway safety. Maintaining shoulders in a proper and timely manner is a primary goal of transportation agencies. In an effort to assist agencies in meeting this goal, the Shaping and Shoulders training provides information on the maintenance of both paved and unpaved shoulders, including specific details on the maintenance of gravel shoulders. This course is primarily intended for inspectors and technicians.

In addition to a discussion of the various types of shoulders, project selection considerations, and key maintenance issues, this training places shoulders and shaping into the context of an overall maintenance and pavement preservation program.

This training was developed as part of the Maintenance Training Series. To access all the trainings in the series, enroll in the 134109 course.

OUTCOMES

Upon completion of the course, participants will be able to:

- Identify desirable characteristics of various types of shoulders
- Identify project selection considerations for shaping and shoulders
- Describe shoulder shaping and blading activities, including equipment requirements and construction activities
- Describe how a shoulder and ditching program forms the core of the overall maintenance and pavement preservation program

TARGET AUDIENCE

This course is designed for State, regional, and county personnel who manage operations programs and deal with oversight and quality assurance across broad geographic areas. This target audience also is involved with handling materials, scheduling, budgeting, and planning.

TRAINING LEVEL: Basic

FEE: 2016: $25 Per Person; 2017: N/A

LENGTH: 1.5 HOURS (CEU: 0 UNITS)

CLASS SIZE: MINIMUM: 0; MAXIMUM: 0

NHI Customer Service: (877) 558-6873 • nhicustomerservice@dot.gov

Course Number

FHWA-NHI-134109C

Course Title

Maintenance Training Series: Thin HMA Overlays and Leveling - WEB-BASED

Thin HMA overlays and leveling are common pavement treatments and can be a central part of a maintenance crew's activities. During the Thin HMA Overlays and Leveling training, participants will be introduced to the characteristics and purposes of thin HMA overlays as well as the placement of leveling courses. Each of these techniques is capable of improving the functionality of an otherwise structurally sound pavement.

The training also covers information on the materials, personnel, and equipment needed for thin HMA overlays; items that should be considered when making project selection decisions; and guidance on proper mixture compaction. This information is designed to help participants improve project planning and execution for thin HMA overlays and leveling treatments.

This training was developed as part of the Maintenance Training Series. To access all the courses in the series, enroll in the 134109 course.

Outcomes

Upon completion of the course, participants will be able to:

- Determine the purpose of thin HMA overlays and leveling courses
- Identify material components of HMA overlays
- Identify personnel and equipment needed for HMA overlays and leveling construction
- Identify project selection considerations for thin HMA overlays and leveling
- Identify how this treatment can be incorporated into an overall system preservation program

Target Audience

This course is designed for State, regional, and county personnel who manage operations programs and deal with oversight and quality assurance across broad geographic areas. This target audience also is involved with handling materials, scheduling, budgeting, and planning.

Training Level: Basic

Fee: 2016: $25 Per Person; 2017: N/A

Length: 1.5 HOURS (CEU: 0 UNITS)

Class Size: MINIMUM: 0; MAXIMUM: 0

NHI Customer Service: (877) 558-6873 • nhicustomerservice@dot.gov

Course Number

FHWA-NHI-134109D

Course Title

Maintenance Training Series: Base and Subbase Stabilization and Repair - WEB-BASED

Before preservation treatments can be applied, localized repairs may be necessary for a pavement's base or subbase. The Base and Subbase Stabilization and Repair course gives participants the knowledge they need to determine if the base or subbase must be stabilized or repaired, to select the appropriate stabilization and repair methods for a given project, and to ensure the repair is performed properly.

This training reviews the failures and distresses that indicate structural deterioration exists in a roadway. The course also covers project selection and trade-off considerations through example roadway projects that give participants the opportunity to evaluate a roadway and determine if it is a candidate for reconstruction or repair. Participants can use this information, as well as guidance on design and construction, to make sound project planning decisions.

This training was developed as part of the Maintenance Training Series. To access all the courses in the series, enroll in the 134109 course.

Outcomes

Upon completion of the course, participants will be able to:

- Identify the symptoms of a localized base or subbase problem, which require greater depth of stabilization and repair than a hot-mix asphalt (HMA) or portland cement concrete (PCC) surface repair patch
- Determine when it is appropriate to employ base or subbase repair on a preventive maintenance project
- Identify the most appropriate repair methods if base or subbase failures are identified in a project

Target Audience

This course is designed for State, regional, and county personnel who manage operations programs and deal with oversight and quality assurance across broad geographic areas. This target audience also is involved with handling materials, scheduling, budgeting, and planning.

Training Level: Basic

Fee: 2016: $25 Per Person; 2017: N/A

Length: 1 HOURS (CEU: 0 UNITS)

Class Size: MINIMUM: 0; MAXIMUM: 0

NHI Customer Service: (877) 558-6873 • nhicustomerservice@dot.gov

Course Number

FHWA-NHI-134109E

Course Title

Maintenance Training Series: Roadway Drainage - WEB-BASED

Shoulder, ditch, and pipe or culvert maintenance activities are performed frequently throughout the year. These activities are critical for avoiding hazardous roadway conditions and extending the life of pavements by controlling water flow along maintainable pathways. This course, Roadway Drainage, provides information on the purpose, function, and components of roadway drainage systems.

This course reviews the components of shoulders and ditches, the purpose of a roadway drainage inventory, and the permits used in roadway drainage maintenance. Examples of existing drainage inventories are provided. In addition, the benefits of proper water removal are discussed through examples of drainage system issues, such as ponding and washouts, in order to emphasize the connection between good drainage and roadway safety.

This training was developed as part of the Maintenance Training Series. To access all the courses in the series, enroll in the 134109 course.

Outcomes

Upon completion of the course, participants will be able to:

- Identify the purpose and function of roadway drainage systems
- Identify eight components of roadway drainage systems
- Identify the purpose of a roadway drainage inventory
- Identify the purpose of permits in roadway drainage maintenance
- Identify the components of shoulders and ditches

Target Audience

This course is designed for State, regional, and county personnel who manage operations programs and deal with oversight and quality assurance across broad geographic areas. This target audience also is involved with handling materials, scheduling, budgeting, and planning.

Training Level: Basic

Fee: 2016: $25 Per Person; 2017: N/A

Length: 1 HOURS (CEU: 0 UNITS)

Class Size: MINIMUM: 0; MAXIMUM: 0

NHI Customer Service: (877) 558-6873 • nhicustomerservice@dot.gov

Course Number

FHWA-NHI-134109F

Course Title

Maintenance Training Series: Outdoor Advertising and Litter Control - WEB-BASED

The Highway Beautification Act (HBA) of 1965 mandated a state program, based on Federal rules and regulations, for improving motorists' visual experiences on the roadway. The HBA affects billboards and advertisements along State roadways. The Outdoor Advertising and Litter Control course familiarizes maintenance personnel with the rules and regulations governing placement and control of outdoor advertising along highway rights-of-way to ensure they are in compliance with the standards stipulated in the HBA. Additionally, the course covers litter control safety for public groups assisting State DOTs in litter pickup.

Participants learn about the rules and regulations for maintaining and controlling outdoor advertising, guidance on administering an outdoor advertising program, the steps involved in the permitting process, and appropriate actions for non-compliance by sign owners. Additionally, participants are encouraged to compare the standards outlined in the HBA to their State's rules and regulations, which may include stricter provisions than those in the HBA.

This training was developed as part of the Maintenance Training Series. To access all the courses in the series, enroll in the 134109 course.

Outcomes

Upon completion of the course, participants will be able to:

- Identify Federal and State regulations, laws, ordinances, guidelines, and policies governing outdoor advertisement placement
- Describe the permit process
- Describe the role of the maintenance supervisor in outdoor advertising control

Target Audience

This course is designed for State, regional, and county personnel who manage operations programs and deal with oversight and quality assurance across broad geographic areas. This target audience also is involved with handling materials, scheduling, budgeting, and planning.

Training Level: Basic

Fee: 2016: $25 Per Person; 2017: N/A

Length: .5 HOURS (CEU: 0 UNITS)

Class Size: MINIMUM: 0; MAXIMUM: 0

NHI Customer Service: (877) 558-6873 • nhicustomerservice@dot.gov

Course Number

FHWA-NHI-134109G

WBT

Course Title

Maintenance Training Series: Roadside Vegetation Management - WEB-BASED

Vegetation management is much more than routine mowing of grass and trimming of bushes and trees. The Roadside Vegetation Management course explains the need for and purpose of good vegetation management. The course also underscores why vegetation management is a critical part of a roadway maintenance program.

Participants learn about equipment and herbicides used for vegetation management, including an overview of mechanical vegetation control and the environmental controls and precautions needed when using herbicides as part of a noxious weed control program.

This training was developed as part of the Maintenance Training Series. To access all the courses in the series, enroll in the 134109 course.

Outcomes

Upon completion of the course, participants will be able to:

- Describe why vegetation control is important to roadway safety and performance
- Identify the types of equipment used for mechanical vegetation control
- Identify types of herbicide vegetation management methods, their use, environmental control, and precautions
- Describe the requirements of a noxious weed control program

Target Audience

This course is designed for State, regional, and county personnel who manage operations programs and deal with oversight and quality assurance across broad geographic areas. This target audience also is involved with handling materials, scheduling, budgeting, and planning.

Training Level: Basic

Fee: 2016: $25 Per Person; 2017: N/A

Length: 1 HOURS (CEU: 0 UNITS)

Class Size: MINIMUM: 0; MAXIMUM: 0

NHI Customer Service: (877) 558-6873 • nhicustomerservice@dot.gov

Course Number

FHWA-NHI-134109H

Course Title

Maintenance Training Series: Weather-related Operations - WEB-BASED

Storm control is a major component of roadway maintenance in many areas of the country. State, municipal, and county agencies are responsible for providing safe, passable roadways even in severe weather. While the majority of the Weather-related Operations course concentrates on snow and ice storms, many of the elements apply to other weather events as well. Tornadoes, hurricanes, and flooding all require coordination and dedication of maintenance personnel. In any weather event, agencies need to restore roadways and bridges and to ensure they are safe for motorists.

Participants learn about the planning requirements for an effective storm response, including scheduling and training personnel, identifying equipment needs, executing dry runs, and the additional requirements posed by a multi-day storm event. This training assists participants with planning and responding effectively to all weather-related operations.

This training was developed as part of the Maintenance Training Series. To access all the courses in the series, enroll in the 134109 course.

Outcomes

Upon completion of the course, participants will be able to:

- Identify the elements of an effective storm response plan
- Identify factors involved in scheduling personnel needs
- Identify safety and training considerations for maintenance personnel who are involved in weather-related operations
- Identify the types of equipment used in a snow and ice removal plan and their uses
- Describe how to identify equipment needs for a particular storm

Target Audience

This course is designed for State, regional, and county personnel who manage operations programs and deal with oversight and quality assurance across broad geographic areas. This target audience also is involved with handling materials, scheduling, budgeting, and planning.

Training Level: Basic

Fee: 2016: $25 Per Person; 2017: N/A

Length: 1 HOURS (CEU: 0 UNITS)

Class Size: MINIMUM: 0; MAXIMUM: 0

NHI Customer Service: (877) 558-6873 • nhicustomerservice@dot.gov

Course Number

FHWA-NHI-134109I

Course Title

Maintenance Training Series: Basics of Work Zone Traffic Control - WEB-BASED

Meeting the national requirements for work zone traffic control is a critically important responsibility of maintenance personnel. The national requirements, found in Part 6 of the Manual on Uniform Traffic Control Devices (MUTCD), promote driver and worker safety during roadway maintenance projects. This training, Basics of Work Zone Traffic Control, provides an introduction to the requirements outlined in Part 6 of the 2009 MUTCD. The course also offers an overview of the manual's structure and requirements regarding traffic control devices and their applications, flagging operations and procedures, and pedestrian and worker safety.

Through a series of work zone scenarios, this training uses the MUTCD Part 6 to review fundamental concepts of setting up work zones, including proper signage, taper lengths, and flagging procedures. Participants are encouraged to compare their State's standards, if available, to the guidance established in the MUTCD and determine what additional requirements may need to be met to establish safe, compliant work zones.

This training was developed as part of the Maintenance Training Series. To access all the courses in the series, enroll in the 134109 course.

Outcomes

Upon completion of the course, participants will be able to:

- Describe the content and use of The Manual on Uniform Traffic Control Devices (MUTCD) Part 6
- Use the MUTCD to correctly answer questions about the basics of work zone traffic control
- Differentiate among standard, guidance, and option conditions in the MUTCD
- Differentiate among standard, guidance, and option conditions in the MUTCD for work zone traffic control in rural and urban areas

Target Audience

This course is designed for State, regional, and county personnel who manage operations programs and deal with oversight and quality assurance across broad geographic areas. This target audience also is involved with handling materials, scheduling, budgeting, and planning.

Training Level: Basic

Fee: 2016: $25 Per Person; 2017: N/A

Length: 1 HOURS (CEU: 0 UNITS)

Class Size: MINIMUM: 0; MAXIMUM: 0

NHI Customer Service: (877) 558-6873 • nhicustomerservice@dot.gov

Course Number

FHWA-NHI-134109J

Course Title

Maintenance Training Series: Underground Storage Tanks - WEB-BASED

The Nation's underground storage tank (UST) systems consist of underground tanks and piping that store petroleum and other hazardous materials. This course, Underground Storage Tanks, addresses the procedures to install, operate, and remove USTs.

Developed specifically for maintenance personnel, this course provides participants with an understanding of the Federal laws and regulations that govern UST systems. During the course, participants acquire the knowledge needed to successfully oversee UST installations and closures. Specifically, the course explores the requirements of industry installation and closure codes, leakage detection, spill and overfill prevention, corrosion protection, and ensuring a "clean" closure.

This training was developed as part of the Maintenance Training Series. To access all the trainings in the series, enroll in the 134109 course.

Outcomes

Upon completion of the course, participants will be able to:

- Describe the regulatory framework governing the operation of underground storage tanks
- Describe UST operations
- Describe the process that must be followed to obtain satisfactory "clean closure" from the appropriate oversight agency
- Describe UST cleanup and removal operations

Target Audience

This course is designed for State, regional, and county personnel who manage operations programs and deal with oversight and quality assurance across broad geographic areas. This target audience also is involved with handling materials, scheduling, budgeting, and planning.

Training Level: Basic

Fee: 2016: $25 Per Person; 2017: N/A

Length: 1 HOURS (CEU: 0 UNITS)

Class Size: MINIMUM: 0; MAXIMUM: 0

NHI Customer Service: (877) 558-6873 • nhicustomerservice@dot.gov

COURSE NUMBER

FHWA-NHI-134109K

COURSE TITLE

Maintenance Training Series: Cultural and Historic Preservation - WEB-BASED

Cultural and historic sites are often located within an area where maintenance activities are scheduled to be completed. This training, Cultural and Historic Preservation, is teaches participants about regulations and concerns related to safeguarding cultural and historic sites from the potential impacts of highway maintenance activities. Examples of maintenance activities that can impact cultural or historic sites include slope stabilization, shoulder or pavement widening, and vegetation control. Additional examples are presented during the course.

This course assists participants with recognizing potential historic or cultural resources, verifying a site's cultural or historic status, and avoiding impacts to sites when carrying out maintenance activities. Since completing these tasks often requires additional expertise, resources for obtaining needed assistance are provided. In addition, participants learn how maintenance activities can enhance cultural and historic sites through utilization of Context Sensitive Solutions (CSS).

This training was developed as part of the Maintenance Training Series. To access all the courses in the series, enroll in the 134109 course.

OUTCOMES

Upon completion of the course, participants will be able to:

- Identify governing bodies and registries that should be consulted prior to commencing maintenance activities on sites of cultural and historic importance
- Recognize what sorts of structures, landmarks, and properties could pose potential cultural and historic preservation issues
- Describe how to avoid impacts to historic sites
- Describe the role of DOT in maintaining and enhancing cultural resources

TARGET AUDIENCE

This course is designed for State, regional, and county personnel who manage operations programs and deal with oversight and quality assurance across broad geographic areas. This target audience also is involved with handling materials, scheduling, budgeting, and planning.

TRAINING LEVEL: Basic

FEE: 2016: $25 Per Person; 2017: N/A

LENGTH: 1 HOURS (CEU: 0 UNITS)

CLASS SIZE: MINIMUM: 0; MAXIMUM: 0

NHI Customer Service: (877) 558-6873 • nhicustomerservice@dot.gov

Course Number

FHWA-NHI-134112

Course Title

Principles and Practices for Enhanced Maintenance Management Systems - WEB-CONFERENCE

Is your agency in the process of enhancing its maintenance management capabilities?

Are you interested in learning more about developing effective performance measures for maintenance activities?

If so, join us for a blended training course that features both independent study material and facilitated Web-conferences. You will be introduced to the methods and practices used in an enhanced maintenance management system (MMS) to effectively maintain and operate a highway network. You will explore the principles and practices of using MMS to effectively examine efficient maintenance and operation of a highway network. Throughout the course, you will learn by participating in activities and assignments specific to using MMS.

The course materials rely heavily on the AASHTO Guidelines for Maintenance Management Systems, Transportation Asset Management Guide, and several other recent publications on the topic. To illustrate the application of the principles, the course materials are supplemented with examples from State and local highway agencies.

Participant Responsibilities:

- 7 Web-based lessons (Duration: 1- 1.5 hrs each)
- 3 Web-conferences (Duration: 2 hours each)

To obtain your certificate, you must complete all Web-based lessons and Web-conferences. To receive Continuing Education Units (CEUs), you must also pass the online test at the end of the course. You will need your own computer with an Internet connection as well as a telephone line in order to participate.

Outcomes

Upon completion of the course, participants will be able to:

- Compare and contrast a first generation MMS with an enhanced MMS
- Describe the terms "outcome-based" and "performance-based" and how they pertain to an enhanced MMS
- Describe the use of service levels to support the programming and budgeting activities incorporated into an MMS
- Identify the types of systems that should be integrated with an MMS and provide several examples of the types of data that should interface between each system
- List the potential benefits to be realized by fully integrating an enhanced MMS
- Identify several steps that will advance an agency's current maintenance management practices now and in the future

Target Audience

The target audience for this course includes State and local maintenance engineers, maintenance supervisors, asset managers, and their industry counterparts. The course is specifically for individuals who are responsible for directing and managing maintenance operations and budgets, maintenance project and treatment selection, and/or the monitoring of system conditions.

Training Level: Basic

Fee: 2016: $300 Per Person; 2017: N/A

Length: 15 HOURS (CEU: 1.5 UNITS)

Class Size: MINIMUM: 20; MAXIMUM: 30

NHI Customer Service: (877) 558-6873 • nhicustomerservice@dot.gov

COURSE NUMBER

FHWA-NHI-134114

COURSE TITLE

Inspector Training for Cold In-Place Recycling (CIR) - WEB-BASED

Cold In-place Recycling (CIR) is a method of reconstructing any flexible pavement where the need arises from structural failures. These failures include: transverse cracking, wheel rutting, potholes, surface irregularities, or a combination of the above.

The proper selection of a CIR process in conjunction with good specifications and quality construction are all equally important in the long-term performance of the pavement rehabilitation.

This series on CIR will introduce each method and provide a background on when, how, and why that method is selected/used.

This training is meant to provide an overview of CIR, including an explanation of the pre-production inspection, completing the control strip, full production of the mix, mix placement, curing and maintenance, acceptance testing, and measurement and payment. This course contains 3 modules:

Module 1: Introduction to Cold In-Place Recycling

Module 2: Cold In-Place Recycling Full Production

Module 3: Cold In-Place Recycling Post Production

This course will provide the inspector with a background and proper inspection procedures when placing cold-in-place hot mix asphalt.

OUTCOMES

Upon completion of the course, participants will be able to:

- Explain what Cold In-Place Recycling (CIR) is, and why it is used
- Describe what happens during pre-production
- Explain how the control strip helps determine compaction procedures and why it is needed
- Identify the factors that can influence a CIR mix
- Describe important considerations during placement, compaction, and finishing
- Explain the importance of curing and maintenance on the quality of a CIR surface
- Describe what happens once the surface is finished

TARGET AUDIENCE

This training is designed for Local, County, and State owner agency technicians/inspectors. It is also useful for individuals who need awareness or basic understanding of cold in-place recycling. This training was developed by the Transportation Curriculum Coordination Council (TCCC) in partnership with AASHTO, NHI, and is recommended for TCCC levels II through IV.

TRAINING LEVEL: Basic

FEE: 2016: $25 Per Person; 2017: N/A

LENGTH: 4 HOURS (CEU: 0 UNITS)

CLASS SIZE: MINIMUM: 1; MAXIMUM: 1

NHI Customer Service: (877) 558-6873 • nhicustomerservice@dot.gov

Course Number

FHWA-NHI-380108

Course Title

Maintenance of Drainage Features for Safety - WEB-BASED

The purpose of this training is to highlight common roadway drainage problems that can cause an unsafe condition and suggest inspection methods and corrective action. Maintaining roadway drainage is important for safety and for ensuring the long life of the roadway by preventing erosion of the roadway, saturation of the subbase, and damage to roadway structures. The training is broken into two modules:

Module 1: Effects of Drainage describes common roadway safety hazards and how to recognize drainage problems.

Module 2: Safe Drainage Features and Work Zones covers solutions to common roadway safety issues and work zone safety.

This training is not intended to be a design guide. Participants may want to contact their State Local Technical Assistance Program (LTAP) for more details on drainage design.

Outcomes

Upon completion of the course, participants will be able to:

- Identify problems created by ponding and standing water on the roadway
- Describe safety issues related to ditches and side slopes
- Describe how drainage features can become safety hazards
- Identify methods for identifying drainage problems
- Recall conditions to look for during field inspections
- Explain how to fix or prevent common roadway side slope problems
- Describe work zone safety procedures

Target Audience

This training is intended to help local road agency maintenance workers understand the importance of maintaining and upgrading drainage features on their road system to avoid an unsafe condition.

Training Level: Basic

Fee: 2016: $25 Per Person; 2017: N/A

Length: 1 HOURS (CEU: 0 UNITS)

Class Size: MINIMUM: 1; MAXIMUM: 1

NHI Customer Service: (877) 558-6873 • nhicustomerservice@dot.gov

COURSE NUMBER

FHWA-NHI-381004

COURSE TITLE

CDL Series - General Knowledge - WEB-BASED

This training was prepared by the Transportation Curriculum Coordination Council (TCCC) in partnership with NHI has been designed for someone interested in commercial driver's license (CDL) general knowledge. This training is recommended for the Transportation Curriculum Coordination Council levels I and II or anyone interested in obtaining a CDL. This course is primarily intended for inspectors and technicians.

This training contains the general knowledge and safe driving information that all commercial drivers should know. It is broken into three modules:

Module 1 reviews vehicle control, shifting gears, seeing the road, communicating, speed control, and space management.

Module 2 covers night driving, driving in cold and hot weather, mountain driving, and railroad crossings.

Module 3 discusses seeing hazards, driving and road emergencies, staying alert and fit to drive, and transporting hazards.

This general knowledge training does not have specific information on air brakes or pre-trip inspection. You may complete other training in the CDL series to learn more about them.

For more information on the CDL examination and requirements that apply to your State, contact your State license agencies.

OUTCOMES

Upon completion of the course, participants will be able to:

- Describe the procedures in controlling your vehicle and shifting gears
- Define the steps to seeing the road in various situations
- Recognize the importance of signaling and communicating your presence
- Identify the important components of speed control and space management
- Describe the proper ways to drive at night
- Identify the correct practices for driving in cold weather and hot weather
- Describe the procedures for driving on a mountain
- Recognize the proper way to cross a railroad
- Describe the procedures in responding to driving emergencies and emergencies on the road
- Identify the guidelines to staying alert and fit to drive
- Define the proper way to transport hazardous materials

TARGET AUDIENCE

This course is designed for any individuals wanting to learn more about commercial driver's license (CDL) general information.

TRAINING LEVEL: Basic

FEE: 2016: $25 Per Person; 2017: N/A

LENGTH: 3 HOURS (CEU: 0 UNITS)

CLASS SIZE: MINIMUM: 1; MAXIMUM: 1

NHI Customer Service: (877) 558-6873 • nhicustomerservice@dot.gov

COURSE NUMBER

FHWA-NHI-381005

COURSE TITLE

CDL Series - Air Brakes - WEB-BASED

This training was prepared by the Transportation Curriculum Coordination Council (TCCC) in partnership with NHI has been designed for someone interested in commercial driver's license (CDL) air brake systems. This training is recommended for the Transportation Curriculum Coordination Council levels I and II or anyone interested in obtaining a CDL. This course is primarily intended for technicians.

In this training we'll discuss the parts of an air brake system, dual air brake systems, how to inspect your air brake system, and how to effectively use your air brake system. If you want to drive a truck or bus with air brakes, or pull a trailer with air brakes, you'll need to take a test on this material.

This training contains information on air brakes system that all commercial drivers should know. It is broken into two modules:

Module 1 consists of air brake system parts and dual air brakes systems.

Module 2 consists of inspecting air brakes and using air brakes.

This air brakes training does not have specific information on general knowledge or pre-trip inspection. You may complete other training in the CDL series to learn more about them.

For more information on the CDL examination and requirements that apply to your State, contact your State license agencies.

OUTCOMES

Upon completion of the course, participants will be able to:

- Identify the important parts of the air brake system
- Define dual air brakes
- Recognize key elements in the air flow process of the dual air brake system
- Identify the important components of air brakes inspection
- Recognize the proper ways to use air brakes

TARGET AUDIENCE

This course is designed for any individuals wanting to learn more about commercial driver's license (CDL) air brake systems.

TRAINING LEVEL: Basic

FEE: 2016: $25 Per Person; 2017: N/A

LENGTH: 1.5 HOURS (CEU: 0 UNITS)

CLASS SIZE: MINIMUM: 1; MAXIMUM: 1

NHI Customer Service: (877) 558-6873 • nhicustomerservice@dot.gov

Course Number

FHWA-NHI-135010

Course Title

River Engineering for Highway Encroachments

The course provides training in the theory and application of alluvial channel flow, fluvial geomorphology, sediment transport, and river mechanics to the planning, location, design, construction, maintenance, and operation of highways. Material for this course comes from "Hydraulic Design Series 6 (HDS-6): River Engineering for Highway Encroachments Highways in the River Environment." The course includes detailed coverage on how to estimate rates of sediment transport by selecting appropriate equations for use in the computations. Additional topics include sediment properties and sediment measurement techniques. Case histories provide practical examples of problems that occur at highway crossings and encroachments of streams and rivers. A computer generated 360-degree virtual tour site visit is used for a comprehensive workshop. Example problems in sediment transport will be worked by the course participants.

Prior to the beginning of the course, participants are strongly encouraged to enroll in the Web-based training entitled, 135091 Basic Hydraulic Principles Review. Mastery of the concepts covered in this WBT is important to successful completion of the Instructor-led training.

Outcomes

Upon completion of the course, participants will be able to:

- Apply open channel flow equations and concepts to flow in alluvial channels
- Determine resistance to flow and sediment transport at highway crossings
- Apply sediment transport and sediment continuity relationships for the analysis of streambed degradation and aggradation
- Evaluate the inter-relationships between fluvial (river) geomorphology and highway hydraulic design

Target Audience

Engineers who are responsible for the evaluation of stream stability and the design of highway hydraulic structures. The course is designed for graduate engineers (BS) who have been trained in basic hydraulics of rigid-boundary, open channel flow.

Training Level: Intermediate

Fee: 2016: $925 Per Person; 2017: N/A

Length: 3 DAYS (CEU: 2.1 UNITS)

Class Size: MINIMUM: 20; MAXIMUM: 30

NHI Customer Service: (877) 558-6873 • nhicustomerservice@dot.gov

Course Number

FHWA-NHI-135027

Course Title

Urban Drainage Design (3-Day)

This course provides a detailed introduction to urban roadway drainage design. Design guidance for solving basic problems encountered in urban roadway drainage design is provided. The topics are hydrology including rational equation, soil conservation method, regression equations, and synthetic hydrographs; and highway drainage including gutter flow, roadway inlet interception, storm drain systems, energy and hydraulic grade lines, detention ponds, and stormwater management.

The 4-day course includes the basic 3-day course, plus presentation of the 1-day course FHWA-NHI-135028 Stormwater Pump Station Design.

Outcomes

Upon completion of the course, participants will be able to:

- Determine runoff (peak flows and volumes) from urban watersheds
- Apply basic hydraulic principles to urban drainage design
- Perform roadway drainage designs using various roadway inlets
- Size and/or analyze storm drain conveyance systems
- Establish the energy and hydraulic grade lines for storm drains
- Design and/or analyze detention basins
- Perform hydraulic design of pumping stations (with optional day four)

Target Audience

Highway designers with limited experience in drainage design, but familiar with mathematical concepts such as algebra and geometry and have some working background in hydrology and hydraulics.

Training Level: Intermediate

Fee: 2016: $900 Per Person; 2017: N/A

Length: 3 DAYS (CEU: 1.8 UNITS)

Class Size: MINIMUM: 20; MAXIMUM: 30

NHI Customer Service: (877) 558-6873 • nhicustomerservice@dot.gov

Course Number

FHWA-NHI-135027A

Course Title

Urban Drainage Design (4-Day)

This course provides a detailed introduction to urban roadway drainage design. Design guidance for solving basic problems encountered in urban roadway drainage design is provided. The topics are hydrology including rational equation, soil conservation method, regression equations, and synthetic hydrographs; and highway drainage including gutter flow, roadway inlet interception, storm drain systems, energy and hydraulic grade lines, detention ponds, and stormwater management.

The 4-day course includes the basic 3-day course, plus presentation of the 1-day course FHWA-NHI-135028 Stormwater Pump Station Design.

Outcomes

Upon completion of the course, participants will be able to:

- Determine runoff (peak flows and volumes) from urban watersheds
- Apply basic hydraulic principles to urban drainage design
- Perform roadway drainage designs using various roadway inlets
- Size and/or analyze storm drain conveyance systems
- Establish the energy and hydraulic grade lines for storm drains
- Design and/or analyze detention basins
- Perform hydraulic design of pumping stations (with optional day four)

Target Audience

Highway designers with limited experience in drainage design, but familiar with mathematical concepts such as algebra and geometry and have some working background in hydrology and hydraulics.

Training Level: Intermediate

Fee: 2016: $1025 Per Person; 2017: N/A

Length: 4 DAYS (CEU: 2.4 UNITS)

Class Size: MINIMUM: 20; MAXIMUM: 30

NHI Customer Service: (877) 558-6873 • nhicustomerservice@dot.gov

COURSE NUMBER

FHWA-NHI-135028

COURSE TITLE

Highway Stormwater Pump Station Design

This course provides detailed instruction in the design and analysis of highway stormwater pump stations including guidance on location and type selection. A major portion of the course is devoted to recommended hydraulic design procedures for sizing and optimizing the performance of stormwater pump stations and includes solving of classroom example problems. This course is also offered as a 1-day add-on to FHWA-NHI-135027 Urban Drainage Design. Topics to be discussed include, site considerations, hydrology, storage, pump configuration, mass curve routing, pump selection, sump dimensions, and mechanical and electrical considerations.

OUTCOMES

Upon completion of the course, participants will be able to:

- Describe what a pump station is and where they are used
- Define the drainage area for a pump station and construct the resulting mass inflow curve
- Calculate the storage volume required for a pump station and discuss ways to acquire that volume
- Determine pump operational schedule and perform mass curve routing of the inflow hydrograph
- Calculate the size of the discharge line and select required pump size
- Define dimensions of the wet well and perform system evaluation
- Describe basic mechanical and electrical concepts important in pump station design
- Describe available pump station software

TARGET AUDIENCE

Highway designers or hydraulic engineers who have responsibility for the design and analysis of highway stormwater pumping stations, and managers who review pump station design projects. To derive the most benefit from this training, course participants should have knowledge of the fundamentals of highway hydrology, pavement drainage, stormdrain design, and open channel flow.

TRAINING LEVEL: Intermediate

FEE: 2016: $350 Per Person; 2017: N/A

LENGTH: 1 DAYS (CEU: .6 UNITS)

CLASS SIZE: MINIMUM: 20; MAXIMUM: 30

NHI Customer Service: (877) 558-6873 • nhicustomerservice@dot.gov

Course Number

FHWA-NHI-135041

Course Title

One-Dimensional Modeling of River Encroachments with HEC-RAS

The host is responsible for providing a minimum of one computer for each pair of participants. The computers shall have the following minimum specifications:

Intel Based Pentium processor higher (Pentium III or higher is recommended), Microsoft Windows 95, 98, ME with 212 MB of RAM (1 GB recommended) or Window NT 4.0, 2000, Vista, XP, 7, or 8 with 1 GB of RAM (1 GB recommended), including the .NET framework, a hard drive with at least 60 megabytes of free space (100 MB or more is recommended), CD-ROM drive, and 1024 x 768 color video display.

The course focuses on the use and application of HEC-RAS software, developed by the Hydrologic Engineering Center of the U.S. Army Corps of Engineers. Modeling principles and techniques will be presented using the latest version of HEC-RAS.

HEC-RAS, River Analysis System, solves the conservation of energy equation for one-dimensional steady flow analysis to determine water surface elevations for a given discharge. The Standard Step solution scheme is used combined with Manning's equation to compute cross section conveyance which allows for the construction of backwater and forewater profiles under subcritical, supercritical, and mixed flow regimes. HEC-RAS is capable of simulating structures in natural waterways and constructed channels. Specifically, it has built-in functionality to simulate a variety of bridge types, culverts, roadway approaches/embankments, and roadway encroachments.

Prior to the beginning of the course, participants are strongly encouraged to enroll in the Web-based training entitled, 135091 Basic Hydraulic Principles Review. Mastery of the concepts covered in this WBT is important to successful completion of the Instructor-led training.

Outcomes

Upon completion of the course, participants will be able to:

- Manage HEC-RAS files.
- Navigate the HEC-RAS windows.
- Describe the types of hydraulic modeling situations for which one-dimensionalapplication of HEC-RAS is appropriate.
- Describe one-dimensional hydraulic modeling principles used in HEC-RAS
- including conservation of energy, mass, and momentum.
- Build input data files for use with HEC-RAS for steady state applications with and without roadway encroachments including bridges, culverts, and multiple openings.
- Develop one-dimensional water surface elevations and velocity estimates using the HEC-RAS computer program.
- View and manipulate the output from the HEC-RAS computer program.
- Evaluate hydraulic conditions using HEC-RAS modeling program through
- various transportation related hydraulic structures including weirs, culverts, and bridges.
- Identify and troubleshoot modeling problems, including those indicated by errors,warnings, and notes.

Target Audience

Federal, State, and local hydraulic engineers who have responsibility for the design and analysis of river systems and stream crossings. Participants should have experience in using the Windows environment and knowledge of the fundamentals of open channel flow, including basic understanding of HEC-2 or WSPRO.

TRAINING LEVEL: Intermediate

FEE: 2016: $825 Per Person; 2017: N/A

LENGTH: 3 DAYS (CEU: 1.7 UNITS)

CLASS SIZE: MINIMUM: 20; MAXIMUM: 30

NHI Customer Service: (877) 558-6873 • nhicustomerservice@dot.gov

Course Number

FHWA-NHI-135046

Course Title

Stream Stability and Scour at Highway Bridges

The National Highway Institute's (NHI) 3-day Stream Stability and Scour at Highway Bridges course provides participants with comprehensive training in the prevention of hydraulic-related bridge failures. Course participants will receive training in conducting a stream stability classification and qualitative analysis of stream response and make estimates of scour at a bridge opening.

Material for the course comes primarily from two Hydraulic Engineering Circulars (HEC), "Evaluating Scour at Bridges" (HEC-18), 5th Edition (2012), and "Stream Stability at Highway Structures" (HEC-20), 4th Edition (2012). The effects of stream instability, scour, erosion, and stream aggradation and degradation are covered. Quantitative techniques are provided for estimating long-term degradation and for calculating the magnitude of contraction scour in a bridge opening. Procedures for estimating local scour at bridge piers and abutments for simple and complex substructures are also provided. A comprehensive workshop integrates qualitative analysis and analytical techniques to determine the need for a Scour Plan of Action for correcting stream instability and scour problems. For this 3-day course, the host agency will need to select 3 optional topics (out of 8 possible topics). Course instructors will contact the host prior to the course to complete a pre-course questionnaire, determine optional topics to be taught, and discuss the course schedule.

This comprehensive training provides preventive techniques for identifying, analyzing, and calculating various hydraulic factors that impact bridge stability. Public and private sector engineers responsible for maintaining the integrity of highway bridges will find it invaluable.

Prior to the beginning of the course, participants are strongly encouraged to enroll in the following Web-based training (WBT) courses: 135091 Basic Hydraulic Principles Review, 135086 Stream Stability Factors and Concepts, and 135087 Scour at Highway Bridges: Concepts and Definitions. Mastery of the concepts covered in these WBTs will enhance participation in the Instructor-led training.

Outcomes

Upon completion of the course, participants will be able to:

- Identify indicators of stream instability that can threaten bridges
- Identify stream types and their potential for instability problems
- Describe open-channel hydraulics concepts in bridge scour and stream instability analyses
- Define types of scour that can occur at bridge crossings
- Describe aggradation, degradation, and contraction scour
- Calculate contraction scour for live bed and clear water conditions
- Describe factors that influence scour at piers
- Calculate pier scour for three typical case studies
- Describe the factors that influence scour at abutments
- Describe how HEC-18, HEC-20, and HEC-23 provide analysis procedures for stream instability and bridge scour
- Perform Level I and II analyses
- Classify a stream using two different classification systems
- Conduct a qualitative analysis of stream responses
- Apply the HEC-18 scour equations to determine total scour at a bridge
- Determine the need for a Scour Plan of Action at a scour-critical bridge

Target Audience

Federal, State, and local highway hydraulic, structural, and geotechnical engineers as well as bridge inspectors responsible for maintaining the integrity of highway bridges against possible hydraulic-related problems. Consultants who perform bridge engineering work are encouraged to attend.

TRAINING LEVEL: Intermediate

FEE: 2016: $975 Per Person; 2017: N/A

LENGTH: 3 DAYS (CEU: 2 UNITS)

CLASS SIZE: MINIMUM: 20; MAXIMUM: 30

NHI Customer Service: (877) 558-6873 • nhicustomerservice@dot.gov

COURSE NUMBER

FHWA-NHI-135047

COURSE TITLE

Stream Stability and Scour at Highway Bridges for Bridge Inspectors

This course is an abbreviated presentation of 135046 Stream Stability and Scour at Highway Bridges. The course provides an understanding of and assistance in detecting hydraulic-related problems at highway bridges. The effects of steam instability, scour, erosion, and stream aggradation and degradation are covered. Countermeasures to these problems are discussed. This course concentrates on visual keys to detecting scour and stream instability problems and provides an introduction to portable scour monitoring instrumentation. The course emphasizes inspection guidelines to complete the hydraulic and scour-related coding requirements of the National Bridge Inspection Standards (NBIS). This course can be offered as a 1-day module in conjunction with the 3-day 135046 or as a stand-alone presentation.

NHI Courses 135086 and 135087 are Web-based trainings and are prerequisites for NHI Hydraulics courses 135047 and 135048.

OUTCOMES

Upon completion of the course, participants will be able to:

- Identify stream instability and scour problems at bridges
- Conduct field evaluations for scour and stream instability problems and properly code the results in the National Bridge Inventory
- Recognize countermeasures for stream instability and scour

TARGET AUDIENCE

Federal, State, and local highway bridge inspectors responsible for detecting possible hydraulic-related problems that may threaten the integrity of highway bridges. Consultants who do bridge inspection work for the States may attend if space is available.

TRAINING LEVEL: Basic

FEE: 2016: $350 Per Person; 2017: N/A

LENGTH: 1 DAYS (CEU: .6 UNITS)

CLASS SIZE: MINIMUM: 20; MAXIMUM: 30

NHI Customer Service: (877) 558-6873 • nhicustomerservice@dot.gov

COURSE NUMBER

FHWA-NHI-135048

COURSE TITLE

Countermeasure Design for Bridge Scour and Stream Instability (2.5-Day)

This course provides an overview of countermeasures to highway related failures from the effects of stream instability, scour, erosion, and stream aggradation and degradation problems. Material for the 2.5-day course comes primarily from Hydraulic Engineering Circular (HEC) "Bridge Scour and Stream Instability Countermeasures - Experience, Selection, and Design Guidance" (HEC-23).

Given a stream instability and scour problem, participants will select appropriate countermeasures to correct the problem. The course provides training in recommended strategies for developing a plan that includes appropriate countermeasures, including alternatives to conventional riprap and filter design.

Participants will apply hydraulics analysis techniques to countermeasure design for seven design guideline workshops. The course provides an introduction to fixed and portable instrumentation for scour monitoring using slides and video demonstrations. Participants will receive training in designing a monitoring program to reduce the risk from scour.

NHI Course 135046 provides training in identifying and analyzing stream instability and scour problems at highway bridges and is recommended as a prerequisite for this course.

NHI Courses #135086 and #135087 are Web-based training module and are prerequisites for NHI Hydraulics courses 135047 and 135048.

OUTCOMES

Upon completion of the course, participants will be able to:

- Develop a plan of action for a scour critical bridge
- Propose countermeasures for stream instability and scour problems
- Identify countermeasures for bridge scour and stream instability using the HEC-23 countermeasures matrix
- Design selected countermeasures with HEC-23 design guidelines

TARGET AUDIENCE

Federal, State, and local highway hydraulic, structural, and geotechnical engineers and bridge inspectors responsible for maintaining the integrity of highway bridges against possible hydraulic-related problems. Consultants who do bridge engineering work are also encouraged to attend.

TRAINING LEVEL: Intermediate

FEE: 2016: $900 Per Person; 2017: N/A

LENGTH: 2.5 DAYS (CEU: 1.5 UNITS)

CLASS SIZE: MINIMUM: 20; MAXIMUM: 30

NHI Customer Service: (877) 558-6873 • nhicustomerservice@dot.gov

COURSE NUMBER

FHWA-NHI-135056

COURSE TITLE

Culvert Design

The National Highway Institute's (NHI) 3-day Culvert Design course provides participants with an in-depth, hands-on understanding of how to hydraulically size and design a highway culvert. The course covers a range of design topics, including allowable headwater at the inlet, permissible outlet velocity, energy dissipation measures, aquatic organism passage, mechanisms of culvert failures, and repair and rehabilitation options.

Material for this 3-day course is primarily derived from the Hydraulic Design Series No. 5 (HDS 5), Hydraulic Design of Highway Culverts textbook, which is provided to participants. Additional references used throughout this course include Hydraulic Engineering Circular No. 14 (HEC-14); Hydraulic Design of Energy Dissipators for Culverts and Channels; HEC-26, Culvert Design for Aquatic Organism Passage; and HEC-9, Debris Control Structures, Evaluation, and Countermeasures. Course topics include culvert design principles and procedures and debris control structures. Throughout the course, participants engage in a number of workshops where problems are completed, both long-hand and with a computer using the FHWA HY-8 Culvert Hydraulic Analysis and Design Program. Additionally, a portable hydraulic flume is set up in the classroom for the participants to observe hydraulic principles associated with various culvert configurations, aquatic organism passage features, and culvert linings.

At the end of this course, participants will be able to apply fundamental engineering concepts, methods, and the HY-8 computer program to analyze and design culvert crossings meeting a variety of hydraulic and environmental design criteria.

Prior to taking this course, participants are strongly encouraged to enroll in the Web-based training (WBT) entitled, 135091 Basic Hydraulic Principles Review. Mastery of the concepts covered in this WBT is important to successful completion of the Instructor-led training.

OUTCOMES

Upon completion of the course, participants will be able to:

- Justify the importance of culvert design
- Explain the overall culvert design process
- Summarize basic hydraulic concepts
- Discuss factors influencing hydraulic performance and design of culverts
- Explain how to calculate culvert outlet velocity
- Apply nomographs and computer methods to design a roadway culvert
- Design culverts that meet aquatic organism passage (AOP) requirements
- Assess impacts of repair and rehabilitation of culverts on hydraulic performance
- Design energy dissipator and debris control structures for culverts
- Design culverts for various situations
- Discuss culvert failures and how they can be prevented

TARGET AUDIENCE

This intermediate-level training course is intended for hydraulic engineers, transportation engineers, and highway designers involved with roadway drainage and culvert design. Environmental scientists with an interest in aquatic organism passage may also benefit from participation in this course.

TRAINING LEVEL: Intermediate

FEE: 2016: $900 Per Person; 2017: N/A

LENGTH: 3 DAYS (CEU: 2 UNITS)

CLASS SIZE: MINIMUM: 20; MAXIMUM: 30

NHI Customer Service: (877) 558-6873 • nhicustomerservice@dot.gov

COURSE NUMBER

FHWA-NHI-135065

COURSE TITLE

Introduction to Highway Hydraulics

This course is based on Hydraulic Design Series No. 4 (HDS-4), "Introduction to Highway Hydraulics." The objective of the course is to provide a broad overview of basic highway drainage concepts. Fundamental hydraulic concepts are discussed, followed by open-channel flow principles and design applications of open-channel flow in highway drainage, including the design of stable channels, and pavement drainage. Closed-conduit concepts and applications in highway drainage include the application of culvert and storm drainage design. The presentation concludes with an introduction to concepts and design of energy dissipators. Detailed design criteria are drawn from other Hydraulic Design Series manuals and Hydraulic Engineering Circulars (HECs), providing a broad overview of all components of highway drainage design with an emphasis on practical applications. A portable hydraulic flume is set up in the classroom for the participants to observe numerous hydraulic principles. The participants take velocity and discharge measurements from the flume while in various setups and use the information to make design calculations.

OUTCOMES

Upon completion of the course, participants will be able to:

- Calculate design discharge using the rational method or regression equation procedures
- Apply the continuity and energy equation to solve practical design problems
- Use the Weir equation to calculate the flow overtopping a roadway embankment
- Use Manning's equation to calculate velocity or flow depth in simple or compound channels and recognize when this equation cannot be appropriately applied
- Evaluate channel flow conditions (subcritical, critical, or supercritical) using the Froude number
- Design a stable channel using basic hydraulic concepts and Hydraulic Engineering Circular HEC-15
- Apply basic pavement drainage concepts in calculation procedures described in HEC-22
- Design a simple culvert crossing using the procedures in HDS-5
- Design a simple storm drain and calculate the Hydraulic Grade Line (HGL) using the energy equation and HEC-22
- Describe which energy dissipaters are useful for culvert or storm drain applications based on HEC-14

TARGET AUDIENCE

Entry-level engineers or engineering technicians who are performing highway drainage calculations on transportation facilities. It will also be useful as a refresher course on hydraulic fundamentals for experienced personnel.

TRAINING LEVEL: Basic

FEE: 2016: $900 Per Person; 2017: N/A

LENGTH: 3 DAYS (CEU: 1.8 UNITS)

CLASS SIZE: MINIMUM: 20; MAXIMUM: 30

NHI Customer Service: (877) 558-6873 • nhicustomerservice@dot.gov

COURSE NUMBER

FHWA-NHI-135067

COURSE TITLE

Practical Highway Hydrology

The course provides engineers and designers with the background and skills necessary for the practical application of hydrologic principles to highway design. Participants will be required to work example problems that stress actual design situations. The course is based on the Hydraulic Design Series (HDS) No. 2, "Highway Hydrology" which is also used in the course as a reference manual.

Participants will learn how to select and effectively implement techniques for estimating peak flows and flood hydrographs in gaged and ungaged streams for watersheds of the size typically encountered in highway drainage design. Through a series of optional modules, additional topics including channel routing, wetland hydrology, arid lands hydrology, and snowmelt hydrology are available given host agency preferences.

The overall course objectives enhance the understanding of basic hydrologic concepts and principles as they pertain to highways, and enable application of appropriate hydrologic concepts and tools in the design of drainage facilities and hydraulic structures.

OUTCOMES

Upon completion of the course, participants will be able to:

- Identify which peak flow design methods are suitable for given watershed characteristics and design requirements
- Estimate times of concentration
- Apply the SCS, regression and rational methods for peak flows
- Analyze gage flows using Log-Pearson III Frequency Analysis
- Develop hydrographs using the unit hydrograph and other techniques
- Perform storage routing calculations
- Design a storm water management facility

TARGET AUDIENCE

Highway engineers and designers who are responsible for designing channels, storm drains, and stormwater detention, as well as those involved in the hydraulic design of bridges and culverts. Attendees will benefit from, but are not required to have, a basic knowledge of hydrologic science. The course is a useful primer for those new to the subject and a thorough review for experienced hydrologic and hydraulic engineers.

TRAINING LEVEL: Intermediate

FEE: 2016: $850 Per Person; 2017: N/A

LENGTH: 3 DAYS (CEU: 1.8 UNITS)

CLASS SIZE: MINIMUM: 20; MAXIMUM: 30

NHI Customer Service: (877) 558-6873 • nhicustomerservice@dot.gov

COURSE NUMBER

FHWA-NHI-135080

COURSE TITLE

Hydrologic Analysis and Modeling with WMS

This course is designed as a hands-on, application-oriented training course using the Watershed Modeling System (WMS) to make hydrologic estimates using a variety of techniques. It will provide attendees with the knowledge and tools necessary to use data derived from geographical information systems (GIS) to develop hydrologic estimates and model runoff from watersheds. The course also teaches how to use digital terrain data for the development of watershed parameters that are required by most commonly used hydrologic analysis programs.

The WMS is a comprehensive environment for hydrologic analysis. It is developed by the Environmental Modeling Research Laboratory (EMRL) of Brigham Young University, and has been licensed for use by all State and Federal highway agencies. WMS makes it possible to take advantage of the wealth of digital terrain, land use, soil, and other GIS data readily available from government and private agencies. This data can then be used for preparing input files for several commonly used hydrologic models. Models supported by the interface include HEC-1 (HMS), TR-20, TR-55, and the Rational Method. This course also includes instruction in use of the regional regression equations contained in the National Flood Frequency (NFF) database. This course teaches the techniques and methods necessary to locate and use GIS data so that labor intensive processes such as delineating watershed boundaries and calculating modeling parameters from paper maps can be avoided when computing design flows and developing flow hydrographs at bridges and culverts.

Participants will receive a notebook that includes course materials, a WMS User's Manual, and copies of the software, workshops, and tutorials used in the course. Non-State highway agency course participants will receive a demonstration version of the proprietary WMS computer program.

OUTCOMES

Upon completion of the course, participants will be able to:

- Automate basin delineation in WMS with GIS vector data, DEMs, and TINs
- Efficiently use digital watershed data for hydrologic modeling parameter development
- Locate and obtain digital data sources for watershed delineation and hydrologic model development
- Use WMS to build hydrologic input data files for use with HEC-1 (HMS), TR-20, TR-55, regional regression equations, and Rational Method programs, including instruction on how to graphically view the output

TARGET AUDIENCE

Federal, State, and local hydrologic/hydraulic engineers who have responsibility for the design and analysis of highway stream crossings. In order to derive the most benefit from this training, course participants should have knowledge of the fundamentals of hydrology and hydrologic modeling. Experience with one of the aforementioned hydrologic modeling computer programs would be helpful.

TRAINING LEVEL: Intermediate

FEE: 2016: $725 Per Person; 2017: N/A

LENGTH: 3 DAYS (CEU: 1.8 UNITS)

CLASS SIZE: MINIMUM: 20; MAXIMUM: 25

NHI Customer Service: (877) 558-6873 • nhicustomerservice@dot.gov

COURSE NUMBER

FHWA-NHI-135082

COURSE TITLE

Highways in the Coastal Environment

Over 60,000 miles of roads in the United States are occasionally exposed to coastal surge and waves. Due to these unique design conditions, many highways and bridges suffer damage during coastal storms, including hurricanes and El Nino events. The purpose of this course is to teach important concepts and terminology of coastal science and engineering to highway engineers for use in the planning and design of coastal roads. The course is based on the Hydraulic Engineering Circular (HEC) No. 25, "Highways in the Coastal Environment" (2nd Edition), which is also used in the course as a reference manual.

The course includes the use of a portable flume for demonstration of key concepts and for hands-on participant activities. In addition to the presentation of materials and the flume demonstrations, the course incorporates various workshops and exercises to reinforce key concepts. Topics covered in the course include:

1. Introduction to highways in the coastal environment
2. Waves
3. Tide and water levels
4. Revetment design for coastal embankments
5. Wave loads on bridge decks
6. Coastal geology and sediments
7. Shoreline change and stabilization
8. Road overwash
9. Tidal inlets and coastal bridges

OUTCOMES

Upon completion of the course, participants will be able to:

- Describe coastal engineering design issues related to highways using standard terminology with an understanding of the physical processes unique to this design environment
- Identify appropriate planning, analysis, and design methods for highways and bridges exposed to coastal surge and waves
- Describe differing levels of complexity involving coastal engineering and appropriate qualifications of engineers and coastal engineering consultants to address this complexity in design.

TARGET AUDIENCE

Participants are adult learners with (1) a general civil engineering education and background who currently work in highway planning and design and (2) coastal engineers with some experience in transportation engineering.

TRAINING LEVEL: Intermediate

FEE: 2016: $875 Per Person; 2017: N/A

LENGTH: 3 DAYS (CEU: 1.8 UNITS)

CLASS SIZE: MINIMUM: 14; MAXIMUM: 24

NHI Customer Service: (877) 558-6873 • nhicustomerservice@dot.gov

COURSE NUMBER

FHWA-NHI-135085

COURSE TITLE

Plan of Action (POA) for Scour Critical Bridges - WEB-BASED

This course supports an FHWA-wide priority and is brought to you at no cost by the Office of Bridges and Structures.

This seminar provides guidance on developing a Plan of Action (POA) for scour critical bridges. The seminar highlights the history of the POA requirement and recommends management and inspection strategies for POA development. The seminar introduces the FHWA POA Standard Template and illustrates the use of the POA via a case study of a scour critical bridge in a riverine setting.

OUTCOMES

Upon completion of the course, participants will be able to:

- Describe the purpose of a Plan of Action (POA) for a scour critical bridge
- Identify strategies for developing and implementing a POA
- Describe the sections of the POA Standard Template

TARGET AUDIENCE

Federal, State, and local bridge owners responsible for developing Plan of Actions (POA) for scour critical bridges.

TRAINING LEVEL: Basic

FEE: 2016: $0 Per Person; 2017: N/A

LENGTH: 1 HOURS (CEU: 0 UNITS)

CLASS SIZE: MINIMUM: 1; MAXIMUM: 1

NHI Customer Service: (877) 558-6873 • nhicustomerservice@dot.gov

Course Number

FHWA-NHI-135086

Course Title

Stream Stability Factors and Concepts (Prerequisite) WEB-BASED

This training is a prerequisite of another NHI training and is offered at no cost.

This course is intended to help participants understand river processes and stream stability factors and concepts as a prerequisite for NHI Courses 135046, 135047, 135048. Participants will also be introduced to the concepts of water and sediment continuity.

Outcomes

Upon completion of the course, participants will be able to:

- After completing this course participants will be able to describe the factors influencing stream stability that are important to a bridge scour evaluation, and define water and sediment continuity concepts.

Target Audience

Federal, State, and local highway hydraulic, structural, and geotechnical engineers and bridge inspectors responsible for maintaining the integrity of highway bridges against possible hydraulic-related problems. Consultants who do bridge engineering work are also encouraged to take this prerequisite.

Training Level: Basic

Fee: 2016: $0 Per Person; 2017: N/A

Length: 1 HOURS (CEU: 0 UNITS)

Class Size: MINIMUM: 0; MAXIMUM: 0

NHI Customer Service: (877) 558-6873 • nhicustomerservice@dot.gov

COURSE NUMBER

FHWA-NHI-135087

COURSE TITLE

Scour at Highway Bridges: Concepts and Definitions (Prerequisite) WEB-BASED

This training is a prerequisite of another NHI training and is offered at no cost.

This course has been designed to provide an introduction to scour as a prerequisite for NHI courses 135046, 135047, and 135048.

OUTCOMES

Upon completion of the course, participants will be able to:

- Define scour
- Define total scour and each of it's three components
- Characterize the time dependency of scour
- Distinguish between live-bed and clear-water scour

TARGET AUDIENCE

Federal, State, and local highway hydraulic, structural, and geotechnical engineers and bridge inspectors responsible for maintaining the integrity of highway bridges against possible hydraulic-related problems. Consultants who do bridge engineering work are also encouraged to take this prerequisite.

TRAINING LEVEL: Basic

FEE: 2016: $0 Per Person; 2017: N/A

LENGTH: 1 DAYS (CEU: 0 UNITS)

CLASS SIZE: MINIMUM: 0; MAXIMUM: 0

NHI Customer Service: (877) 558-6873 • nhicustomerservice@dot.gov

COURSE NUMBER

FHWA-NHI-135090

COURSE TITLE

Hydraulic Design of Safe Bridges

The National Highway Institute's (NHI) 3-day 135090 Hydraulic Design of Safe Bridges course provides participants with an intensive training on the hydraulic analysis and design of bridges. The goal of this course is to provide information needed to safely build bridges, while optimizing costs and limiting the impact to property and the environment.

This engaging course includes 12 mandatory lessons that are standard to the course and 3 optional lessons that allow the host agency to customize the course to their particular needs. The optional lessons are: a lesson intended for coastal states with bridges crossing tidal waterways; a lesson that supplements the Unsteady Flow Modeling Concepts lesson and provides additional knowledge of the requirements for one-dimensional unsteady flow modeling; and a lesson that supplements the Scour and Stream Instability Concepts lesson, which enables participants to identify situations requiring sediment transport computations as part of the bridge hydraulics analysis. The host agency will select two optional lessons for the delivery of this course.

Material for this 3-day course is primarily derived from the Hydraulic Design Series No. 7 (HDS 7), Hydraulic Design of Safe Bridges, which is provided to course participants. The course covers significant aspects of bridge hydraulic design including: regulatory topics, specific approaches for bridge hydraulic modeling, hydraulic model selection, bridge design impacts on scour and stream instability, and sediment transport.

Prior to the beginning of the course, participants are strongly encouraged to enroll in the Web-based training (WBT) entitled, 135091 Basic Hydraulic Principles Review. Mastery of the concepts covered in this WBT is important to successful completion of this course.

OUTCOMES

Upon completion of the course, participants will be able to:

- Describe the ways hydraulic design affects bridge performance and public safety
- Describe hydraulic conditions that occur in the vicinity of bridges
- Identify regulatory requirements and design constraints important to bridge projects
- Describe the input requirements for one-dimensional models
- Identify conditions when one-dimensional modeling is adequate to develop accurate hydraulic results for safe bridge design
- Describe the effects of atypical bridge hydraulic conditions on bridge design
- Perform a qualitative risk assessment for a bridge replacement project
- Describe the properties and input requirements for two-dimensional models
- Distinguish conditions requiring two-dimensional modeling to develop accurate hydraulic results for safe bridge design
- Define the types of scour and stream instability that affect bridge design
- Identify how hydraulic variables are obtained from one- and two-dimensional models
- Assess whether a replacement bridge design alternative will have adequate hydraulic capacity to meet design criteria
- Distinguish conditions requiring unsteady flow modeling to develop accurate hydraulic results for safe bridge design
- Describe additional analyses that contribute to the hydraulic aspects of safe bridge design
- Determine the minimum required foundation depth based on scour conditions
- Assess the likelihood of a bridge project causing adverse hydraulic impacts downstream
- Demonstrate strategies for communicating hydraulic recommendations to various stakeholders

TARGET AUDIENCE

The target audience for 135090 Hydraulic Design of Safe Bridges is primarily members of Federal or State departments of transportation. This typically includes hydraulic engineers with a wide range of experience; however, structural and geotechnical engineers would benefit from an understanding of many of the topics in this course. The complexity of some of the engineering decisions made can have significant impacts on structural and geotechnical designs. Additionally, many other segments of the national and international engineering community may find this course valuable. Federal, State, and local highway hydraulic engineers responsible for maintaining the integrity of highway bridges against possible hydraulic related problems will rely on this course and HDS 7 for guidance. Consultants who perform bridge engineering work are also encouraged to attend.

TRAINING LEVEL: Intermediate

FEE: 2016: $950 Per Person; 2017: N/A

LENGTH: 3 DAYS (CEU: 1.8 UNITS)

CLASS SIZE: MINIMUM: 20; MAXIMUM: 30

NHI Customer Service: (877) 558-6873 • nhicustomerservice@dot.gov

Course Number

FHWA-NHI-135091

Course Title

Basic Hydraulic Principles Review (WBT)

Basic Hydraulic Principles Review is designed to familiarize participants with the background concepts, theories, and equations associated with basic hydraulic principles routinely used in highway engineering. NHI strongly suggests that participants complete this self-paced Web-based training (WBT) before attending any Instructor-led hydraulics courses. To fully understand the material presented in NHI hydraulics courses, participants must have an understanding of the basic hydraulic principles presented in this course.

In this course, "hydraulics" is considered to be the determination of various properties and characteristics of flowing water. Such determinations are essential for quantifying the nature of water flow under various conditions. This includes natural features such as streams and rivers, as well as man-made structures such as: bridges, drainage ditches, pipes, culverts, weirs, and spillways.

This WBT consists of three main lessons: Fundamental Concepts, Steady Uniform Flow, and Steady Non-Uniform Flow. After each lesson, knowledge check questions will be presented to test participants' understanding of the material presented in the lesson. The fundamental principles of hydraulics are used as a basis for designing new structures, as well as determining the flow capacity of existing structures.

135091 Basic Hydraulic Review (WBT) is a prerequisite for ILT courses 135010 (River Engineering), 135041 and 041A (HEC-RAS), 135046 (Stream Stability and Scour), and 135056 (Culvert Design). Mastery of the concepts covered in this WBT is important to successful completion of the Instructor-led training.

Outcomes

Upon completion of the course, participants will be able to:

- Define fundamental hydraulic concepts of open-channel flow
- Identify steady uniform flow conditions
- Describe the equations used for steady uniform flow
- Identify steady non-uniform flow conditions
- Describe the equations used for steady non-uniform flow

Target Audience

The primary target audience includes Federal and State Department of Transportation Hydraulic Engineering Units and consultants. The course is relevant to anyone involved in bridge designs over waterways, regardless of their technical discipline or whether they are in the private, municipal, State or Federal sectors. This course is designed primarily for entry-level engineers or engineering technicians who deal with hydraulics. It is also beneficial for experienced personnel as a refresher course on hydraulic fundamentals.

Training Level: Basic

Fee: 2016: $0 Per Person; 2017: N/A

Length: 1 HOURS (CEU: 0 UNITS)

Class Size: MINIMUM: 0; MAXIMUM: 0

NHI Customer Service: (877) 558-6873 • nhicustomerservice@dot.gov

COURSE NUMBER

FHWA-NHI-135092

COURSE TITLE

Highway Hydrology: Basic Concepts and Methods Web-Based

NHI Web-based Training Course #135092 Highway Hydrology: Basic Concepts and Methods provides training on basic hydrologic concepts that will enable users to determine peak flow for transportation hydraulic structures. For engineers, the course teaches basic hydrologic concepts as a review before taking more advanced hydraulic courses. For non-engineers the course enables learners to better understand hydrologic concepts used by engineers.

The Web-based training uses a range of text, graphics, animations, and problem solving in its three lessons. The first lesson focuses on the hydrologic cycle, associated terms, and the relationship of risk to return period and probability of exceedance. The second lesson explains the variability of storms based on three general types of storms, how variations in storm duration and intensity impact runoff, and the watershed characteristics that influence runoff. The third lesson discusses the Rational Method, the NRCS Graphical Method, and Regression Equations as methods to determine peak flow quantities. At the end of the training, learners will be able to apply basic hydrologic concepts to fundamental methods to determine peak flow for highway drainage and hydraulic structures.

OUTCOMES

Upon completion of the course, participants will be able to:

- Identify the hydrologic cycle processes most important to transportation hydraulic engineering.
- Define the relationship between return period and probability of exceedance in hydraulic design.
- Define the temporal and spatial variations observed in precipitation patterns.
- List watershed characteristics that affect peak flows.

TARGET AUDIENCE

Highway Hydrology: Basic Concepts and Methods is a Web-based training course designed for Federal, state, and local hydraulic engineers, highway designers, design consultants, and environmental specialists who have responsibility for the analysis, design, and permitting of roadway drainage features and stream crossings (both culverts and bridges). Designers and reviewers of erosion and sediment control plans may also benefit from the course.

TRAINING LEVEL: Basic

FEE: 2016: $0 Per Person; 2017: N/A

LENGTH: 2 HOURS (CEU: .2 UNITS)

CLASS SIZE: MINIMUM: 1; MAXIMUM: 1

NHI Customer Service: (877) 558-6873 • nhicustomerservice@dot.gov

Course Number

FHWA-NHI-135095

Course Title

Two-Dimensional Hydraulic Modeling of Rivers at Highway Encroachments

THIS AN UPDATE OF COURSE NO. 135071.

The course provides a well-balanced mix of lessons, demonstrations, and exercises for a comprehensive introduction to two-dimensional modeling concepts, including; background data necessary to support a model, hydraulic modeling parameters, mesh development, model simulation parameters, model calibration, hydraulic structures, and reviewing two-dimensional model results. Extracting hydraulic parameters for use in bridge scour evaluation is also discussed. Each concept is demonstrated and participants are given hands-on exercises to apply what they have learned. Once all modeling concepts are covered a comprehensive exercise is provided for participants to apply their skills on a project from start to finish.

Participants will receive a participant workbook that includes hard copies of presentation slides and step-by-step exercises. Electronic data needed for the exercises will also be provided.

Following completion of this course, participants should recognize situations where two-dimensional modeling is preferred and use SMS to successfully compile, execute, and review results for a SRH-2D model on a bridge or other hydraulic structure project.

PREREQUISITE NOTE: Course participants should have knowledge of the fundamentals of open channel flow hydraulics. The free web-based training course, NHI 135091 "Basic Hydraulic Principles Review" is available for those wishing to refresh their knowledge.

HOST NOTE: The host is responsible for providing a minimum of one computer for each pair of students. The computers shall have the following minimum specifications: Microsoft Windows XP with 512 MB of RAM (2 GB recommended) or Windows Vista, Windows 7, or Windows 8 with 1 GB of RAM (4 GB recommended), graphics card (OpenGL 1.5 or higher must be supported). The use of a dedicated graphics card is strongly recommended, display resolution of 1024 x 768 or greater.

Outcomes

Upon completion of the course, participants will be able to:

- Recognize the differences between 1D and 2D hydraulic models
- Use background data in SMS for 2D modeling projects
- Use SMS to setup and run 2D models
- Visualize and review 2D model results
- Add structures to 2D models
- Evaluate 2D hydraulic parameters for use in bridge scour analysis

Target Audience

The target audience for this course is FHWA and state Department of Transportation hydraulicspersonnel and other Federal, state, local or consulting engineers who have responsibility for, or desire to work with, the hydraulic analysis and design of highway river crossings.

Training Level: Intermediate

Fee: 2016: $775 Per Person; 2017: N/A

Length: 3 DAYS (CEU: 2.1 UNITS)

Class Size: MINIMUM: 20; MAXIMUM: 26

NHI Customer Service: (877) 558-6873 • nhicustomerservice@dot.gov

Course Number

FHWA-NHI-134001

Course Title

Principles and Applications of Highway Construction Specifications

Well-written highway construction specifications are those that can be interpreted accurately to minimize confusion and reduce owner-contractor disputes. Across the country, current practices, standards, and requirements for writing specifications are changing. Agencies also are using effective specifications to manage risk and support alternative contracting methods.

NHI 134001 Principles of Writing Highway Construction Specifications is a highly engaging, two-day, instructor-led training session. It includes content that highlights the role of specifications as contract documents and tools for assigning risk. Course participants engage in lessons and practice sessions to identify types of specifications, select the most appropriate type for a given project, and generate an original, effective highway construction specification.

This is not a grammar course; however, adequate course content emphasizes the use of basic grammar and writing style so that the learners can generate specifications that are correct, consistent, clear, complete, and concise.

Outcomes

Upon completion of the course, participants will be able to:

- Explain the purposes of a specification.
- Explain how specifications are used to assign risk and influence the behavior of different parties, within a given a scenario.
- Compare the functions of Standard and Supplemental Specifications with the functions of Special Provisions.
- Explain how the "order of precedence" affects writing specifications and preparing plans.
- Describe the purpose of the General Provisions.
- Explain how a consistent writing style can affect the interpretation of specifications.
- Complete a checklist of the information needed before writing or revising a specification.
- Explain the potential benefits of writing in the active voice.
- Rewrite passive voice sentences into the active voice.
- Evaluate specifications to determine the need for imperative or indicative mood.
- State the five Cs used in specification writing. (Note: the five Cs include: correct; consistent; clear; complete; concise.)
- Explain each element of the AASHTO five-part format.
- Identify potential ambiguities in the wording, given a sample specification.
- Identify the potential benefits of each of the five Cs, given a sample specification.
- Apply the five Cs and the host agency's preferred format to revise the specification, given a sample specification.
- Write a new specification to a given set of criteria using the five Cs and the host agency's preferred format, given a sample specification.
- Compare method versus end-result specifications.
- Relate the type of specification to the allocation of risk.
- Write an end-result specification to replace a method specification, given an excerpt from a method specification.

Target Audience

This course is designed primarily for individuals who write, review, and implement an agency's contract specifications. Participants might represent Federal, State, and local transportation agencies; other public agencies; contractors; and consultant firms.Individuals who do not write specifications but may contribute to their development, as well as those who use specifications, could also benefit from this course and the interaction with their classmates. Such participants might include personnel from environmental, materials, or construction sections or units; legal departments; work zone and safety professionals; contractor personnel; and any others involved with the design and construction of transportation facilities.

TRAINING LEVEL: Intermediate

FEE: 2016: $475 Per Person; 2017: $475 Per Person

LENGTH: 2 DAYS (CEU: 1.2 UNITS)

CLASS SIZE: MINIMUM: 20; MAXIMUM: 30

NHI Customer Service: (877) 558-6873 • nhicustomerservice@dot.gov

Course Number

FHWA-NHI-134064

Course Title

Transportation Construction Quality Assurance (1.5-Day)

The Federal Highway Administration (FHWA) identified the need for transportation construction and materials personnel to increase their knowledge of the fundamentals of effective transportation construction Quality Assurance (QA). This course was developed to ensure that agency, contractor, producer, and consultant personnel responsible for interpreting and applying quality assurance specifications in transportation construction are properly qualified. The course will utilize a Quality Assurance Reference Manual, adapted from the current NETTCP manual.

This one and a half-day version of the course covers Chapters 1 through 6 of the course materials and will be available to, and appropriate for, all audiences including management level personnel. The content covered in this first day includes how quality assurance is featured in a transportation construction quality assurance program, quality assurance program elements, the evolution of quality assurance specifications, measuring quality, and the roles and responsibilities of both contractor and agency personnel.

Outcomes

Upon completion of the course, participants will be able to:

- Consistently apply fundamental Quality Assurance concepts, terminology, and definitions
- Differentiate QA specifications from other specifications
- Explain each of the six core elements of a QA program and how each is essential to successful implementation of Quality Assurance
- Describe the respective roles and responsibilities of the project decision makers (Contractor QC and Agency Acceptance personnel) and how their interaction contributes to construction quality

Target Audience

This is an intermediate-level course for personnel who are implementing QA specifications on construction projects. Necessary background knowledge for participants is 3-5 years minimum in transportation construction specifications inspections. The suggested list of personnel that may consider attending, if they have the requisite background knowledge are Contractor/Consultant Personnel (QC managers/QC Plan Administrators, Senior Production Facility QC Technician/Inspectors, Senior QC Laboratory Personnel, and Senior Field QC Technicians/Inspectors) and Agency Personnel (Project Managers/Resident Engineers, Senior Production Facility Acceptance Technicians/Inspectors, Senior Acceptance Laboratory Personnel, and Senior Field Acceptance Technicians/Inspectors.

Training Level: Intermediate

Fee: 2016: $250 Per Person; 2017: $250 Per Person

Length: 1.5 DAYS (CEU: 1.1 UNITS)

Class Size: MINIMUM: 20; MAXIMUM: 30

NHI Customer Service: (877) 558-6873 • nhicustomerservice@dot.gov

Course Number

FHWA-NHI-134064A

Course Title

Transportation Construction Quality Assurance (3-Day)

The Federal Highway Administration (FHWA) identified the need for transportation construction and materials personnel to increase their knowledge of the fundamentals of effective transportation construction Quality Assurance (QA). This course was developed to ensure that agency, contractor, producer, and consultant personnel responsible for interpreting and applying quality assurance specifications in transportation construction are properly qualified. The course will utilize a Quality Assurance Reference Manual, adapted from the current NETTCP manual.

This three-day version of the course covers Chapters 1 through 10 of the course materials and will be available to, and appropriate for, production, laboratory, and field QC and Acceptance technicians and inspectors. This version contains mathematical terms and principles used in QA sampling, testing, and decision-making. The content also includes how quality assurance is featured in a transportation construction quality assurance program, quality assurance program elements, the evolution of quality assurance specifications, measuring quality, and the roles and responsibilities of both contractor and agency personnel.

Outcomes

Upon completion of the course, participants will be able to:

- Consistently apply fundamental Quality Assurance concepts, terminology, and definitions
- Differentiate QA specifications from other specifications
- Explain each of the six core elements of a QA program and how each is essential to successful implementation of Quality Assurance
- Describe the respective roles and responsibilities of the project decision makers (Contractor QC and Agency Acceptance personnel) and how their interaction contributes to construction quality
- Apply the mathematical concepts of variability, statistical distribution, and sampling protocols to measure construction quality
- Describe the primary components of inspection, properly document the results of inspection, and utilize inspection data to quantify quality of workmanship

Target Audience

This is an intermediate-level course for personnel who are implementing QA specifications on construction projects. Necessary background knowledge for participants: 3-5 years minimum in transportation construction specifications inspections, basic statistical knowledge/training, some usage of tools necessary to the Quality Assurance process (contractor test results). The suggested list of personnel that may consider attending, if they have the requisite background knowledge are Contractor/Consultant Personnel (QC managers/QC Plan Administrators, Senior Production Facility QC Technician/Inspectors, Senior QC Laboratory Personnel, and Senior Field QC Technicians/Inspectors) and Agency Personnel (Project Managers/Resident Engineers, Senior Production Facility Acceptance Technicians/Inspectors, Senior Acceptance Laboratory Personnel, andSenior Field Acceptance Technicians/Inspectors).

Training Level: Intermediate

Fee: 2016: $350 Per Person; 2017: $350 Per Person

Length: 3 DAYS (CEU: 1.8 UNITS)

Class Size: MINIMUM: 20; MAXIMUM: 30

NHI Customer Service: (877) 558-6873 • nhicustomerservice@dot.gov

Course Number

FHWA-NHI-134070

Course Title

SpecRisk Quality Assurance Specification Development and Validation Course - WEB-BASED

This course will provide an introduction to statistical analysis and the development of statistically valid quality assurance specifications, introducing general guidelines established and put forth by the Federal Government and FHWA policy. The course also provides participants with an introduction to SpecRisk, the resource that is necessary to successfully develop statistically valid specifications. The course is designed and delivered to motivate members of the target audience to use SpecRisk software to develop their specifications. Although the course demonstrates basic functions of the software, it is not intended to be an in-depth training on how to use SpecRisk.

This course requires a solid foundation in basic statistics. Completion of FHWA-NHI 134042, or equivalent training, is also recommended. NHI 134042 trains participants to identify the importance of organizing data and how to plot frequency histograms. It explains how a sample relates to the population, the relationship between single and multiple samples, and the use of random stratified sampling tables. This knowledge provides an excellent foundation for this course.

Outcomes

Upon completion of the course, participants will be able to:

- Recognize key concepts to develop an effective, statistically valid Quality Assurance (QA) specification.
- Make an informed selection among available options when developing an acceptance plan.
- Develop QA specifications in alignment with best practices, Federal regulations, and FHWA policy.
- Apply SpecRisk software to understand risks and develop statistically valid specifications.

Target Audience

Personnel involved in specification development: Federal, State, and local highway agency engineers and technicians in materials, construction, and research. The training is also appropriate for industry personnel that are involved in reviewing and providing input to the specification development process.

Training Level: Basic

Fee: 2016: $50 Per Person; 2017: $50 Per Person

Length: 8 HOURS (CEU: 0 UNITS)

Class Size: MINIMUM: 1; MAXIMUM: 1

NHI Customer Service: (877) 558-6873 • nhicustomerservice@dot.gov

Course Number

FHWA-NHI-134112

Course Title

Principles and Practices for Enhanced Maintenance Management Systems - WEB-CONFERENCE

Is your agency in the process of enhancing its maintenance management capabilities?

Are you interested in learning more about developing effective performance measures for maintenance activities?

If so, join us for a blended training course that features both independent study material and facilitated Web-conferences. You will be introduced to the methods and practices used in an enhanced maintenance management system (MMS) to effectively maintain and operate a highway network. You will explore the principles and practices of using MMS to effectively examine efficient maintenance and operation of a highway network. Throughout the course, you will learn by participating in activities and assignments specific to using MMS.

The course materials rely heavily on the AASHTO Guidelines for Maintenance Management Systems, Transportation Asset Management Guide, and several other recent publications on the topic. To illustrate the application of the principles, the course materials are supplemented with examples from State and local highway agencies.

Participant Responsibilities:

- 7 Web-based lessons (Duration: 1- 1.5 hrs each)

- 3 Web-conferences (Duration: 2 hours each)

To obtain your certificate, you must complete all Web-based lessons and Web-conferences. To receive Continuing Education Units (CEUs), you must also pass the online test at the end of the course. You will need your own computer with an Internet connection as well as a telephone line in order to participate.

Outcomes

Upon completion of the course, participants will be able to:

- Compare and contrast a first generation MMS with an enhanced MMS
- Describe the terms "outcome-based" and "performance-based" and how they pertain to an enhanced MMS
- Describe the use of service levels to support the programming and budgeting activities incorporated into an MMS
- Identify the types of systems that should be integrated with an MMS and provide several examples of the types of data that should interface between each system
- List the potential benefits to be realized by fully integrating an enhanced MMS
- Identify several steps that will advance an agency's current maintenance management practices now and in the future

Target Audience

The target audience for this course includes State and local maintenance engineers, maintenance supervisors, asset managers, and their industry counterparts. The course is specifically for individuals who are responsible for directing and managing maintenance operations and budgets, maintenance project and treatment selection, and/or the monitoring of system conditions.

Training Level: Basic

Fee: 2016: $300 Per Person; 2017: N/A

Length: 15 HOURS (CEU: 1.5 UNITS)

Class Size: MINIMUM: 20; MAXIMUM: 30

NHI Customer Service: (877) 558-6873 • nhicustomerservice@dot.gov

COURSE NUMBER

FHWA-NHI-136065

COURSE TITLE

Risk Management

During planning for a transportation improvement, thorough consideration of potential risks can help reduce or eliminate them, ensuring a smoother project life cycle. In the same way, accounting for risk in program management and strategic planning can help transportation agencies minimize or avoid unexpected costs or delays. Recent legislation reflects the growing importance of understanding risk in government programs. Congress included requirements for risk management in the latest highway authorization, the Moving Ahead for Progress in the 21st Century Act (MAP-21).

This course provides an understanding of risk management concepts and processes, including terminology, benefits of use, risk management planning, and a framework for implementation. The course presents a cyclical risk management framework through classroom training and an independent study assignment, focusing on gathering information; identifying risk; analyzing and prioritizing risk events; documenting risk; identifying risk response strategies; incorporating response strategies into a plan; and monitoring, evaluating, and adjusting strategies. The 1-hour independent study assignment is completed by participants before attending the 2-day classroom session.

Participants use tools and methods from each step of the framework in a series of exercises that provide realistic, job-relevant practice in applying the risk management process. In order to maximize the impact of the training and increase the likelihood of participants' mastery of the risk management process, the agency can select active agency issues (project, program, or network) to be used during the exercises. In addition, the agency can provide problem statements and pre-select the teams for the exercises. Contact the NHI Training Program Manager for further information on making such arrangements.

OUTCOMES

Upon completion of the course, participants will be able to:

- Explain the overall organizational context, framework, and importance of risk management to others.
- Follow a consistent process for managing risk.
- Utilize standard risk terminology, tools, and methods.
- Apply the risk management process to a project or program within established time-constraints.
- Develop effective plans for internal and external communication and consultation about risk and risk-related issues.

TARGET AUDIENCE

The target audience for this course includes Federal, State and local highway employees who are responsible for directing and managing all aspects of highway related programs and projects such as planning, environment, project development, design, construction, operations, maintenance, and finance. Audience experience, background, knowledge, skills and abilities are expected to vary widely. No previous experience with risk management is required.

TRAINING LEVEL: Basic

FEE: 2016: $375 Per Person; 2017: N/A

LENGTH: 2 DAYS (CEU: 1.2 UNITS)

CLASS SIZE: MINIMUM: 20; MAXIMUM: 30

NHI Customer Service: (877) 558-6873 • nhicustomerservice@dot.gov

Course Number

FHWA-NHI-136106

Course Title

Introduction to Transportation Asset Management

Asset management principles are becoming increasingly important to help agencies manage their assets as they face fewer available resources, higher expectations for customer service, and increased demand for more transparency in the decision process. In an asset management environment, investment decisions are linked to targeted performance levels that have been established based on current and expected asset conditions. Trade-offs between investments in different types of assets and different investment priorities can be assessed because of the availability of reliable data and a clear set of performance metrics that the agency hopes to achieve. As a result, agencies are better able to use their funding effectively and to defend their need for additional resources.

The Transportation Asset Management course introduces a strategic approach to managing physical transportation infrastructure. This 1-day course covers the principles of asset management and introduces the five core questions every agency should be able to answer about its assets. The course contains modules on the following topics: asset management principles, performance measures, long-term financial planning, and risk assessment. Through a series of workshops, the course material introduces the participants to the application of asset management principles in real life situations.

This course can be delivered with the addition of a half-day workshop. During the workshop, participants review self-assessments to identify agency gaps between the desired and actual application of asset management principles. The workshop also includes recommendations for advancing the implementation of asset management practices within the agency. Refer to 136106A "Introduction to Transportation Asset Management with Workshop" on the NHI Web site for additional information.

This course is the first in a series and previously cataloged under 131106. The other courses in this series are 136106B "Development of a Transportation Asset Management Plan" and 136106C "Introduction to Transportation Asset Management Plans", which is Web-based. See the NHI Web site for additional information on each of these courses.

Outcomes

Upon completion of the course, participants will be able to:

- Champion the use of asset management principles and concepts within the organization.
- Define their role in supporting the agency's asset management efforts.
- Describe the maturity of the agency's asset management program.

Target Audience

This training is intended for senior-level and mid-level managers from State departments of transportation and other transportation agencies, who typically have the responsibility for decision-making in one or more areas addressed by transportation asset management. Participants should represent a number of organizational units, including (but not limited to) planning, engineering (e.g., facility management, design, construction), capital programming, maintenance and operations, financial management, traffic and safety engineering, system operation and management, and information technology. The course is also intended for individuals who manage or provide critical information to senior managers, or who have direct responsibility for meeting specific transportation system performance or program delivery targets.

Training Level: Basic

Fee: 2016: $450 Per Person; 2017: N/A

Length: 1 DAYS (CEU: 0 UNITS)

Class Size: MINIMUM: 20; MAXIMUM: 30

NHI Customer Service: (877) 558-6873 • nhicustomerservice@dot.gov

Course Number

FHWA-NHI-136106A

Course Title

Introduction to Transportation Asset Management with Workshop

Asset management principles are becoming increasingly important to help agencies manage their assets as they face fewer available resources, higher expectations for customer service, and increased demand for more transparency in the decision process. In an asset management environment, investment decisions are linked to targeted performance levels that have been established based on current and expected asset conditions. Trade-offs between investments in different types of assets and different investment priorities can be assessed because of the availability of reliable data and a clear set of performance metrics that the agency hopes to achieve. As a result, agencies are better able to use their funding effectively and to defend their need for additional resources.

The Transportation Asset Management course introduces a strategic approach to managing physical transportation infrastructure. This 1.5-day course covers the principles of asset management and introduces the five core questions every agency should be able to answer about its assets. The course contains modules on the following topics: asset management principles, performance measures, long-term financial planning, risk assessment, and implementation. Through a series of workshops, the course material introduces the participants to the application of asset management principles in real life situations. It also provides a review of a self-assessment that can be used to identify agency gaps between the desired and actual use of these principles. The course concludes with recommendations for advancing the implementation of asset management practices within the agency.

This course can be delivered without the half-day workshop with the approval of the FHWA Technical Lead. See 136106 "Introduction to Transportation Asset Management" on the NHI Web site for further information.

This course is the first in a series and previously cataloged under 131106A. It serves as a prerequisite for 136106B "Development of a Transportation Asset Management Plan." The other course in the series is 136106C "Introduction to Transportation Asset Management Plans," which is Web-based. See the NHI Web site for additional information on each of these courses.

Outcomes

Upon completion of the course, participants will be able to:

- Champion the use of asset management principles and concepts within the organization.
- Define their role in supporting the agency's asset management efforts.
- Describe the maturity of the agency's asset management program.
- Draft a plan for using existing resources to advance the agency's use of asset management principles.

Target Audience

This training is designed for senior-level and mid-level managers from State departments of transportation and other transportation agencies, who typically have the responsibility for decision-making in one or more areas addressed by transportation asset management. Participants should represent a number of organizational units, including (but not limited to) planning, engineering (e.g., facility management, design, construction), capital programming, maintenance and operations, financial management, traffic and safety engineering, system operation and management, and information technology. The course is also intended for individuals who manage or provide critical information to senior managers, or who have direct responsibility for meeting specific transportation system performance or program delivery targets.

Training Level: Basic

Fee: 2016: $575 Per Person; 2017: N/A

Length: 1.5 DAYS (CEU: 0 UNITS)

Class Size: MINIMUM: 20; MAXIMUM: 30

NHI Customer Service: (877) 558-6873 • nhicustomerservice@dot.gov

Course Number

FHWA-NHI-136106B

Course Title

Developing a Transportation Asset Management Plan

"Developing a Transportation Asset Management Plan" is blend of online and classroom training that introduces the role of the Transportation Asset Management Plan (TAMP) as a planning, communication, and accountability tool. This course also provides the information necessary for an agency to develop or enhance a TAMP. This course is the second in the series of courses on transportation asset management and previously cataloged under 131106B.

It is required that participants take the Web-based training 136106C and attend 136106 or 136106A prior to attending this course, or have a solid background in transportation asset management principles and planning.

The course focuses on three primary components to the TAMP, including strategic performance measurement, risk assessment, and financial management. It also provides templates and guidelines for use in developing a TAMP.

The course is 1.5-days in length. Prior to attending the first day of class, participants must complete the online, 1-hour training module NHI-136106C "Introduction to a Transportation Asset Management Plan." Information on how participants access this training is provided to the host by the instructor once the session is confirmed.

Outcomes

Upon completion of the course, participants will be able to:

- Describe the role of a Transportation Asset Management Plan in a transportation agency.
- Identify strategies for incorporating risk into investment decisions.
- Explain how to determine whether an agency is making sustainable, long-term investments in its assets.
- Develop a Transportation Asset Management Plan that matches the amount of data and the sophistication of the analysis tools available.

Target Audience

The course is intended for senior-level and mid-level managers from State departments of transportation and other transportation agencies, who have the responsibility for decision-making in one or more areas addressed by transportation asset management. Course participants should represent a broad range of organizational units, such as (but not limited to) planning, engineering (facility management, design, and construction), capital programming, maintenance and operations, financial management, traffic and safety engineering, system operation and management, and information technology. If the agency has an Asset Management Steering Committee, its members would benefit from this course. In addition, individuals who manage individual assets or provide critical information to senior managers, or who have direct responsibility for meeting specific transportation system performance or program delivery targets, are also excellent candidates for attending the course.

Training Level: Intermediate

Fee: 2016: $575 Per Person; 2017: N/A

Length: 1.5 DAYS (CEU: 0 UNITS)

Class Size: MINIMUM: 20; MAXIMUM: 30

NHI Customer Service: (877) 558-6873 • nhicustomerservice@dot.gov

ASSET MANAGEMENT

Course Number

FHWA-NHI-136106C

Course Title

Introduction to a Transportation Asset Management Plan (WEB-BASED)

This training is a prerequisite of another NHI training and is offered at no cost.

A Transportation Asset Management Plan (TAMP) can be considered a business plan that builds the case for making infrastructure investments and explains how resources will be used. This course, "Introduction to a Transportation Asset Management Plan," is a 1-hour, Web-based training (WBT) that introduces the content and organization of a TAMP and the typical TAMP development process. This course was previously cataloged under 131106C

This training includes the following topics.

- The use of a TAMP in transportation agencies
- The typical content of a TAMP (including a comparison with requirements in MAP-21)
- Key components, including performance projections and the financial summary
- Examples of TAMPs at various levels of maturity
- Existing and anticipated use of a TAMP in state highway agencies
- The expected involvement of agency personnel in developing and updating a TAMP

This training includes audio clips from leaders in state highway agencies that convey the anticipated benefits from the development of a TAMP and the way they expect to use their TAMP. In addition, the WBT highlights the use of existing documentation to develop the TAMP and plans for enhancing the content of future TAMPs.

This training serves as a prerequisite for NHI-136106B "Developing a Transportation Asset Management Plan", which describes the role of a TAMP in a transportation agency and explores in some detail three important components: strategic performance management, risk assessment and management, and financial management.

Outcomes

Upon completion of the course, participants will be able to:

- Describe the role of a TAMP as a communication tool with internal and external stakeholders.
- List the typical content of a TAMP.
- Identify several sources of information that will contribute to the development of a TAMP.

Target Audience

The course is intended for senior-level and mid-level managers from State departments of transportation and other transportation agencies, who typically have the responsibility for decision-making in one or more areas addressed by transportation asset management. Participants should represent a number of organizational units, including (but not limited to) planning, engineering (e.g., facility management, design, construction), capital programming, maintenance and operations, financial management, traffic and safety engineering, system operation and management, and information technology. The course is also intended for individuals who manage or provide critical information to senior managers, or who have direct responsibility for meeting specific transportation system performance or program delivery targets.

Training Level: Intermediate

Fee: 2016: $0 Per Person; 2017: N/A

Length: 1 HOURS (CEU: 0 UNITS)

Class Size: MINIMUM: 0; MAXIMUM: 0

NHI Customer Service: (877) 558-6873 • nhicustomerservice@dot.gov

COURSE NUMBER

FHWA-NHI-136113

COURSE TITLE

Transportation Asset Management Overview - WEB-BASED

This training was developed by the Transportation Curriculum Coordination Council (TCCC) in partnership with AASHTO and NHI. This training explains the basics of asset management and why asset management is important. After you complete this training, you'll have new terms, and new ways of thinking about what you're already doing. More importantly, you'll understand why it's so important to be strategic and systematic when you're responsible for managing huge numbers of assets.

This training contains the following lessons:

Lesson 1: What is Transportation Asset Management? This lesson will explain the concept of asset management; give examples of how asset management is used in the planning process; and explain how current asset management practices have been impacted by past transportation needs.

Lesson 2: Asset Management Principles and Practices. This lesson lists the categories of activity that inform spending decisions; explain how policy goals and objectives impact asset management; relate planning and programming to managing assets; describe how asset management principles apply to program delivery; explain why system monitoring is necessary; and explain how quality data and analysis impact asset management.

OUTCOMES

Upon completion of the course, participants will be able to:

- Explain what transportation asset management is and why it is important
- Describe the asset management principles and practices used to make informed spending decisions

TARGET AUDIENCE

This training was developed by the Transportation Curriculum Coordination Council (TCCC) in partnership with AASHTO, NHI, and is recommended for TCCC levels II through IV.

TRAINING LEVEL: Basic

FEE: 2016: $25 Per Person; 2017: N/A

LENGTH: 2 HOURS (CEU: 0 UNITS)

CLASS SIZE: MINIMUM: 1; MAXIMUM: 1

NHI Customer Service: (877) 558-6873 • nhicustomerservice@dot.gov

Course Number

FHWA-NHI-137046

WBT

Course Title

ITS Deployment Analysis System (IDAS) - WEB-BASED

This course is a Web-based training session on the newly developed ITS Deployment Analysis System (IDAS) software. IDAS provides ITS sketch planning capability to calculate the relative costs and benefits of ITS investments. IDAS incorporates a cost module, a benefit module and an internal travel demand model to generate cost/benefit comparisons for alternative ITS deployment scenarios. IDAS uses the output from an existing transportation planning model to establish a best-case scenario on which the user can deploy ITS services on specific links in the regional transportation network model.

The participant's computer should have the following recommended requirements: 500 MHz Intel Pentium II Processor or equivalent with 128 MB of RAM, Windows 2000, Windows NT, or Windows XP, color monitors, 2 GB of available disk space.

Outcomes

Upon completion of the course, participants will be able to:

- Explain the importance of integrating operations/ITS into the planning and decision-making processes
- Explain that IDAS software can be used to link operations to the planning process
- Demonstrate how IDAS uses the network and output from an existing regional travel demand model
- Employ IDAS to screen ITS alternatives and produce a cost/benefit analysis.
- Interpret IDAS results
- Review and refine IDAS defaults

Target Audience

FHWA, State DOT, metropolitan planning organization, and local government transportation planning staff members who are involved in the day-to-day elements of transportation planning and modeling would benefit for this course. Operations Engineers, ITS Project Managers, and Transit Agency Personnel (this includes individuals who: 1. develop inputs for, set up, and carry out analyses of operations/ITS alternatives and/or 2. examine results, conduct sensitivity analyses, and explore tradeoffs of such analyses created by others) would also benefit for this course.

Training Level: Basic

Fee: 2016: $50 Per Person; 2017: N/A

Length: 5 HOURS (CEU: .5 UNITS)

Class Size: MINIMUM: 1; MAXIMUM: 1

NHI Customer Service: (877) 558-6873 • nhicustomerservice@dot.gov

COURSE NUMBER

FHWA-NHI-137048

COURSE TITLE

Turbo Architecture-Web-Based

This course is based upon Turbo Architecture Version 5.0. The current version is 7.0 (as Turbo Architecture jumped directly from Version 5.0 to Version 7.0 to align the Turbo Architecture version with the corresponding version of the National ITS Architecture.) There are some minor differences in the versions but the information presented in this course is still applicable to Version 7.0. For more information on the differences in version 7.0, see the Turbo Architecture page on The National ITS Architecture 7.0 website, http://www.iteris.com/itsarch/html/turbo/turbomain.htm.

This training is provided to you at no cost by the ITS Joint Program Office.

Turbo Architecture is an interactive software program that assists transportation planners and system integrators in the development of regional and project architectures. This Web-based training (WBT) provides ITS professionals with a hands-on experience using the Turbo software. Participants will work with simulated examples and practice exercises to create, maintain, and use regional and project ITS architectures.

At the end of the training, participants will be able to use the Turbo software to create and modify a regional or project architecture including providing a link to planning, entering stakeholders, entering inventory data, selecting ITS services, creating operational concepts, tailoring functional requirements, building and customizing interfaces, customizing standards mappings, entering agreements, creating outputs, and applying features to new projects.

OUTCOMES

Upon completion of the course, participants will be able to:

- Recall training objective and delivery elements
- Verify the correct installation of Turbo
- Explain the use and importance of Turbo
- Explain Turbo's support of the ITS project life cycle

TARGET AUDIENCE

The Turbo Architecture WBT is designed for ITS professionals employed by MPOs, transit agencies, municipalities, State DOTs, FHWA Division Offices, or consultants and system integrators who use and/or maintain an ITS architecture and are involved with ITS planning, deployment, and operations.

TRAINING LEVEL: Basic

FEE: 2016: $0 Per Person; 2017: N/A

LENGTH: 6 HOURS (CEU: 0 UNITS)

CLASS SIZE: MINIMUM: 1; MAXIMUM: 1

NHI Customer Service: (877) 558-6873 • nhicustomerservice@dot.gov

Course Number

FHWA-NHI-138001

Course Title

Transportation Performance Management Awareness - Federal Aid Version - WEB-BASED

This course provides an introduction to performance management and covers the definition of performance management and basic performance management concepts, explains the critical role that the planning process plays in implementing a performance management program, and addresses what performance management means to the Federal Highway Administration.

Outcomes

Upon completion of the course, participants will be able to:

- Describe performance management
- Describe the basic elements of a performance management program
- Explain the critical role planning plays in implementing a performance management program

Target Audience

The target audience for this training will be all FHWA employees.

Training Level: Basic

Fee: 2016: $25 Per Person; 2017: N/A

Length: 1 HOURS (CEU: 0 UNITS)

Class Size: MINIMUM: 1; MAXIMUM: 1

NHI Customer Service: (877) 558-6873 • nhicustomerservice@dot.gov

Course Number

FHWA-NHI-138003

Course Title

Introduction to Performance Measurement - WEB-BASED

This training is one in a series of introductory modules on the topic of Transportation Performance Management. This training will cover developing performance measures.

The training will give an overview of performance measures. It will go into FHWA's role in developing performance measures. It will go over the criteria for developing effective performance measures. Next, it will discuss the importance of data in developing performance measures, tools available to collect that data and considerations for analysis of the data. Lastly, we will look at how performance measurement information can be used.

Outcomes

Upon completion of the course, participants will be able to:

- Explain why performance measures are important
- Explain the FHWA's role in developing performance measures
- Describe the criteria an effective performance measure must meet
- Recognize the importance of data in developing performance measures
- Explain how performance measurement information is used

Target Audience

The target audience for this training will be all FHWA employees.

Training Level: Basic

Fee: 2016: $25 Per Person; 2017: N/A

Length: 2 HOURS (CEU: 0 UNITS)

Class Size: MINIMUM: 1; MAXIMUM: 1

NHI Customer Service: (877) 558-6873 • nhicustomerservice@dot.gov

Course Number

FHWA-NHI-138004

Course Title

Overview of MAP-21 (101) Transportation Performance Management

The MAP-21/FAST ACT Transportation Performance Management Overview is a one-day course that presents an overview of the Transportation Performance Management (TPM) provisions of MAP-21 and FAST Act, describes the responsibilities that agencies at several levels - Federal, State, Metropolitan Planning Organization (MPO), Regional Transportation Planning Organizations (RTPOs) and Transit - have in delivering these requirements, highlights the importance of data in meeting the performance management provisions and captures noteworthy practices in use today that may help agencies get started on meeting these requirements. This course is highly recommended for participants interested in taking the more detailed system specific MAP-21 implementation training.

The aim of this course is to familiarize transportation professionals with the TPM provisions of MAP-21 and FAST and how it can impact existing products and processes. The course introduces key points of overlap between the transit and highway performance measurement provisions in MAP-21, but the primary focus is on the highway provisions. Course material is organized around the TPM elements of goals, measures, targets, plans, accountability and transparency and introduces the related MAP-21/FACT Act provisions for each element. The course provides an introduction on how TPM provisions impact individual performance areas (i.e., safety, pavement/bridge, operations) and how these provisions come together under planning and programming. The course allows participants to interact and to learn from their peers through open discussions and course exercises.

Outcomes

Upon completion of the course, participants will be able to:

- Identify the transportation performance management provisions of MAP-21, how they are tied together, and the associated products and delivery timelines.
- List the roles and responsibilities different agencies (Federal, state, and MPO) have in meeting the TPM requirements.
- Describe noteworthy practices in use at other agencies that may be helpful to begin the process of implementing MAP-21 transportation performance management requirements.
- Explain the importance of having good quality data in meeting the MAP-21 TPM provisions.

Target Audience

Overview of MAP-21 (101) Transportation Performance Management is a one-day course for primarily intended for state DOT, MPO and FHWA staff who have a role in meeting the MAP21 TPM requirements. Executives and senior decision makers are a secondary audience.

Training Level: Basic

Fee: 2016: $425 Per Person; 2017: N/A

Length: 1 DAYS (CEU: .6 UNITS)

Class Size: MINIMUM: 20; MAXIMUM: 30

NHI Customer Service: (877) 558-6873 • nhicustomerservice@dot.gov

Course Number

FHWA-NHI-139004

Course Title

Principles of Effective Commercial Motor Vehicle (CMV) Size and Weight Enforcement

Principles of Effective Commercial Motor Vehicle Size and Weight Enforcement is a two-day course intended to provide advanced, in-depth, understanding of federal motor vehicle size and weight regulations and the importance of state level vehicle size and weight enforcement programs. This course targets transportation professionals responsible for overseeing the preservation of federal and state highway assets through annual VSW enforcement planning and federal certification, as well as personnel directly involved in commercial VSW enforcement. The course provides techniques and strategies designed for those individuals working to implement VSW enforcement programs.

In summary, this course will provide participants with the knowledge and tools to implement effective programs for enforcing federal VSW regulations.

Outcomes

Upon completion of the course, participants will be able to:

- Demonstrate the importance of Commercial VSW Enforcement
- Describe Federal VSW Regulations
- Discuss current issues, trends and technologies related to VSW Enforcement
- Use the SEP and Annual Certification to measure progress for implementing an effective VSW Enforcement Program

Target Audience

The target audience for the course are: FHWA Division Office Staff; State DOTs/Public Works Policymakers; Public employees from State and Local transportation agencies with Commercial Vehicle Operations (CVO) size and weight responsibilities; Personnel from state and local law enforcement agencies; Federal Motor Carrier Safety Administration (FMCSA) field office personnel; Trucking company managers; Trucking association officials; Law enforcement associations; and Training staff from state transportation agencies.

Training Level: Basic

Fee: 2016: $525 Per Person; 2017: N/A

Length: 2 DAYS (CEU: 1.2 UNITS)

Class Size: MINIMUM: 20; MAXIMUM: 30

NHI Customer Service: (877) 558-6873 • nhicustomerservice@dot.gov

FREIGHT AND TRANSPORTATION LOGISTICS

Course Number

FHWA-NHI-139006

Course Title

Integrating Freight in the Transportation Planning Process - WBT-Standard Version

This training is a prerequisite of another NHI training and is offered at no cost.

Freight transportation issues can be complex and involve many different stakeholders, all of whom have different perspectives on the freight transportation system. The challenge faced by many public-sector transportation planners is how to best incorporate these freight perspectives into the transportation planning process in a way that results in a safe and efficient transportation system for both people and goods. This Web-based training course will provide a greater understanding of freight trends, its stakeholders, and its issues, so that public-sector transportation planners are better able to incorporate freight into their respective transportation planning processes and programs.

This is a prerequisite course for other Freight courses.

In accordance with the Rehabilitation Act of 1973, as amended, this WBT is also available in an accessible 508 compliant version. See course number FHWA-NHI-139006W for more information.

Outcomes

Upon completion of the course, participants will be able to:

- Upon completion of the course, participants will be able to:
- Identify the stakeholders involved in freight transportation
- Explain the role of different modes in freight transportation
- Describe some trends affecting freight transportation, and their impact on a State's transportation system and communities
- Discuss some of the common issues that prevent freight from being fully incorporated into the planning process
- Identify key resources to help guide statewide and metropolitan freight planning effort

Target Audience

Transportation planners and freight transportation planners from State DOTs, MPOs, local governments, and Federal agencies.

Training Level: Basic

Fee: 2016: $0 Per Person; 2017: N/A

Length: 6 HOURS (CEU: .6 UNITS)

Class Size: MINIMUM: 1; MAXIMUM: 1

NHI Customer Service: (877) 558-6873 • nhicustomerservice@dot.gov

COURSE NUMBER

FHWA-NHI-139006W

COURSE TITLE

Integrating Freight in the Transportation Planning Process - WBT-Accessible 508 Version

This training is a prerequisite of another NHI training and is offered at no cost.

Freight transportation issues can be complex and involve many different stakeholders, all of whom have different perspectives on the freight transportation system. The challenge faced by many public-sector transportation planners is how to best incorporate these freight perspectives into the transportation planning process in a way that results in a safe and efficient transportation system for both people and goods. This Web-based training course will provide a greater understanding of freight trends, its stakeholders, and its issues, so that public-sector transportation planners are better able to incorporate freight into their respective transportation planning processes and programs.

This is a prerequisite course for other Freight courses.

In accordance with the Rehabilitation Act of 1973, as amended, this WBT was developed as an accessible 508 compliant version. See course number FHWA-NHI-139006 for the standard WBT version.

OUTCOMES

Upon completion of the course, participants will be able to:

- Upon completion of the course, participants will be able to:
- Identify the stakeholders involved in freight transportation
- Explain the role of different modes in freight transportation
- Describe some trends affecting freight transportation, and their impact on a State's transportation system and communities
- Discuss some of the common issues that prevent freight from being fully incorporated into the planning process
- Identify key resources to help guide statewide and metropolitan freight planning effort

TARGET AUDIENCE

Transportation planners and freight transportation planners from State DOTs, MPOs, local governments, and Federal agencies.

TRAINING LEVEL: Basic

FEE: 2016: $0 Per Person; 2017: N/A

LENGTH: 6 HOURS (CEU: .6 UNITS)

CLASS SIZE: MINIMUM: 1; MAXIMUM: 1

NHI Customer Service: (877) 558-6873 • nhicustomerservice@dot.gov

Course Number

FHWA-NHI-139008

Course Title

Freight and Land Use Workshop

The Freight and Land Use Workshop is a one-day training session intended to provide an in-depth, hands-on understanding of the land use needs of freight-generating facilities; the potential impacts freight land uses can have on the transportation system, communities, and the natural environment; strategies that can reduce or mitigate the impacts of freight land uses; and the roles stakeholder groups can play in implementing those strategies.

This workshop provides specific guidance to planning practitioners on how to integrate freight and land use considerations within the transportation planning and programming process, and offers the tools and resources to assess the impacts of land use decisions on freight movements, as well as the impacts of freight development and growth on land use planning goals.

Outcomes

Upon completion of the course, participants will be able to:

- Discuss the importance of coordinating land use policies and freight transportation planning
- Describe methods for coordinating freight land use with local, regional, and state comprehensive and transportation planning processes
- Describe methods to avoid or mitigate impacts of freight land uses on surrounding land uses and the environment
- Identify examples of sustainable freight land use strategies and initiatives
- Locate the resources and tools available to address freight and land use considerations within the transportation and comprehensive planning processes

Target Audience

The target audience for this course includes a variety of transportation and planning professionals involved in transportation or land use planning, with a particular focus on municipal and county land use and transportation planners; and local, regional, or state economic development officials. This course is also relevant for State DOT planners; environmental specialists; engineers and specialists; MPO staff; Federal transportation employees, particularly FHWA and FRA staff; private sector stakeholders, particularly representatives from industrial real estate development, trucking, rail, and other industry associations, and port and port authority staff. Secondary audiences include Federal and state resource agency staff, EPA staff, elected officials or MPO board members, shippers or industry leaders, consultants, and citizen activists.

Training Level: Basic

Fee: 2016: $350 Per Person; 2017: N/A

Length: 1 DAYS (CEU: 0 UNITS)

Class Size: MINIMUM: 20; MAXIMUM: 30

NHI Customer Service: (877) 558-6873 • nhicustomerservice@dot.gov

Course Number

FHWA-NHI-139009

Course Title

Engaging the Private Sector in Freight Planning

Engaging the Private Sector in Freight Planning is a one-day workshop intended to provide techniques and strategies to help practitioners establish and strengthen relationships with the private sector. It is designed for practitioners addressing freight issues at metropolitan planning organizations (MPOs), State departments of transportation, local governments, and economic development agencies.

Building knowledge of both private sector business needs and the public sector transportation planning process supports the effective integration of freight into many statewide and metropolitan transportation plans and investment decisions. Developing and sustaining relationships, either formally or informally, with key private sector stakeholders is critical to effective freight transportation planning. Upon completion of this workshop participants will be able to develop an action plan that will engage freight stakeholders in their agencies' planning processes.

Outcomes

Upon completion of the course, participants will be able to:

- Describe the value that private sector freight stakeholders can add to the planning process
- Apply tools and resources for identifying freight stakeholders
- Construct an action plan for engaging freight stakeholders in their agencies' planning processes

Target Audience

The target audience for this course includes practitioners and others involved in transportation planning, project development, and project implementation specifically, State DOT Planners, Freight Transportation Specialists, MPO Staff, City/County Engineers and Planners, Economic Development Agency staff, Quasi-public Agencies (such as ports), FHWA Employees, and other Federal Employees. Private sector entities are considered a secondary audience; however, their participation in the workshop can be extremely beneficial to them and the public sector participants.

Training Level: Basic

Fee: 2016: $350 Per Person; 2017: N/A

Length: 1 DAYS (CEU: 0 UNITS)

Class Size: MINIMUM: 20; MAXIMUM: 30

NHI Customer Service: (877) 558-6873 • nhicustomerservice@dot.gov

Course Number

FHWA-NHI-141029

Course Title

Basic Relocation under the Uniform Act

The course is designed for the beginning relocation agent or for those persons interested in a basic knowledge of the Uniform Relocation Assistance and Real Property Acquisition Policies Act of 1970 (Uniform Act). The purpose is to answer questions, meet technical needs, and broaden the knowledge of those engaged in the relocation of persons displaced as a result of a Federal or Federally-funded project. The course covers all functional areas of the relocation assistance program, with emphasis on residential displacements.

This course is part of the Certificate of Accomplishment in Relocation under the Uniform Act. To learn more about how you can achieve a certificate in Relocation visit the NHI Web site at http://www.nhi.fhwa.dot.gov/training/cert_programs.aspx.

Outcomes

Upon completion of the course, participants will be able to:

- Explain the principles of the Uniform Act and implementing regulations
- Describe the Uniform Act planning requirements
- Describe an agency's advisory services responsibilities
- Describe the elements of comparable replacement housing
- Calculate replacement housing payments for owners and tenants
- Explain replacement housing of last resort
- Compute residential and non-residential moving costs

Target Audience

Federal, State, and local public agencies, FHWA personnel, and other interested persons.

Training Level: Basic

Fee: 2016: $650 Per Person; 2017: N/A

Length: 3 DAYS (CEU: 1.8 UNITS)

Class Size: MINIMUM: 20; MAXIMUM: 30

NHI Customer Service: (877) 558-6873 • nhicustomerservice@dot.gov

Course Number

FHWA-NHI-141030

Course Title

Advanced Relocation under the Uniform Act

This training goes beyond the basic functional areas of relocation assistance and concentrates on areas of specific concern, such as mortgage differential payments, settlement costs, and partial acquisitions. Other topics, including comparability, last resort housing, multiple use, tenants, and nonresidential moves -- including businesses, are also covered. The training is designed to allow flexibility in adjusting course materials to meet the needs of the requesting agency.

Prerequisites: Completion of FHWA-NHI-141029 Basic Relocation and the Web-based training FHWA-NHI-141045 Real Estate Acquisition Under the Uniform Act: An Overview or approximately one year of experience working in the relocation program. The training is peppered with interesting case study exercises, so bring an HP12 C calculator to class.

This training is part of the Certificate of Accomplishment in Relocation under the Uniform Act. To learn more about how you can achieve a certificate in Relocation visit the NHI Web site at http://www.nhi.fhwa.dot.gov/training/cert_programs.aspx.

Outcomes

Upon completion of the course, participants will be able to:

- Explain the principles that govern relocation provisions of the Uniform Relocation and Real Property Acquisition Policies Act of 1970 (Uniform Act) and implementing regulations
- Describe at least three factors involved in difficult relocation subject areas
- Describe issues that may arise when developing advisory assistance plans for difficult relocation areas
- Determine eligibility for certain relocation payments in difficult relocation cases
- Determine challenging issues when calculating complex nonresidential moving costs
- Calculate complex nonresidential moving costs

Target Audience

Federal, State, and local public agencies, FHWA personnel, right-of-way contractors, and other interested persons.

Training Level: Intermediate

Fee: 2016: $650 Per Person; 2017: N/A

Length: 3 DAYS (CEU: 1.8 UNITS)

Class Size: MINIMUM: 20; MAXIMUM: 30

NHI Customer Service: (877) 558-6873 • nhicustomerservice@dot.gov

Course Number

FHWA-NHI-141031

Course Title

Business Relocation under the Uniform Act

This course provides comprehensive information on the various aspects of business relocation and is designed to address the relocation of businesses, farms and nonprofit organizations. The main topics include eligibility, moving payments and benefits, advisory services, actual direct loss of tangible personal property, substitute personal property payments, reestablishment expenses, and fixed payment in lieu of (ILO) payments. A module about the move process includes the move option available to a business, as well as the need for an inventory and move specifications.

This course is part of the Certificate of Accomplishment in Relocation under the Uniform Act. To learn more about how you can achieve a certificate in Relocation visit the NHI Web site at http://www.nhi.fhwa.dot.gov/training/cert_programs.aspx.

Outcomes

Upon completion of the course, participants will be able to:

- Provide advisory services for businesses
- Determine moving and related expense payments for businesses, farms and non-profit organizations
- Determine reestablishment expenses for small businesses
- Determine fixed payments for businesses, farms and non-profit organizations
- Evaluate the move process for businesses
- Determine how to move hazardous materials for businesses

Target Audience

State departments of transportation, local public agencies, FHWA personnel, and other Federal agency personnel. Suggest that participants have at least two years general relocation experience.

Training Level: Accomplished

Fee: 2016: $650 Per Person; 2017: N/A

Length: 3 DAYS (CEU: 1.8 UNITS)

Class Size: MINIMUM: 20; MAXIMUM: 35

NHI Customer Service: (877) 558-6873 • nhicustomerservice@dot.gov

Course Number

FHWA-NHI-141043

Course Title

Appraisal for Federal-Aid Highway Programs

Please note that this training has been approved for Continuing Education Credits in several states by their respective appraisal licensing boards. As part of our training delivery, we will assist in preparing the documents required for course approval in your state. However, any fees associated with the application process are the responsibility of the requestor. Additionally, this course counts toward IRWA's SR/WA designation and R/W-AC re-certification. Participants should bring an HP 12c calculator for the classroom exercises.

The Uniform Relocation Assistance and Real Property Acquisition Policies Act of 1970, as amended (Uniform Act) and its implementing regulations require the uniform and equitable treatment of persons displaced from their homes, businesses, or farms and establish uniform and equitable land acquisition policies for public programs using Federal funds. Title III of the Uniform Act addresses real property acquisition policies, including appraisal requirements.

The training is designed to help transportation professionals understand and conform with the appraisal requirements of the Uniform Act and 49 CFR Part 24. It is intended for experienced appraisal personnel and focuses on preparing, presenting, and understanding appraisal reports in conformance with the Uniform Act. In addition, the training addresses the appraiser's role in the overall project development process and how an appraiser's expertise can assist in completing a transportation project effectively and efficiently.

Outcomes

Upon completion of the course, participants will be able to:

- Explain basic eminent domain principles
- Apply Federal-aid appraisal requirements
- Use partial acquisition appraisal techniques
- Explain the use and application of the waiver valuation process
- Apply appraisal techniques to unique situations within highway programs
- Describe the role of the appraiser in the land acquisition process

Target Audience

State departments of transportation (DOTs), local public agencies (LPAs), city and county attorneys, consultants; FHWA and other Federal agency staff involved in the appraisal process. Prerequisite: A course in the basic practices and principles of real estate appraisal (e.g. International Right of Way Association course 400, the Appraisal Institute's courses 110 and 120) or a college-level course in appraisal.

Training Level: Intermediate

Fee: 2016: $650 Per Person; 2017: N/A

Length: 2 DAYS (CEU: 1.2 UNITS)

Class Size: MINIMUM: 20; MAXIMUM: 35

NHI Customer Service: (877) 558-6873 • nhicustomerservice@dot.gov

Course Number

FHWA-NHI-141044

Course Title

Appraisal Review for Federal-Aid Highway Programs

Please note that this training has been approved for Continuing Education Credits in several States by their respective appraisal licensing boards. As part of our course delivery, we will assist in preparing the documents required for course approval in your State. However, any fees associated with the application process are the responsibility of the requestor. Additionally, this training counts toward IRWA's SR/WA designation and R/W-AC re-certification. Participants should bring an HP 12c calculator for the classroom exercises.

The Uniform Relocation Assistance and Real Property Acquisition Policies Act of 1970 as amended (Uniform Act) ensures that persons whose real property is acquired or who are displaced as a result of a Federal or Federally-assisted project are treated fairly and consistently. This course focuses on the application of appraisal review principles and how they fit within the Uniform Act and 49 CFR Part 24 as related to transportation project development. Focusing on larger parcel, uneconomic remnants, cost to cure, and severance damages, the course discusses the qualifications, roles, and responsibilities of the review appraiser from pre- to post-appraisal activities.

Outcomes

Upon completion of the course, participants will be able to:

- Explain basic eminent domain principles
- Apply Federal-Aid appraisal review requirements
- Apply appraisal review techniques to unique situations within Federal-Aid highway programs
- Describe the role of the review appraiser in the land acquisition process

Target Audience

State departments of transportation (DOTs), local public agencies (LPAs), city and county attorneys, consultants; FHWA and other Federal agency staff involved in the appraisal process. Prerequisite: A course in the basic practices and principles of real estate appraisal (e.g. International Right of Way Association course 400, the Appraisal Institute's courses 110 and 120) or a college-level course in appraisal.

Training Level: Accomplished

Fee: 2016: $450 Per Person; 2017: N/A

Length: 1 DAYS (CEU: .6 UNITS)

Class Size: MINIMUM: 20; MAXIMUM: 35

NHI Customer Service: (877) 558-6873 • nhicustomerservice@dot.gov

Course Number

FHWA-NHI-141045

Course Title

Real Estate Acquisition under the Uniform Act: An Overview - WEB-BASED

The Uniform Relocation Assistance and Real Property Acquisition Policies Act of 1970 (Uniform Act) is the basis for Federally-funded real estate acquisition programs. This self-paced training provides an overview of the Uniform Act's three key elements: valuation, acquisition, and relocation. This course underscores the importance of following Uniform Act requirements when acquiring property for a Federally-funded transportation project.

Outcomes

Upon completion of the course, participants will be able to:

- Provide a basic overview of the Uniform Relocation Assistance and Real Property Acquisition Policies Act of 1970 (Uniform Act)
- Discuss the three key elements of the Uniform Act: valuation/appraisal, acquisition and relocation
- Explain how to develop an estimate of just compensation using the appraisal process or appraisal waiver procedure(s)
- Identify relocation benefits and services required by the Uniform Act
- List places to obtain relevant resource documents and materials

Target Audience

Federal, State, and local government employees and consultants who acquire real estate for Federally-funded transportation projects. This includes acquisition and relocation agents; program or project managers; grant administrators or grant recipients; appraisers, realty specialists, attorneys, engineers, planners, and others involved with real property acquisition.

Training Level: Basic

Fee: 2016: $0 Per Person; 2017: N/A

Length: 6 HOURS (CEU: 0 UNITS)

Class Size: MINIMUM: 0; MAXIMUM: 0

NHI Customer Service: (877) 558-6873 • nhicustomerservice@dot.gov

COURSE NUMBER

FHWA-NHI-141047

COURSE TITLE

Local Public Agency Real Estate Acquisition - WEB-BASED

141047 Local Public Agency Real Estate Acquisition training is designed for those who are unfamiliar with Federal requirements when acquiring real property for Federally-assisted transportation projects. This training provides participants with a working knowledge of these Federal requirements when acquiring real property, including relocation guidance related to individuals and businesses.

Comprised of seven distinct learning modules, this self-paced Web-based training (WBT) provides an overview on real estate acquisition authority and the Uniform Relocation Assistance and Real Property Acquisition Policies Act of 1970 (Uniform Act) and related regulations. Additional modules include project development and administrative matters; valuation; acquisition and negotiation; relocation; and property management. This training also includes case studies, important resources, and suggestions for other companion courses.

Failure to comply with the Uniform Act when acquiring real property for a Federally-funded transportation project can put this funding at risk and may lead to project delays.

OUTCOMES

Upon completion of the course, participants will be able to:

- Explain the statutory basis for Federal requirements and relate these to State and local laws, regulations and procedures
- Explain the intent of the Uniform Act and describe what States and LPAs must do to comply
- Describe how a typical project is developed and strategies for enhancing project delivery
- Describe the LPA role in the appraisal process and determine the appropriate valuation format for specific situations
- Describe the sequence for land acquisition and options available to the negotiator
- Explain what relocation advisory services are to be provided to property owners and tenants and differentiate the residential and nonresidential relocation processes
- Summarize various property management activities and evaluate property management actions using specific case studies

TARGET AUDIENCE

Those within local public agencies who are responsible for acquiring right-of-way for federally-funded projects, as well as those responsible for oversight of LPAs, in addition to FHWA personnel, consultants, Federal and State staff and other interested parties.

TRAINING LEVEL: Basic

FEE: 2016: $50 Per Person; 2017: N/A

LENGTH: 6 HOURS (CEU: 0 UNITS)

CLASS SIZE: MINIMUM: 0; MAXIMUM: 0

NHI Customer Service: (877) 558-6873 • nhicustomerservice@dot.gov

Course Number

FHWA-NHI-141048

Course Title

Outdoor Advertising Control: Bonus States - WEB-BASED

Outdoor Advertising Control (OAC) limits the location, size, spacing, and lighting of signs adjacent to the interstate, National Highway System, and other Federal-Aid primary routes. Regulators are responsible for enforcing these requirements. The material in this training applies to all States and will help participants interpret major legislation and make effective decisions in support of OAC.

There are two related OAC Web-based trainings (WBTs): one designed for Bonus States (23 states) and the other for the remaining non-Bonus States. 141048 Outdoor Advertising Control: Bonus States includes one additional lesson addressing unique requirements these States must follow. Please refer to 141049 for information on this companion course.

Comprised of eight distinct learning modules, this self-paced WBT addresses Federal laws and regulations regarding signs adjacent to the right-of-way; zoning and related programs; commercial advertising signs adjacent to the right-of-way; recognized Federal sign classifications; implementation; maintenance and illegal sign removal; acquisition of signs on highway projects under the Uniform Act, as amended; and bonus state requirements.

This training provides participants an overview on laws and requirements related to Outdoor Advertising Control.

Outcomes

Upon completion of the course, participants will be able to:

- Apply Federal laws and regulations to assist in interpreting State and local laws and regulations for effective control
- Identify major Federal outdoor advertising legislation and regulations, and their importance for effective control
- Implement the process of effective control

Target Audience

State department of transportation employees; county, city town, and township staff involved with outdoor advertising; FHWA staff; and consultants assisting governmental entities with their Outdoor Advertising Control program.

Training Level: Basic

Fee: 2016: $0 Per Person; 2017: N/A

Length: 6 DAYS (CEU: 0 UNITS)

Class Size: MINIMUM: 1; MAXIMUM: 1

NHI Customer Service: (877) 558-6873 • nhicustomerservice@dot.gov

COURSE NUMBER

FHWA-NHI-141049

COURSE TITLE

Outdoor Advertising Control: Non-Bonus States - WEB-BASED

Outdoor Advertising Control (OAC) limits the location, size, spacing, and lighting of signs adjacent to the interstate, National Highway System, and other Federal-Aid primary routes. Regulators are responsible for enforcing these requirements. The material in this training applies to all States and will help participants interpret major legislation and make effective decisions in support of OAC.

There are two related OAC Web-based trainings (WBTs): one designed for Bonus States (23 states) and the other for the remaining non-Bonus States. 141049 Outdoor Advertising Control: Non-Bonus States and 141048 Outdoor Advertising Control: Bonus States. Please refer to 141048 for information on this companion course.

The course follows recommended FHWA specifications and practices for drilled shaft construction but may be modified to follow local agency specifications and practices.

Comprised of seven distinct learning modules, this self-paced WBT addresses Federal laws and regulations regarding signs adjacent to the right-of-way; zoning and related programs; commercial advertising signs adjacent to the right-of-way; recognized Federal sign classifications; implementation; maintenance and illegal sign removal; and acquisition of signs on highway projects, under the Uniform Act, as amended.

This training provides participants an overview on laws and requirements related to Outdoor Advertising Control.

OUTCOMES

Upon completion of the course, participants will be able to:

- Apply Federal laws and regulations to assist in interpreting State and local laws and regulations for effective control
- Identify major Federal outdoor advertising legislation and regulations, and their importance for effective control
- Implement the process of effective control

TARGET AUDIENCE

State department of transportation employees; county, city town, and township staff involved with outdoor advertising; FHWA staff; and consultants assisting governmental entities with their Outdoor Advertising Control program.

TRAINING LEVEL: Basic

FEE: 2016: $0 Per Person; 2017: N/A

LENGTH: 6 DAYS (CEU: 0 UNITS)

CLASS SIZE: MINIMUM: 1; MAXIMUM: 1

NHI Customer Service: (877) 558-6873 • nhicustomerservice@dot.gov

Course Number

FHWA-NHI-141050

Course Title

Introduction to Federal-Aid Right of Way Requirements for Local Public Agencies

This two-day introductory course provides Local Public Agencies (LPAs) with a working knowledge of Federal requirements and procedures for acquiring property for Federally-assisted transportation projects. The course focuses on applying the Uniform Act and related Federal Regulations to specific situations and issues. Designed as a hands-on, highly interactive learning experience, instructors guide participants through a series of right-of-way (ROW) problem solving exercises and large group discussions. We encourage those with limited ROW knowledge to register for the free NHI 141045 web-based training course in advance of this instructor-led course session.

Outcomes

Upon completion of the course, participants will be able to:

- Explain the legal basis for land acquisition by a governmental entity
- Assess the impact of a roadway improvement as it relates to the Uniform Act
- Sequence the right-of-way process (ROW) within the overall project development process
- Determine the appropriate valuation process for ROW acquisition
- Apply the Uniform Act requirements for ROW acquisition
- Apply the Uniform Act requirements to relocation assistance
- Determine the agency's responsibilities for managing real property

Target Audience

Those within local public agencies who are responsible for acquiring right-of-way for federally-funded projects, as well as those responsible for oversight of LPAs, in addition to FHWA personnel, consultants, Federal and State staff and other interested parties.

Training Level: Basic

Fee: 2016: $650 Per Person; 2017: N/A

Length: 2 DAYS (CEU: 1.2 UNITS)

Class Size: MINIMUM: 20; MAXIMUM: 35

NHI Customer Service: (877) 558-6873 • nhicustomerservice@dot.gov

Course Number

FHWA-NHI-141052

Course Title

Successful Acquisition under the Uniform Act

This course will provide the knowledge and skills that a public agency negotiator needs to complete acquisitions that comply with the Uniform Act.

Outcomes

Upon completion of the course, participants will be able to:

- Explain the legal basis for land acquisition by a governmental entity
- Identify the pre-acquisition materials necessary for property acquisition
- Explain the basics of the valuation process
- Describe the acquisition process under the Uniform Act
- Formulate effective negotiation skills, using best practices
- Discuss legal aspects of real property acquisition
- Discuss the role and limitations of consultants in the acquisition process

Target Audience

Federal, State, and local public agencies, FHWA personnel, contractors, and other interested persons.

Training Level: Basic

Fee: 2016: $800 Per Person; 2017: N/A

Length: 3 DAYS (CEU: 1.8 UNITS)

Class Size: MINIMUM: 20; MAXIMUM: 30

NHI Customer Service: (877) 558-6873 • nhicustomerservice@dot.gov

Course Number

FHWA-NHI-134109J

Course Title

Maintenance Training Series: Underground Storage Tanks - WEB-BASED

The Nation's underground storage tank (UST) systems consist of underground tanks and piping that store petroleum and other hazardous materials. This course, Underground Storage Tanks, addresses the procedures to install, operate, and remove USTs.

Developed specifically for maintenance personnel, this course provides participants with an understanding of the Federal laws and regulations that govern UST systems. During the course, participants acquire the knowledge needed to successfully oversee UST installations and closures. Specifically, the course explores the requirements of industry installation and closure codes, leakage detection, spill and overfill prevention, corrosion protection, and ensuring a "clean" closure.

This training was developed as part of the Maintenance Training Series. To access all the trainings in the series, enroll in the 134109 course.

Outcomes

Upon completion of the course, participants will be able to:

- Describe the regulatory framework governing the operation of underground storage tanks
- Describe UST operations
- Describe the process that must be followed to obtain satisfactory "clean closure" from the appropriate oversight agency
- Describe UST cleanup and removal operations

Target Audience

This course is designed for State, regional, and county personnel who manage operations programs and deal with oversight and quality assurance across broad geographic areas. This target audience also is involved with handling materials, scheduling, budgeting, and planning.

Training Level: Basic

Fee: 2016: $25 Per Person; 2017: N/A

Length: 1 HOURS (CEU: 0 UNITS)

Class Size: MINIMUM: 0; MAXIMUM: 0

NHI Customer Service: (877) 558-6873 • nhicustomerservice@dot.gov

Course Number

FHWA-NHI-134109K

Course Title

Maintenance Training Series: Cultural and Historic Preservation - WEB-BASED

Cultural and historic sites are often located within an area where maintenance activities are scheduled to be completed. This training, Cultural and Historic Preservation, is teaches participants about regulations and concerns related to safeguarding cultural and historic sites from the potential impacts of highway maintenance activities. Examples of maintenance activities that can impact cultural or historic sites include slope stabilization, shoulder or pavement widening, and vegetation control. Additional examples are presented during the course.

This course assists participants with recognizing potential historic or cultural resources, verifying a site's cultural or historic status, and avoiding impacts to sites when carrying out maintenance activities. Since completing these tasks often requires additional expertise, resources for obtaining needed assistance are provided. In addition, participants learn how maintenance activities can enhance cultural and historic sites through utilization of Context Sensitive Solutions (CSS).

This training was developed as part of the Maintenance Training Series. To access all the courses in the series, enroll in the 134109 course.

Outcomes

Upon completion of the course, participants will be able to:

- Identify governing bodies and registries that should be consulted prior to commencing maintenance activities on sites of cultural and historic importance
- Recognize what sorts of structures, landmarks, and properties could pose potential cultural and historic preservation issues
- Describe how to avoid impacts to historic sites
- Describe the role of DOT in maintaining and enhancing cultural resources

Target Audience

This course is designed for State, regional, and county personnel who manage operations programs and deal with oversight and quality assurance across broad geographic areas. This target audience also is involved with handling materials, scheduling, budgeting, and planning.

Training Level: Basic

Fee: 2016: $25 Per Person; 2017: N/A

Length: 1 HOURS (CEU: 0 UNITS)

Class Size: MINIMUM: 0; MAXIMUM: 0

NHI Customer Service: (877) 558-6873 • nhicustomerservice@dot.gov

Course Number

FHWA-NHI-142005

Course Title

NEPA and the Transportation Decisionmaking Process

This comprehensive, three-day, instructor-led course presents the historical background and evolution of the National Environmental Policy Act (NEPA) and related environmental laws. It discusses their influence on FHWA's policies and procedures for the transportation project development and decisionmaking process. The course examines how the framework of laws, regulations, policies, and guidance integrate social, environmental, and economic factors in making transportation project decisions that are in the best overall public interest.

The course emphasizes the Council on Environmental Quality and FHWA's regulations; FHWA policy and guidance for implementing NEPA, Section 4(f) of the Department of Transportation Act, and related environmental requirements. It discusses the NEPA Essential Elements in detail, including purpose and need, alternatives, impacts, mitigation, public involvement, interagency coordination and documentation. The course presents the requirements and considerations used to decide whether to prepare an environmental impact statement, an environmental assessment or determination that a project is categorically excluded from either. While this is not a course in environmental document writing, it presents the key principles for preparing high quality environmental documents, including the core principles from the FHWA/AASHTO/American Council of Engineering Companies Improving the Quality of Environmental Documents and the FHWA IQED initiative. The course also includes group exercises that allow participants to apply the course concepts to a realistic project scenario involving several transportation, social and environmental considerations.

Outcomes

Upon completion of the course, participants will be able to:

- Describe the NEPA principles in relation to transportation project development.
- Describe how the NEPA umbrella concept influences the transportation decisionmaking process.
- Explain the roles and responsibilities of participants in the NEPA process.
- Describe the imporatnce of a reasoned, collaborative process when developing and evaluating alternatives.
- Discuss balancing an array of interests and values in making transportation decisions.
- List the milestones in transportation planning that link to the NEPA project development process.
- Describe the documentation requirements of the NEPA process.
- Discuss environmental streamlining, stewardship, and leadership in managing the NEPA process.

Target Audience

The target audience for this course includes FHWA, State departments of transportation (including consultants acting on behalf of the State), Federal and State environmental resource agencies, local governments, and metropolitan planning organizations who participate in the transportation decisionmaking process. We strongly encourage the sponsoring organization to invite a mix of planning and environmental staff from these agencies.

Training Level: Intermediate

Fee: 2016: $725 Per Person; 2017: N/A

Length: 3 DAYS (CEU: 1.8 UNITS)

Class Size: MINIMUM: 20; MAXIMUM: 30

NHI Customer Service: (877) 558-6873 • nhicustomerservice@dot.gov

Course Number

FHWA-NHI-142036

Course Title

Public Involvement in the Transportation Decisionmaking Process

Public involvement is much more than public hearings. It involves creative thinking as well as the willingness and ability to interact openly and sensitively to the public's preferred forms of communication and participation. Public involvement is about reaching out to and involving the public in transportation decisionmaking. The public should have a role in every phase of decisionmaking, including the design of the participation plan itself. Successful public involvement addresses the public's procedural, psychological, and substantive needs while gathering useful information. By focusing on interests--rather than positions--public involvement can become more meaningful as well as useful.

Outcomes

Upon completion of the course, participants will be able to:

- Describe U.S. DOT transportation decisionmaking processes, including those that trigger the National Environmental Policy Act
- Describe the relationship between public involvement and decisionmaking
- Develop a public involvement plan with stakeholder assistance that includes attention to non-traditional populations as an evaluation component
- Describe interest-based problem solving and the values that underlie it
- Identify ways to enhance public involvement plans

Target Audience

Federal, State, and local transportation agency staff, metropolitan planning organization personnel, transit operators, consultants, and others who are responsible for planning, implementing, or participating in any phase of the public involvement process.

Training Level: Basic

Fee: 2016: $1050 Per Person; 2017: N/A

Length: 3 DAYS (CEU: 1.8 UNITS)

Class Size: MINIMUM: 20; MAXIMUM: 30

NHI Customer Service: (877) 558-6873 • nhicustomerservice@dot.gov

Course Number

FHWA-NHI-142042

Course Title

Fundamentals of Environmental Justice

Addressing environmental justice applies to every stage of transportation decisionmaking. The U.S. Department of Transportation (USDOT) and its partners are committed to integrating the nondiscrimination principles of environmental justice into all Federal-aid programs. Through these and other transportation programs, many opportunities exist to establish partnerships with other public and private organizations to create livable communities that meet the needs of all people. This course presents participants with a framework for using a variety of approaches and tools for accomplishing environmental justice goals in Federal-aid programs and other transportation projects.

Outcomes

Upon completion of the course, participants will be able to:

- Define environmental justice and describe its relationship to Title VI
- Explain the fundamental principles of environmental justice
- Apply the principles of environmental justice to transportation decisions
- Identify how environmental justice applies to each stage of transportation decisionmaking
- Describe the benefits of environmental justice in transportation decisionmaking
- Develop proactive strategies, methods, and techniques to implement environmental justice in transportation programs and projects

Target Audience

We strongly encourage the sponsoring organization to invite a mix of civil rights, environmental, engineering, and planning staff from the following agencies: Federal, State, and regional organizations, local transit agencies, and consultants who act on their behalf. Others who interact with minority and low-income communities; provide community services; and elected officials and their representatives.

Training Level: Basic

Fee: 2016: $650 Per Person; 2017: N/A

Length: 2 DAYS (CEU: 1.2 UNITS)

Class Size: MINIMUM: 20; MAXIMUM: 30

NHI Customer Service: (877) 558-6873 • nhicustomerservice@dot.gov

Course Number

FHWA-NHI-142045

Course Title

Pedestrian Facility Design

To emphasize the importance of planning for pedestrians, the course focuses on case examples involving corridor and intersection design issues. Participants are engaged through lecture, discussion, video demonstrations of problem areas in corridors and intersections, small group problem identification, and the development of design alternatives. This training was developed to provide information and application opportunities to those involved in the design of pedestrian facilities. The Americans with Disabilities Act (ADA) requires newly constructed and altered sidewalks to be accessible and usable by people with disabilities, and accessibility improvements need to be implemented for existing facilities.

Outcomes

Upon completion of the course, participants will be able to:

- List the characteristics of pedestrians and motorized traffic that influence pedestrian facility design
- Apply the concepts of universal design and applicable design reference material to redesigning an existing location and/or designing a new location that meets the needs of motorized and nonmotorized users
- Given a case example, identify potential conflicts between pedestrians and other traffic and propose design options that improve access and safety
- Given a case example, analyze the network for improvement options to meet the needs of pedestrian and other traffic

Target Audience

Engineers with planning, design, construction, or maintenance responsibilities; pedestrian and bicycle specialists, disability and orientation specialists, transportation planners, architects, landscape architects, as well as decisionmakers at the project planning level.

Training Level: Intermediate

Fee: 2016: $700 Per Person; 2017: N/A

Length: 1.5 DAYS (CEU: .9 UNITS)

Class Size: MINIMUM: 20; MAXIMUM: 30

NHI Customer Service: (877) 558-6873 • nhicustomerservice@dot.gov

Course Number

FHWA-NHI-142046

Course Title

Bicycle Facility Design

This training will assist planners and designers in learning how to apply the existing standards and how to deal with other technical issues involved. The availability of Federal, State, and local transportation funding for bicycle facilities that serve transportation and recreational users is resulting in a dramatic increase in the number of bicycling (and shared use) facilities being planned and built. Although there are no Federal design standards for bicycle facilities, the AASHTO Guide for the Development of Bicycle Facilities, or a modification thereof, serves as a design guide. As with most guides, the AASHTO guide cannot address every possible scenario so designers often need to apply engineering judgment where specific information is not provided. The training fee includes a copy of the AASHTO Guide for the Development of Bicycle Facilities.

Outcomes

Upon completion of the course, participants will be able to:

- List the needs of bicyclists as transportation facility users
- Identify common roadway and traffic conditions that affect bicyclists
- Describe the characteristics of a roadway and a shared-use path that are designed to accommodate bicyclists
- List the benefits to the transportation system of accommodating bicyclists with different abilities
- Recognize opportunities to accommodate bicyclists during the planning, design, construction, and operational phases of a project

Target Audience

Federal, State, or local engineers with planning, design, construction, or maintenance responsibilities; bicycle specialists, transportation planners, landscape architects, as well as decisionmakers at the project planning level.

Training Level: Accomplished

Fee: 2016: $700 Per Person; 2017: N/A

Length: 1.5 DAYS (CEU: 1 UNITS)

Class Size: MINIMUM: 20; MAXIMUM: 30

NHI Customer Service: (877) 558-6873 • nhicustomerservice@dot.gov

COURSE NUMBER

FHWA-NHI-142047

COURSE TITLE

Water Quality Management of Highway Runoff

In reaction to the impact of human activity on water quality, the Clean Water Act was passed in 1972 in order to restore and maintain the chemical, physical, and biological integrity of the Nation's waters. The act regulates discharges to U.S. waters through permits issued under the National Pollutant Discharge Elimination System permitting program and places requirements on State transportation agencies for managing runoff water quality. Understanding the legal responsibilities, terminology, and the general roles of players in the regulatory process is critical in order to properly plan for, budget, and implement water quality management.

The intent of this course is to provide a basic understanding of water quality parameters, processes, requirements, and best management practices (BMPs) in order to provide the transportation community with guidance on how to mitigate impacts and protect water quality. The course shares approaches and technologies for the water quality management of highway stormwater runoff, including the effective maintenance, inspection, and performance evaluation of BMPs.

OUTCOMES

Upon completion of the course, participants will be able to:

- Identify and characterize the quantity and quality of highway runoff
- Describe how highway runoff can affect ecosystems
- List major Federal requirements that apply to management of highway runoff
- Explain how to select a mitigation strategy from a watershed perspective
- Describe design concepts and considerations in selecting and siting appropriate BMPs for controlling highway runoff
- Develop conceptual designs for various BMPs considering treatment targets, design requirements, BMP performance goals, siting and maintenance considerations, etc.
- Explain how to integrate mitigation of highway runoff impacts into the project development process
- Discuss the importance of BMP inspection, performance evaluation, monitoring, and maintenance

TARGET AUDIENCE

This course is designed for State department of transportation staff who negotiate permit conditions with the appropriate State agency; design engineers who must be cognizant of permit requirements; construction personnel who implement the highway designs; inspectors who ensure that water quality management features (BMPs) are functioning as designed; biologists who identify habitat for wildlife and potential ecosystem impacts; landscape architects and botanists who ensure that vegetation is preserved to the maximum extent practicable and that appropriate vegetation is used to provide water quality benefits after construction; and environmental scientists who monitor and evaluate water quality.

TRAINING LEVEL: Intermediate

FEE: 2016: $775 Per Person; 2017: N/A

LENGTH: 2 DAYS (CEU: 1.2 UNITS)

CLASS SIZE: MINIMUM: 20; MAXIMUM: 30

NHI Customer Service: (877) 558-6873 • nhicustomerservice@dot.gov

COURSE NUMBER

FHWA-NHI-142049

COURSE TITLE

Beyond Compliance: Historic Preservation in Transportation Project Development

Section 106 of the National Historic Preservation Act requires Federal agencies to take into account the effects of their undertakings on properties listed in, or eligible for listing in the National Register of Historic Places. This is accomplished through consultation with resource agencies, stakeholders, tribes, Native Hawaiian organizations, and the public. The regulation implementing Section 106 strongly encourages close coordination between the Section 106 process and National Environmental Policy Act (NEPA) requirements. The regulation also gives agencies great flexibility in how they fulfill their Section 106 responsibilities.

This training is designed to help transportation professionals meet the requirements of Section 106 and take advantage of the flexibility offered by the Section 106 regulation. The training focuses on the fundamentals of Section 106, placing it in context with other environmental requirements, including NEPA and Section 4(f) of the Department of Transportation Act. The course presents a number of innovative and programmatic approaches to Section 106 compliance that streamline and enhance environmental reviews and project delivery. The training emphasizes practical, real-world approaches for completing the Section 106 process, with the goal of balancing historic preservation concerns with the needs of transportation projects.

OUTCOMES

Upon completion of the course, participants will be able to:

- Identify key historic preservation laws and other authorities
- Describe the Section 106 process
- Define the roles and responsibilities of all parties in the Section 106 process
- Describe the NEPA transportation decisionmaking process
- Describe the relationship among Section 106, NEPA project development, and Section 4(f)
- Identify principles and opportunities for environmental streamlining and stewardship

TARGET AUDIENCE

Those involved in or affected by the Federal-Aid Highway program, including staff of State DOTs, MPOs, FHWA headquarters and field offices, city and county governments, tribal governments, consultants, State and tribal Historical Preservation Offices (SHPO/THPO), and other Federal and State resource agencies that deal with transportation issues.

TRAINING LEVEL: Basic

FEE: 2016: $1000 Per Person; 2017: N/A

LENGTH: 3 DAYS (CEU: 1.8 UNITS)

CLASS SIZE: MINIMUM: 20; MAXIMUM: 30

NHI Customer Service: (877) 558-6873 • nhicustomerservice@dot.gov

Course Number

FHWA-NHI-142051

Course Title

Highway Traffic Noise

This course was recently updated to reflect current FHWA noise regulations and policies: Title 23 CFR Part 772: Procedures for Abatement of Highway Traffic Noise and Construction Noise, and related State highway agency noise policies.

We recommend that participants registering for this course take the NHI 142063 Highway Traffic Noise: Basic Acoustics in advance of their session.

The Interactive Sound Information System (ISIS) demonstrations that are part of this training are deliberately very loud. So, we encourage Session Hosts to secure a training room accordingly.

This course provides an introductory- yet comprehensive- overview of highway traffic noise that will help educate engineers, environmental specialists, designers, planners, and consultants about traffic noise and ways to reduce the impacts. This course is designed not just for those performing noise analyses, but also for those responsible for the review and approval of noise studies and incorporating the results of these studies into design and environmental documents. Among topics covered are the basic principles of acoustics, how to determine when a noise analysis is required, typical strategies to mitigate noise in highway projects, noise study requirements and documentation, noise measurement, construction noise, and public involvement.

Participants will also learn about noise-compatible planning, which encourages local governments and developers to avoid noise-sensitive land uses adjacent to highways. And the course includes an overview of the FHWA Traffic Noise Model (FHWA TNM)--developed to predict noise levels and evaluate noise abatement measures.

Additionally, the NHI commissioned a customized version of the Interactive Sound Information System (ISIS) as part of the curriculum design for this course. ISIS is a noise simulation software program that employs high-quality digital recordings, precise sound control, and graphic imagery to present basic acoustical concepts and sounds from various traffic loads. ISIS also demonstrates the sound level reductions provided by various noise barrier configurations.

Outcomes

Upon completion of the course, participants will be able to:

- Explain the basic principles of acoustics
- Describe all necessary documentation to fulfill FHWA noise requirements, as codified in 23 CFR 772
- Determine when a noise study is required
- Explain applicable Federal noise abatement regulations and policies

Target Audience

FHWA staff; State department of transportation environmental specialists, designers, planners or engineers; city or county environmental engineers, coordinators or specialists; consultants.

Training Level: Basic

Fee: 2016: $1000 Per Person; 2017: N/A

Length: 3 DAYS (CEU: 1.8 UNITS)

Class Size: MINIMUM: 20; MAXIMUM: 30

NHI Customer Service: (877) 558-6873 • nhicustomerservice@dot.gov

Course Number

FHWA-NHI-142052

Course Title

Introduction to NEPA and Transportation Decisionmaking - WEB-BASED

The National Environmental Policy Act of 1969 (NEPA) requires, to the fullest extent possible, that the policies, regulations, and laws of the Federal Government be interpreted and administered in accordance with Federal environmental protection goals. NEPA also requires Federal agencies to use an interdisciplinary approach in planning and decisionmaking for any action that adversely impacts the environment.

This Web-based training is a basic introduction to FHWA's NEPA transportation decisionmaking process that takes into account the potential impacts of transportation projects on the human and natural environment balanced with the public's need for safe and efficient transportation. The training covers NEPA requirements as implemented by the Council on Environmental Quality, as well as FHWA's regulations and guidance for NEPA implementation and project decisionmaking. Topics include purpose and need, alternatives development and analysis, impact analysis, public involvement, interagency coordination, mitigation, and documentation. We strongly recommend completion of this self-paced training prior to enrolling in FHWA-NHI 142005.

Outcomes

Upon completion of the course, participants will be able to:

- Relate the origin, evolution, and context of NEPA
- Describe the intent, goals, and basic requirements of NEPA
- Describe the NEPA umbrella concept in transportation decisionmaking
- Identify the NEPA principles in the development of transportation projects
- Explain the roles and responsibilities of the lead agency, applicant, and cooperating agencies in the NEPA process
- List documentation requirements of the NEPA process

Target Audience

Staff from FHWA, State Departments of Transportation (DOT), Federal and State environmental resource agencies, local government, MPOs who participate in the transportation decisionmaking process, and consultants acting on behalf of State and local governments.

Training Level: Basic

Fee: 2016: $25 Per Person; 2017: N/A

Length: 4 HOURS (CEU: .4 UNITS)

Class Size: MINIMUM: 1; MAXIMUM: 1

NHI Customer Service: (877) 558-6873 • nhicustomerservice@dot.gov

Course Number

FHWA-NHI-142054

Course Title

Design and Implementation of Erosion and Sediment Control

This training is the result of a joint effort between the Federal Highway Administration (FHWA) and the U.S. Environmental Protection Agency (EPA), and reflects the agencies' commitment to providing education and training on planning, design, implementation, enforcement, inspection, and maintenance strategies to control erosion and sediment on highway construction projects. The agencies also are committed to ensuring that regulatory issues are addressed accurately and uniformly. Each discipline involved in a highway construction project has a different set of priorities. Reflecting the National Highway Institute's (NHI) commitment to learner-centered training, the course offers participants opportunities for discussion and joint problem solving, enabling participants to gain information about the roles and responsibilities of other team members.

Outcomes

Upon completion of the course, participants will be able to:

- Describe the components of an erosion and sediment control (ESC) plan
- List the sources of information for the ESC plan
- Identify management practices and related measures that are appropriate for typical situations and for a case example
- List typical construction and inspection problems. Describe both suitable prevention strategies and remedies for failures
- Link Federal and State environmental regulations to the components of the ESC plan

Target Audience

The training targets Federal, State, and local highway design, construction, inspection, and maintenance staff. In addition, environmental agency representatives, as well as consultants and members of the construction industry, are encouraged to attend to provide their perspectives, learn each other's responsibilities, and explore an array of options to control erosion and sedimentation.

Training Level: Intermediate

Fee: 2016: $775 Per Person; 2017: N/A

Length: 2 DAYS (CEU: 1.2 UNITS)

Class Size: MINIMUM: 20; MAXIMUM: 30

NHI Customer Service: (877) 558-6873 • nhicustomerservice@dot.gov

ENVIRONMENT

Course Number

FHWA-NHI-142055

Course Title

Advanced Seminar on Transportation Project Development: Navigating the NEPA Maze

Building upon demonstrated knowledge and understanding of the NEPA project development process, this advanced training provides practical tools and approaches to successfully resolve complex environmental issues and challenges. Designed in seminar format, this training is highly interactive and guides participants through the NEPA decisionmaking process, pointing out potential pitfalls and providing the skills and knowledge to apply critical thinking to reach defensible decisions.

Outcomes

Upon completion of the course, participants will be able to:

- Manage and deliver projects and programs more effectively
- Apply tools and techniques to their jobs
- Apply principles of environmental stewardship and streamlining to complex projects
- Employ integrated coordination of related laws and regulations, as well as coordination among all stakeholders
- Identify strategies to manage controversial projects
- Formulate solutions to complex environmental challenges
- Apply lessons learned from relevant case law
- Build a defensible administrative record
- Identify solutions to emerging issues

Target Audience

Experienced environmental practitioners and project development managers (i.e. planning, design, legal, and technical specialists) involved in the NEPA and transportation decisionmaking process. We encourage a mix of experienced staff from FHWA, State DOTs, resource and permitting agencies, and local governments, as well as consultants.

Training Level: Accomplished

Fee: 2016: $800 Per Person; 2017: N/A

Length: 3 DAYS (CEU: 1.8 UNITS)

Class Size: MINIMUM: 20; MAXIMUM: 30

NHI Customer Service: (877) 558-6873 • nhicustomerservice@dot.gov

Course Number

FHWA-NHI-142060

Course Title

Practical Conflict Management Skills for Environmental Issues

This course is designed to teach basic conflict management skills, including interest-based negotiation, communication, and facilitation skills, as well as leadership behaviors and to examine opportunities for applying these collaborative skills in the context of transportation decisionmaking where there are environmental issues. The purpose of the course is to help transportation and environmental agencies, Tribes, and stakeholders bridge their different agency mandates and diverse interests and offers opportunities for participants to share their varied perspectives on important issues and resources. The course models how to participate in meaningful discussions and demonstrates how to maintain a positive and constructive dialogue, arrive at integrative decisions, make good use of limited time and personnel resources, achieve streamlined processes, and make decisions that serve the public.

Outcomes

Upon completion of the course, participants will be able to:

- Use interpersonal skills to engage productively with individuals within their agency
- Use interpersonal skills to work productively with other agencies, organizations, Tribes, and the general public
- Analyze agency roles and decisionmaking processes with respect to potential conflict
- Apply conflict management strategies to planning, project development, and project implementation
- Apply conflict management strategies to increase the effectiveness of inter-agency and intra-agency working relationships and programmatic initiatives

Target Audience

This training program is intended for Federal, State, Metropolitan Planning Organization, Local Public Agency, and Tribal representatives who are involved in Federally-funded transportation projects. The target audience may include environmental protection specialists; transportation planners, project managers, design engineers, and transportation/ environmental liaisons; legal counsel, and Federal, Tribal, and State resource agency staff who review and implement transportation projects and are involved in transportation planning as well as environmental consultants and non-governmental organizational representatives. Participants should have a working knowledge of the National Environmental Protection Act (NEPA) and the transportation planning and project development processes.

Training Level: Basic

Fee: 2016: $1050 Per Person; 2017: N/A

Length: 3 DAYS (CEU: 1.9 UNITS)

Class Size: MINIMUM: 20; MAXIMUM: 30

NHI Customer Service: (877) 558-6873 • nhicustomerservice@dot.gov

Course Number

FHWA-NHI-142062

Course Title

Administrative Record - WEB-BASED

This web-based training (WBT) course describes the purpose of an Administrative Record, explains what should be included in an Administrative Record, and presents best practices for building a defensible Administrative Record. Please note that this WBT is not meant to take the place of formal legal advice and consultation with FHWA counsel is strongly recommended.

Outcomes

Upon completion of the course, participants will be able to:

- Describe the purpose of an Administrative Record
- Explain the factors involved in determining what should be included in an Administrative Record
- Describe best practices for building a defensible Administrative Record

Target Audience

This course is designed for Federal Highway Administration (FHWA) Division office staff who are responsible for the Administrative Record, as well as for State DOT employees and their contractors who need to build and maintain an Administrative Record.

Training Level: Basic

Fee: 2016: $25 Per Person; 2017: N/A

Length: 1 HOURS (CEU: 0 UNITS)

Class Size: MINIMUM: 1; MAXIMUM: 1

NHI Customer Service: (877) 558-6873 • nhicustomerservice@dot.gov

COURSE NUMBER

FHWA-NHI-142063

COURSE TITLE

Highway Traffic Noise: Basic Acoustics - WEB-BASED

This Web-based training (WBT) course provides an overview of Acoustic Principles related to highway traffic noise. The course incorporates the Interactive Sounds Information System (ISIS) -- a customized noise simulation model -- to demonstrate Acoustic Principles.

The WBT begins with the characteristics of sound and discusses how to apply basic concepts of acoustics to solve sample problems. It proceeds to the propagation of sound: a presentation on the four phenomena of divergence, ground effects, atmospheric effects, and shielding by natural and man-made features. The interaction between highway noise and barriers is described and key noise barrier concepts are presented (i.e. direct, incident, transmittal, reflected, and diffracted). Traffic noise sources are discussed, as well as a brief overview of traffic noise modeling and vehicle classification types.

The goal for the Highway Traffic Noise: Basic Acoustics WBT is to explain the basic principles of highway traffic acoustics.

OUTCOMES

Upon completion of the course, participants will be able to:

- Apply basic concepts of acoustics to solve sample problems
- Add levels in decibels
- Describe characteristics affecting noise propagation
- Describe how noise interacts with barriers
- Determine the increase in level for N sources vs. one source
- Determine the change in level with changing distance for a point source and for a line source

TARGET AUDIENCE

The Basic Acoustics of Highway Traffic Noise WBT is intended for FHWA staff; State department of transportation (DOT) environmental specialists, designers, planners or engineers; city or county environmental engineers, coordinators or specialists, and consultants. The training design for this WBT assumes that participants have basic computer skills (e.g. manipulating windows, using directories, a web browser, etc).

TRAINING LEVEL: Basic

FEE: 2016: $25 Per Person; 2017: N/A

LENGTH: 2 HOURS (CEU: 0 UNITS)

CLASS SIZE: MINIMUM: 1; MAXIMUM: 1

NHI Customer Service: (877) 558-6873 • nhicustomerservice@dot.gov

Course Number

FHWA-NHI-142068

Course Title

Air Quality Planning: Clean Air Act Overview - WEB-BASED

The purpose of this training is to provide participants with an overview of air quality planning, including requirements, processes, interactions with and implications for, transportation planning and project development.

This is the first in a future series of air quality Web-based trainings (WBTs):

142068: Clear Air Act Overview

142069: SIP and TCM Requirements and Policies

142070: SIP Development Process

142071: Transportation Conformity

Outcomes

Upon completion of the course, participants will be able to:

- Define the purpose of the Clean Air Act
- Describe the 1990 Clean Air Act Amendments
- Identify and explain Clean Air Act Amendment provisions relevant to transportation
- Recognize impacts of Clean Air Act

Target Audience

The target audience for the Air Quality Series is transportation and air quality planners and engineers from State and local departments of transportation (DOT), metropolitan planning organizations (MPO), transit agencies, Federal agencies (Federal Highway Administration, Federal Transit Administration, U.S. Environmental Protection Agency, U.S. Department of Energy, etc.), and State and local environmental agencies. Others include transportation and environmental consultants, public officials and staff members, community and interest groups, as well as other stakeholders in the planning process.

Training Level: Basic

Fee: 2016: $0 Per Person; 2017: N/A

Length: 1.5 HOURS (CEU: 0 UNITS)

Class Size: MINIMUM: 1; MAXIMUM: 1

NHI Customer Service: (877) 558-6873 • nhicustomerservice@dot.gov

ENVIRONMENT

COURSE NUMBER

FHWA-NHI-142069

COURSE TITLE

Air Quality Planning: SIP and TCM Requirements and Policies - WEB-BASED

This course covers the different types of SIPs and key CAA SIP requirements general to all SIPs and specific to ozone, CO and PM SIPs; discusses how the EPA processes SIPs; explores the key features of EPA SIP policies and how they differ from CAA requirements; and explains RACM and how it applies to TCMs.

This is the second in a future series of air quality Web-based trainings (WBTs):

142068: Clear Air Act Overview

142069: SIP and TCM Requirements and Policies

142070: SIP Development Process

142071: Transportation Conformity

OUTCOMES

Upon completion of the course, participants will be able to:

- Define SIP
- List different types of SIPs and their purposes
- Identify SIP requirements in Title I of the Clean Air Act
- Describe TCM requirements
- Describe what is meant by Reasonably Available Control Measure, or RACM, and how this applies to TCMs

TARGET AUDIENCE

The target audience for the Air Quality Series is transportation and air quality planners and engineers from State and local departments of transportation (DOT), metropolitan planning organizations (MPO), transit agencies, Federal agencies (Federal Highway Administration, Federal Transit Administration, U.S. Environmental Protection Agency, U.S. Department of Energy, etc.), and State and local environmental agencies. Others include transportation and environmental consultants, public officials and staff members, community and interest groups, as well as other stakeholders in the planning process.

TRAINING LEVEL: Basic

FEE: 2016: $0 Per Person; 2017: N/A

LENGTH: 1 HOURS (CEU: 0 UNITS)

CLASS SIZE: MINIMUM: 1; MAXIMUM: 1

NHI Customer Service: (877) 558-6873 • nhicustomerservice@dot.gov

Course Number

FHWA-NHI-142070

Course Title

Air Quality Planning: SIP Development Process - WEB-BASED

This course provides an overview of the State Implementation Plan (or SIP) development process, focusing on agency roles, with an explanation of the problem definition and solution parts of the process. This course also covers motor vehicle emission budgets that are included in SIPs and used in conformity determinations, as well as describes EPAs procedures in approving and disapproving SIPs.

This is the third in a series of air quality Web-based trainings (WBTs):

142068: Clear Air Act Overview

142069: SIP and TCM Requirements and Policies

142070: SIP Development Process

142071: Transportation Conformity

Outcomes

Upon completion of the course, participants will be able to:

- Describe the SIP development process;
- Identify the various emission sources and describe emission trends;
- List the steps involved in preparing emission inventories;
- Describe how SIP emission reductions are determined; and
- Describe the different types of control measures, emphasizing the role of transportation-related strategies.

Target Audience

The target audience for the Air Quality Series is transportation and air quality planners and engineers from State and local departments of transportation (DOT), metropolitan planning organizations (MPO), transit agencies, Federal agencies (Federal Highway Administration, Federal Transit Administration, U.S. Environmental Protection Agency, U.S. Department of Energy, etc.), and State and local environmental agencies. Others include transportation and environmental consultants, public officials and staff members, community and interest groups, as well as other stakeholders in the planning process.

Training Level: Basic

Fee: 2016: $0 Per Person; 2017: N/A

Length: 2 HOURS (CEU: 0 UNITS)

Class Size: MINIMUM: 1; MAXIMUM: 1

NHI Customer Service: (877) 558-6873 • nhicustomerservice@dot.gov

COURSE NUMBER

FHWA-NHI-142071

COURSE TITLE

Air Quality Planning: Transportation Conformity - WEB-BASED

This course defines transportation conformity and is designed for individuals that are new to transportation conformity, with little to no experience with the Transportation Conformity Rule.

This introductory transportation conformity course will answer questions related to the "what" of transportation conformity. What is transportation conformity? What activities are covered by conformity? What are the major requirements?

This course does not address how transportation conformity regulations are met. That topic is an advanced subject matter area and out of scope for an introductory transportation conformity course.

This is the fourth in a future series of air quality Web-based trainings (WBTs):

142068: Clear Air Act Overview

142069: SIP and TCM Requirements and Policies

142070: SIP Development Process

142071: Transportation Conformity

OUTCOMES

Upon completion of the course, participants will be able to:

- Relate transportation conformity to Transportation Improvement Programs (TIPs) and transportation plans
- Define transportation conformity
- Explain the transportation activities that are subject to conformity in a given timeframe
- Describe transportation conformity requirements for different activities
- Explain stakeholder responsibilities related to transportation conformity

TARGET AUDIENCE

The target audience for the Air Quality Series is transportation and air quality planners and engineers from State and local departments of transportation (DOT), metropolitan planning organizations (MPO), transit agencies, Federal agencies (Federal Highway Administration, Federal Transit Administration, U.S. Environmental Protection Agency, U.S. Department of Energy, etc.), and State and local environmental agencies. Others include transportation and environmental consultants, public officials and staff members, community and interest groups, as well as other stakeholders in the planning process.

TRAINING LEVEL: Basic

FEE: 2016: $0 Per Person; 2017: N/A

LENGTH: 1.5 HOURS (CEU: 0 UNITS)

CLASS SIZE: MINIMUM: 1; MAXIMUM: 1

NHI Customer Service: (877) 558-6873 • nhicustomerservice@dot.gov

ENVIRONMENT

Course Number

FHWA-NHI-142073

Course Title

Applying Section 4(f): Putting Policy into Practice

NHI 142073 Applying Section 4(f): Putting Policy into Practice is a 2-day interactive course that explains the history, purpose, and application of Section 4(f) within the context of the transportation project development process. Lessons include identifying Section 4(f) properties; explanations on types of use; an overview of Section 4(f) approval options; requirements for De Minimis determinations, individual 4(f) evaluations, and nationwide programmatic evaluations; selecting the appropriate approval option; and the relationship of Section 4(f) with NEPA and other environment laws and regulations.

Outcomes

Upon completion of the course, participants will be able to:

- Explain the history and purpose of Section 4(f)
- Identify the agencies subject to Section 4(f) compliance
- Describe the applicability criteria for Section 4(f) properties
- Describe the relationship among Section 4(f), NEPA project development, and other environmental requirements
- Differentiate the roles and responsibilities of participants in the Section 4(f) process
- Apply the Section 4(f) decision-making process within transportation project development
- Describe what is necessary to document Section 4(f) compliance

Target Audience

State Departments of TransportationFHWA Headquarters and Field staff, including Federal LandsConsultantsOfficials with jurisdiction of affected Section 4(f) resources, e.g. State Historic Preservation Offices, Tribal Historic Preservation Offices, park owners, etc.Other Federal agencies involved with environmental resourcesLocal agencies, including project sponsors and transit agenciesPublic/Special interest groups or Non-Governmental Organizations (NGOs) Transportation Planning PartnersTribes

Training Level: Basic

Fee: 2016: $650 Per Person; 2017: N/A

Length: 2 DAYS (CEU: 1.4 UNITS)

Class Size: MINIMUM: 20; MAXIMUM: 30

NHI Customer Service: (877) 558-6873 • nhicustomerservice@dot.gov

Course Number

FHWA-NHI-151043

Course Title

Transportation and Land Use

The course is designed to help practitioners develop a multimodal transportation system that supports desired land uses and helps them shape land uses to support the transportation system. Course lessons include the principles of transportation and land use; the processes through which transportation and land use issues can be jointly addressed; and implementation steps to ensure that transportation and land use systems are designed in a compatible, mutually supportive manner.

Outcomes

Upon completion of the course, participants will be able to:

- Explain how transportation decisions affect land use, growth patterns and related community impacts on both regional and local scales.
- Explain how land use patterns affect peoples' travel patterns and the overall performance of the transportation system.
- Describe the various transportation planning processes--including statewide planning, metropolitan planning, corridor planning/alternatives analysis, the NEPA process, subarea planning, and project development--and how land use considerations can be integrated into these processes.
- Describe local comprehensive planning and land use regulatory activities, and how the process and outcomes of these activities can support local and regional transportation objectives.
- Identify the full range of stakeholders--including public agencies, private and nonprofit organizations, and the general public--who should be involved in transportation and land use planning and decisionmaking, and describe methods for involving these stakeholders.
- Describe methods that are available for implementing coordinated transportation and land use strategies.
- Identify analytical tools that are available for measuring and forecasting the impacts of transportation and land use decisions.

Target Audience

Primary: Mid-level State DOT employees, City and County engineers and planners, MPO staff, transit operators, Federal employees (FHWA, FTA, EPA), resource agency staff, consultants. Secondary: Elected officials, regulatory agency staff, local zoning officials, site designers, citizen activists, developers, media representatives and business leaders.

Training Level: Intermediate

Fee: 2016: $600 Per Person; 2017: N/A

Length: 3 DAYS (CEU: 1.8 UNITS)

Class Size: MINIMUM: 20; MAXIMUM: 30

NHI Customer Service: (877) 558-6873 • nhicustomerservice@dot.gov

Course Number

FHWA-NHI-151050

Course Title

Traffic Monitoring Programs: Guidance and Procedures

Participants are encouraged to bring their own copy of the FHWA Traffic Monitoring Guide 2013 and a calculator. The training room must be large enough to allow for group exercises, as well as room to display local traffic data collection equipment.

Additionally, the FHWA Office of Highway Policy Information offers a complimentary presentation of the Travel Monitoring Analysis System (TMAS) in conjunction with this training course. Please contact Steven Jessberger (Steven.Jessberger@dot.gov) for more information.

Developed in conjunction with the 5th revision of the FHWA Traffic Monitoring Guide (TMG 2013), this new course replaces NHI 151018 and offers guidance on how to manage a successful traffic monitoring program. The training begins with an overview of Federal traffic monitoring regulations and a presentation of the host State's traffic monitoring program. Subsequent lessons introduce federal guidance, best practices, and recommended procedures for developing a data collection framework for traffic volume, speed, classification, weight, and non-motorized programs. The course also incorporates related traffic monitoring elements of transportation management and operations, traffic data needs and uses, traffic data submittal requirements, and relevant traffic monitoring research. The critical importance of quality data collection is emphasized to support project planning, programming, design, and maintenance decisions-- all of which affect the Nation's transportation network.

Outcomes

Upon completion of the course, participants will be able to:

- Describe the appropriate use of the TMG
- Describe the TMG procedures for obtaining traffic monitoring data for Federal and State programs
- Explain how to apply traffic monitoring data to answer specific questions on Federal and State issues
- Explain traffic data reporting requirements
- Explain the value of cooperative and multi-disciplinary approaches to traffic monitoring programs

Target Audience

This Instructor-led training (ILT) course is designed for transportation professionals involved in traffic monitoring programs. Primarily intended for FHWA and State DOT staff, this training is also relevant to regional and local government staff, as well as others whose roles include development and/or oversight of traffic monitoring programs. There are no course pre-requisites or assumed pre-training competencies.

Training Level: Basic

Fee: 2016: $450 Per Person; 2017: N/A

Length: 2 DAYS (CEU: 1.4 UNITS)

Class Size: MINIMUM: 20; MAXIMUM: 30

NHI Customer Service: (877) 558-6873 • nhicustomerservice@dot.gov

Course Number

FHWA-NHI-139006

Course Title

Integrating Freight in the Transportation Planning Process - WBT-Standard Version

This training is a prerequisite of another NHI training and is offered at no cost.

Freight transportation issues can be complex and involve many different stakeholders, all of whom have different perspectives on the freight transportation system. The challenge faced by many public-sector transportation planners is how to best incorporate these freight perspectives into the transportation planning process in a way that results in a safe and efficient transportation system for both people and goods. This Web-based training course will provide a greater understanding of freight trends, its stakeholders, and its issues, so that public-sector transportation planners are better able to incorporate freight into their respective transportation planning processes and programs.

This is a prerequisite course for other Freight courses.

In accordance with the Rehabilitation Act of 1973, as amended, this WBT is also available in an accessible 508 compliant version. See course number FHWA-NHI-139006W for more information.

Outcomes

Upon completion of the course, participants will be able to:

- Upon completion of the course, participants will be able to:
- Identify the stakeholders involved in freight transportation
- Explain the role of different modes in freight transportation
- Describe some trends affecting freight transportation, and their impact on a State's transportation system and communities
- Discuss some of the common issues that prevent freight from being fully incorporated into the planning process
- Identify key resources to help guide statewide and metropolitan freight planning effort

Target Audience

Transportation planners and freight transportation planners from State DOTs, MPOs, local governments, and Federal agencies.

Training Level: Basic

Fee: 2016: $0 Per Person; 2017: N/A

Length: 6 HOURS (CEU: .6 UNITS)

Class Size: MINIMUM: 1; MAXIMUM: 1

NHI Customer Service: (877) 558-6873 • nhicustomerservice@dot.gov

Course Number

FHWA-NHI-139006W

Course Title

Integrating Freight in the Transportation Planning Process - WBT-Accessible 508 Version

This training is a prerequisite of another NHI training and is offered at no cost.

Freight transportation issues can be complex and involve many different stakeholders, all of whom have different perspectives on the freight transportation system. The challenge faced by many public-sector transportation planners is how to best incorporate these freight perspectives into the transportation planning process in a way that results in a safe and efficient transportation system for both people and goods. This Web-based training course will provide a greater understanding of freight trends, its stakeholders, and its issues, so that public-sector transportation planners are better able to incorporate freight into their respective transportation planning processes and programs.

This is a prerequisite course for other Freight courses.

In accordance with the Rehabilitation Act of 1973, as amended, this WBT was developed as an accessible 508 compliant version. See course number FHWA-NHI-139006 for the standard WBT version.

Outcomes

Upon completion of the course, participants will be able to:

- Upon completion of the course, participants will be able to:
- Identify the stakeholders involved in freight transportation
- Explain the role of different modes in freight transportation
- Describe some trends affecting freight transportation, and their impact on a State's transportation system and communities
- Discuss some of the common issues that prevent freight from being fully incorporated into the planning process
- Identify key resources to help guide statewide and metropolitan freight planning effort

Target Audience

Transportation planners and freight transportation planners from State DOTs, MPOs, local governments, and Federal agencies.

Training Level: Basic

Fee: 2016: $0 Per Person; 2017: N/A

Length: 6 HOURS (CEU: .6 UNITS)

Class Size: MINIMUM: 1; MAXIMUM: 1

NHI Customer Service: (877) 558-6873 • nhicustomerservice@dot.gov

COURSE NUMBER

FHWA-NHI-141045

COURSE TITLE

Real Estate Acquisition under the Uniform Act: An Overview - WEB-BASED

The Uniform Relocation Assistance and Real Property Acquisition Policies Act of 1970 (Uniform Act) is the basis for Federally-funded real estate acquisition programs. This self-paced training provides an overview of the Uniform Act's three key elements: valuation, acquisition, and relocation. This course underscores the importance of following Uniform Act requirements when acquiring property for a Federally-funded transportation project.

OUTCOMES

Upon completion of the course, participants will be able to:

- Provide a basic overview of the Uniform Relocation Assistance and Real Property Acquisition Policies Act of 1970 (Uniform Act)
- Discuss the three key elements of the Uniform Act: valuation/appraisal, acquisition and relocation
- Explain how to develop an estimate of just compensation using the appraisal process or appraisal waiver procedure(s)
- Identify relocation benefits and services required by the Uniform Act
- List places to obtain relevant resource documents and materials

TARGET AUDIENCE

Federal, State, and local government employees and consultants who acquire real estate for Federally-funded transportation projects. This includes acquisition and relocation agents; program or project managers; grant administrators or grant recipients; appraisers, realty specialists, attorneys, engineers, planners, and others involved with real property acquisition.

TRAINING LEVEL: Basic

FEE: 2016: $0 Per Person; 2017: N/A

LENGTH: 6 HOURS (CEU: 0 UNITS)

CLASS SIZE: MINIMUM: 0; MAXIMUM: 0

NHI Customer Service: (877) 558-6873 • nhicustomerservice@dot.gov

Course Number

FHWA-NHI-141052

Course Title

Successful Acquisition under the Uniform Act

This course will provide the knowledge and skills that a public agency negotiator needs to complete acquisitions that comply with the Uniform Act.

Outcomes

Upon completion of the course, participants will be able to:

- Explain the legal basis for land acquisition by a governmental entity
- Identify the pre-acquisition materials necessary for property acquisition
- Explain the basics of the valuation process
- Describe the acquisition process under the Uniform Act
- Formulate effective negotiation skills, using best practices
- Discuss legal aspects of real property acquisition
- Discuss the role and limitations of consultants in the acquisition process

Target Audience

Federal, State, and local public agencies, FHWA personnel, contractors, and other interested persons.

Training Level: Basic

Fee: 2016: $800 Per Person; 2017: N/A

Length: 3 DAYS (CEU: 1.8 UNITS)

Class Size: MINIMUM: 20; MAXIMUM: 30

NHI Customer Service: (877) 558-6873 • nhicustomerservice@dot.gov

Course Number

FHWA-NHI-142036

Course Title

Public Involvement in the Transportation Decisionmaking Process

Public involvement is much more than public hearings. It involves creative thinking as well as the willingness and ability to interact openly and sensitively to the public's preferred forms of communication and participation. Public involvement is about reaching out to and involving the public in transportation decisionmaking. The public should have a role in every phase of decisionmaking, including the design of the participation plan itself. Successful public involvement addresses the public's procedural, psychological, and substantive needs while gathering useful information. By focusing on interests--rather than positions--public involvement can become more meaningful as well as useful.

Outcomes

Upon completion of the course, participants will be able to:

- Describe U.S. DOT transportation decisionmaking processes, including those that trigger the National Environmental Policy Act
- Describe the relationship between public involvement and decisionmaking
- Develop a public involvement plan with stakeholder assistance that includes attention to non-traditional populations as an evaluation component
- Describe interest-based problem solving and the values that underlie it
- Identify ways to enhance public involvement plans

Target Audience

Federal, State, and local transportation agency staff, metropolitan planning organization personnel, transit operators, consultants, and others who are responsible for planning, implementing, or participating in any phase of the public involvement process.

Training Level: Basic

Fee: 2016: $1050 Per Person; 2017: N/A

Length: 3 DAYS (CEU: 1.8 UNITS)

Class Size: MINIMUM: 20; MAXIMUM: 30

NHI Customer Service: (877) 558-6873 • nhicustomerservice@dot.gov

Course Number

FHWA-NHI-142042

Course Title

Fundamentals of Environmental Justice

Addressing environmental justice applies to every stage of transportation decisionmaking. The U.S. Department of Transportation (USDOT) and its partners are committed to integrating the nondiscrimination principles of environmental justice into all Federal-aid programs. Through these and other transportation programs, many opportunities exist to establish partnerships with other public and private organizations to create livable communities that meet the needs of all people. This course presents participants with a framework for using a variety of approaches and tools for accomplishing environmental justice goals in Federal-aid programs and other transportation projects.

Outcomes

Upon completion of the course, participants will be able to:

- Define environmental justice and describe its relationship to Title VI
- Explain the fundamental principles of environmental justice
- Apply the principles of environmental justice to transportation decisions
- Identify how environmental justice applies to each stage of transportation decisionmaking
- Describe the benefits of environmental justice in transportation decisionmaking
- Develop proactive strategies, methods, and techniques to implement environmental justice in transportation programs and projects

Target Audience

We strongly encourage the sponsoring organization to invite a mix of civil rights, environmental, engineering, and planning staff from the following agencies: Federal, State, and regional organizations, local transit agencies, and consultants who act on their behalf. Others who interact with minority and low-income communities; provide community services; and elected officials and their representatives.

Training Level: Basic

Fee: 2016: $650 Per Person; 2017: N/A

Length: 2 DAYS (CEU: 1.2 UNITS)

Class Size: MINIMUM: 20; MAXIMUM: 30

NHI Customer Service: (877) 558-6873 • nhicustomerservice@dot.gov

Course Number

FHWA-NHI-142051

Course Title

Highway Traffic Noise

This course was recently updated to reflect current FHWA noise regulations and policies: Title 23 CFR Part 772: Procedures for Abatement of Highway Traffic Noise and Construction Noise, and related State highway agency noise policies.

We recommend that participants registering for this course take the NHI 142063 Highway Traffic Noise: Basic Acoustics in advance of their session.

The Interactive Sound Information System (ISIS) demonstrations that are part of this training are deliberately very loud. So, we encourage Session Hosts to secure a training room accordingly.

This course provides an introductory- yet comprehensive- overview of highway traffic noise that will help educate engineers, environmental specialists, designers, planners, and consultants about traffic noise and ways to reduce the impacts. This course is designed not just for those performing noise analyses, but also for those responsible for the review and approval of noise studies and incorporating the results of these studies into design and environmental documents. Among topics covered are the basic principles of acoustics, how to determine when a noise analysis is required, typical strategies to mitigate noise in highway projects, noise study requirements and documentation, noise measurement, construction noise, and public involvement.

Participants will also learn about noise-compatible planning, which encourages local governments and developers to avoid noise-sensitive land uses adjacent to highways. And the course includes an overview of the FHWA Traffic Noise Model (FHWA TNM)--developed to predict noise levels and evaluate noise abatement measures.

Additionally, the NHI commissioned a customized version of the Interactive Sound Information System (ISIS) as part of the curriculum design for this course. ISIS is a noise simulation software program that employs high-quality digital recordings, precise sound control, and graphic imagery to present basic acoustical concepts and sounds from various traffic loads. ISIS also demonstrates the sound level reductions provided by various noise barrier configurations.

Outcomes

Upon completion of the course, participants will be able to:

- Explain the basic principles of acoustics
- Describe all necessary documentation to fulfill FHWA noise requirements, as codified in 23 CFR 772
- Determine when a noise study is required
- Explain applicable Federal noise abatement regulations and policies

Target Audience

FHWA staff; State department of transportation environmental specialists, designers, planners or engineers; city or county environmental engineers, coordinators or specialists; consultants.

Training Level: Basic

Fee: 2016: $1000 Per Person; 2017: N/A

Length: 3 DAYS (CEU: 1.8 UNITS)

Class Size: MINIMUM: 20; MAXIMUM: 30

NHI Customer Service: (877) 558-6873 • nhicustomerservice@dot.gov

COURSE NUMBER

FHWA-NHI-142068

COURSE TITLE

Air Quality Planning: Clean Air Act Overview - WEB-BASED

The purpose of this training is to provide participants with an overview of air quality planning, including requirements, processes, interactions with and implications for, transportation planning and project development.

This is the first in a future series of air quality Web-based trainings (WBTs):

142068: Clear Air Act Overview

142069: SIP and TCM Requirements and Policies

142070: SIP Development Process

142071: Transportation Conformity

OUTCOMES

Upon completion of the course, participants will be able to:

- Define the purpose of the Clean Air Act
- Describe the 1990 Clean Air Act Amendments
- Identify and explain Clean Air Act Amendment provisions relevant to transportation
- Recognize impacts of Clean Air Act

TARGET AUDIENCE

The target audience for the Air Quality Series is transportation and air quality planners and engineers from State and local departments of transportation (DOT), metropolitan planning organizations (MPO), transit agencies, Federal agencies (Federal Highway Administration, Federal Transit Administration, U.S. Environmental Protection Agency, U.S. Department of Energy, etc.), and State and local environmental agencies. Others include transportation and environmental consultants, public officials and staff members, community and interest groups, as well as other stakeholders in the planning process.

TRAINING LEVEL: Basic

FEE: 2016: $0 Per Person; 2017: N/A

LENGTH: 1.5 HOURS (CEU: 0 UNITS)

CLASS SIZE: MINIMUM: 1; MAXIMUM: 1

NHI Customer Service: (877) 558-6873 • nhicustomerservice@dot.gov

Course Number

FHWA-NHI-142069

Course Title

Air Quality Planning: SIP and TCM Requirements and Policies - WEB-BASED

This course covers the different types of SIPs and key CAA SIP requirements general to all SIPs and specific to ozone, CO and PM SIPs; discusses how the EPA processes SIPs; explores the key features of EPA SIP policies and how they differ from CAA requirements; and explains RACM and how it applies to TCMs.

This is the second in a future series of air quality Web-based trainings (WBTs):

142068: Clear Air Act Overview

142069: SIP and TCM Requirements and Policies

142070: SIP Development Process

142071: Transportation Conformity

Outcomes

Upon completion of the course, participants will be able to:

- Define SIP
- List different types of SIPs and their purposes
- Identify SIP requirements in Title I of the Clean Air Act
- Describe TCM requirements
- Describe what is meant by Reasonably Available Control Measure, or RACM, and how this applies to TCMs

Target Audience

The target audience for the Air Quality Series is transportation and air quality planners and engineers from State and local departments of transportation (DOT), metropolitan planning organizations (MPO), transit agencies, Federal agencies (Federal Highway Administration, Federal Transit Administration, U.S. Environmental Protection Agency, U.S. Department of Energy, etc.), and State and local environmental agencies. Others include transportation and environmental consultants, public officials and staff members, community and interest groups, as well as other stakeholders in the planning process.

Training Level: Basic

Fee: 2016: $0 Per Person; 2017: N/A

Length: 1 HOURS (CEU: 0 UNITS)

Class Size: MINIMUM: 1; MAXIMUM: 1

NHI Customer Service: (877) 558-6873 • nhicustomerservice@dot.gov

Course Number

FHWA-NHI-142070

Course Title

Air Quality Planning: SIP Development Process - WEB-BASED

This course provides an overview of the State Implementation Plan (or SIP) development process, focusing on agency roles, with an explanation of the problem definition and solution parts of the process. This course also covers motor vehicle emission budgets that are included in SIPs and used in conformity determinations, as well as describes EPAs procedures in approving and disapproving SIPs.

This is the third in a series of air quality Web-based trainings (WBTs):

142068: Clear Air Act Overview

142069: SIP and TCM Requirements and Policies

142070: SIP Development Process

142071: Transportation Conformity

Outcomes

Upon completion of the course, participants will be able to:

- Describe the SIP development process;
- Identify the various emission sources and describe emission trends;
- List the steps involved in preparing emission inventories;
- Describe how SIP emission reductions are determined; and
- Describe the different types of control measures, emphasizing the role of transportation-related strategies.

Target Audience

The target audience for the Air Quality Series is transportation and air quality planners and engineers from State and local departments of transportation (DOT), metropolitan planning organizations (MPO), transit agencies, Federal agencies (Federal Highway Administration, Federal Transit Administration, U.S. Environmental Protection Agency, U.S. Department of Energy, etc.), and State and local environmental agencies. Others include transportation and environmental consultants, public officials and staff members, community and interest groups, as well as other stakeholders in the planning process.

Training Level: Basic

Fee: 2016: $0 Per Person; 2017: N/A

Length: 2 HOURS (CEU: 0 UNITS)

Class Size: MINIMUM: 1; MAXIMUM: 1

NHI Customer Service: (877) 558-6873 • nhicustomerservice@dot.gov

COURSE NUMBER

FHWA-NHI-142071

COURSE TITLE

Air Quality Planning: Transportation Conformity - WEB-BASED

This course defines transportation conformity and is designed for individuals that are new to transportation conformity, with little to no experience with the Transportation Conformity Rule.

This introductory transportation conformity course will answer questions related to the "what" of transportation conformity. What is transportation conformity? What activities are covered by conformity? What are the major requirements?

This course does not address how transportation conformity regulations are met. That topic is an advanced subject matter area and out of scope for an introductory transportation conformity course.

This is the fourth in a future series of air quality Web-based trainings (WBTs):

142068: Clear Air Act Overview

142069: SIP and TCM Requirements and Policies

142070: SIP Development Process

142071: Transportation Conformity

OUTCOMES

Upon completion of the course, participants will be able to:

- Relate transportation conformity to Transportation Improvement Programs (TIPs) and transportation plans
- Define transportation conformity
- Explain the transportation activities that are subject to conformity in a given timeframe
- Describe transportation conformity requirements for different activities
- Explain stakeholder responsibilities related to transportation conformity

TARGET AUDIENCE

The target audience for the Air Quality Series is transportation and air quality planners and engineers from State and local departments of transportation (DOT), metropolitan planning organizations (MPO), transit agencies, Federal agencies (Federal Highway Administration, Federal Transit Administration, U.S. Environmental Protection Agency, U.S. Department of Energy, etc.), and State and local environmental agencies. Others include transportation and environmental consultants, public officials and staff members, community and interest groups, as well as other stakeholders in the planning process.

TRAINING LEVEL: Basic

FEE: 2016: $0 Per Person; 2017: N/A

LENGTH: 1.5 HOURS (CEU: 0 UNITS)

CLASS SIZE: MINIMUM: 1; MAXIMUM: 1

NHI Customer Service: (877) 558-6873 • nhicustomerservice@dot.gov

Course Number

FHWA-NHI-142073

Course Title

Applying Section 4(f): Putting Policy into Practice

NHI 142073 Applying Section 4(f): Putting Policy into Practice is a 2-day interactive course that explains the history, purpose, and application of Section 4(f) within the context of the transportation project development process. Lessons include identifying Section 4(f) properties; explanations on types of use; an overview of Section 4(f) approval options; requirements for De Minimis determinations, individual 4(f) evaluations, and nationwide programmatic evaluations; selecting the appropriate approval option; and the relationship of Section 4(f) with NEPA and other environment laws and regulations.

Outcomes

Upon completion of the course, participants will be able to:

- Explain the history and purpose of Section 4(f)
- Identify the agencies subject to Section 4(f) compliance
- Describe the applicability criteria for Section 4(f) properties
- Describe the relationship among Section 4(f), NEPA project development, and other environmental requirements
- Differentiate the roles and responsibilities of participants in the Section 4(f) process
- Apply the Section 4(f) decision-making process within transportation project development
- Describe what is necessary to document Section 4(f) compliance

Target Audience

State Departments of TransportationFHWA Headquarters and Field staff, including Federal LandsConsultantsOfficials with jurisdiction of affected Section 4(f) resources, e.g. State Historic Preservation Offices, Tribal Historic Preservation Offices, park owners, etc.Other Federal agencies involved with environmental resourcesLocal agencies, including project sponsors and transit agenciesPublic/Special interest groups or Non-Governmental Organizations (NGOs) Transportation Planning PartnersTribes

Training Level: Basic

Fee: 2016: $650 Per Person; 2017: N/A

Length: 2 DAYS (CEU: 1.4 UNITS)

Class Size: MINIMUM: 20; MAXIMUM: 30

NHI Customer Service: (877) 558-6873 • nhicustomerservice@dot.gov

Course Number

FHWA-NHI-151043

Course Title

Transportation and Land Use

The course is designed to help practitioners develop a multimodal transportation system that supports desired land uses and helps them shape land uses to support the transportation system. Course lessons include the principles of transportation and land use; the processes through which transportation and land use issues can be jointly addressed; and implementation steps to ensure that transportation and land use systems are designed in a compatible, mutually supportive manner.

Outcomes

Upon completion of the course, participants will be able to:

- Explain how transportation decisions affect land use, growth patterns and related community impacts on both regional and local scales.
- Explain how land use patterns affect peoples' travel patterns and the overall performance of the transportation system.
- Describe the various transportation planning processes--including statewide planning, metropolitan planning, corridor planning/alternatives analysis, the NEPA process, subarea planning, and project development--and how land use considerations can be integrated into these processes.
- Describe local comprehensive planning and land use regulatory activities, and how the process and outcomes of these activities can support local and regional transportation objectives.
- Identify the full range of stakeholders--including public agencies, private and nonprofit organizations, and the general public--who should be involved in transportation and land use planning and decisionmaking, and describe methods for involving these stakeholders.
- Describe methods that are available for implementing coordinated transportation and land use strategies.
- Identify analytical tools that are available for measuring and forecasting the impacts of transportation and land use decisions.

Target Audience

Primary: Mid-level State DOT employees, City and County engineers and planners, MPO staff, transit operators, Federal employees (FHWA, FTA, EPA), resource agency staff, consultants. Secondary: Elected officials, regulatory agency staff, local zoning officials, site designers, citizen activists, developers, media representatives and business leaders.

Training Level: Intermediate

Fee: 2016: $600 Per Person; 2017: N/A

Length: 3 DAYS (CEU: 1.8 UNITS)

Class Size: MINIMUM: 20; MAXIMUM: 30

NHI Customer Service: (877) 558-6873 • nhicustomerservice@dot.gov

COURSE NUMBER

FHWA-NHI-151044

COURSE TITLE

Traffic Monitoring and Pavement Design Programs - WEB-BASED

The goal of this online presentation is to promote interaction and collaboration between traffic monitoring program staff and pavement program staff. The presentation supports implementation of the new Mechanistic Empirical Pavement Design Guide (MEPDG). FHWA's Office of Highway Policy Information, in collaboration with the Design Guide Implementation Team (DIGI Team), created this presentation to help ensure that pavement data needs are met with the existing traffic monitoring program or adjustments to the program.

Please note that the Flash Player must be installed on your computer in order to view the presentation.

OUTCOMES

Upon completion of the course, participants will be able to:

- Describe the traffic monitoring program
- Describe the pavement design program, as it relates to traffic monitoring
- Explain the interconnectivity and interdependency between the traffic monitoring and pavement design programs
- Identify ways to make the best use of available funding to meet users' data needs

TARGET AUDIENCE

Federal and State department of transportation specialists, designers, and administrators who are responsible for traffic monitoring and pavement programs. Local transportation agencies, as well as those who are new to the traffic program and pavement programs, may also find this presentation to be interesting and helpful.

TRAINING LEVEL: Basic

FEE: 2016: $25 Per Person; 2017: N/A

LENGTH: 1 HOURS (CEU: 0 UNITS)

CLASS SIZE: MINIMUM: 1; MAXIMUM: 1

NHI Customer Service: (877) 558-6873 • nhicustomerservice@dot.gov

COURSE NUMBER

FHWA-NHI-151046

COURSE TITLE

FHWA Planning and Research Grants: History, Sources, and Regulations - WEB-BASED

NHI 151046 is a 2-hour WBT that introduces Federal financial assistance and FHWA's planning and research grant regulations. The course covers sources for requirements and funds; sources and hierarchy of Federal grant requirements; and the process for providing grant funding, as it relates to 23 CFR 420 and 450.

This Web-based training (WBT) course is one of a series designed as an introduction to FHWA planning and research grant administration. The series includes four independent WBTs that cover the history of FHWA planning and research grants, the Common Grant Rule, Cost Principles, and Audits:

FHWA-NHI-151046--FHWA Planning and Research Grants: History, Sources, and Regulations;

FHWA-NHI-151047--FHWA Planning and Research Grants: Common Grant Rule;

FHWA-NHI-151048--FHWA Planning and Research Grants: Cost Principles; and

FHWA-NHI-151049--FHWA Planning and Research Grants: Audits.

These WBTs are designed to complement FHWA-NHI-151021 Administration of FHWA Planning and Research Grants, a 2-day Instructor-led training course. They are not intended to replace NHI 151021.

OUTCOMES

Upon completion of the course, participants will be able to:

- Identify basic principles of grant administration
- Describe the hierarchy of laws, regulations, requirements, and the relationship among them
- Explain terminology associated with grant administration
- Explain the purpose and policy for 23 CFR, Part 420

TARGET AUDIENCE

The target audience includes FHWA, State Department of Transportation (State DOT), Metropolitan Planning Organization (MPO), and other agency staff who expend or administer Federal-aid funds, including planning, engineering, and fiscal staff.

TRAINING LEVEL: Basic

FEE: 2016: $25 Per Person; 2017: N/A

LENGTH: 2 HOURS (CEU: 0 UNITS)

CLASS SIZE: MINIMUM: 1; MAXIMUM: 1

NHI Customer Service: (877) 558-6873 • nhicustomerservice@dot.gov

Course Number

FHWA-NHI-151047

Course Title

FHWA Planning and Research Grants: Common Grant Rule - WEB-BASED

NHI 151047 is a 3-hour WBT that discusses the process for providing grant funding and Common Grant Rule requirements found in 49 CFR Part 18 (i.e. pre-award, post award, and after-the-award). The course reviews OMB grant circulars and the general Grant Rule for Institutions of Higher Education, Hospitals and other Non-Profit Organizations.

This Web-based training (WBT) course is one of a series designed as an introduction to FHWA planning and research grant administration. The series includes four independent WBTs that cover the history of FHWA planning and research grants, the Common Grant Rule, Cost Principles, and Audits:

FHWA-NHI-151046--FHWA Planning and Research Grants: History, Sources, and Regulations;

FHWA-NHI-151047--FHWA Planning and Research Grants: Common Grant Rule;

FHWA-NHI-151048--FHWA Planning and Research Grants: Cost Principles; and

FHWA-NHI-151049--FHWA Planning and Research Grants: Audits.

These WBTs are designed to complement FHWA-NHI-151021 Administration of FHWA Planning and Research Grants, a 2-day Instructor-led training course. They are not intended to replace NHI 151021.

Outcomes

Upon completion of the course, participants will be able to:

- Identify laws and implementing regulations (issued by FHWA, U.S. Department of Administration [DOT] and applicable OMB circulars) that apply to the administration of Federal-aid funding
- Describe how to apply Federal laws and regulations to administer funds
- Define the basic principles of Federalism incorporated in the Common Grant Rule
- Describe pre-award, post-award, and after-the-award Common Grant Rule requirements
- Differentiate among requirements at different stages of the Common Grant Rule

Target Audience

The target audience includes FHWA, State Department of Transportation (State DOT), Metropolitan Planning Organization (MPO), and other agency staff who expend or administer Federal-aid funds, including planning, engineering, and fiscal staff.

Training Level: Basic

Fee: 2016: $25 Per Person; 2017: N/A

Length: 3 HOURS (CEU: 0 UNITS)

Class Size: MINIMUM: 1; MAXIMUM: 1

NHI Customer Service: (877) 558-6873 • nhicustomerservice@dot.gov

Course Number

FHWA-NHI-151048

Course Title

FHWA Planning and Research Grants: Cost Principles - WEB-BASED

NHI 151048 is a 2.5-hour WBT that presents general cost principles, selected items of cost, and indirect cost plans (2CFR 225, fomerly OMB Circular A-87).

This Web-based training (WBT) course is one of a series designed as an introduction to FHWA planning and research grant administration. The series includes four independent WBTs that cover the history of FHWA planning and research grants, the Common Grant Rule, Cost Principles, and Audits:

FHWA-NHI-151046--FHWA Planning and Research Grants: History, Sources, and Regulations;

FHWA-NHI-151047--FHWA Planning and Research Grants: Common Grant Rule;

FHWA-NHI-151048--FHWA Planning and Research Grants: Cost Principles; and

FHWA-NHI-151049--FHWA Planning and Research Grants: Audits.

These WBTs are designed to complement FHWA-NHI-151021 Administration of FHWA Planning and Research Grants, a 2-day Instructor-led training course. They are not intended to replace NHI 151021.

Outcomes

Upon completion of the course, participants will be able to:

- Identify basic cost principles
- Identify selected items of cost
- Explain the purpose of indirect cost plans

Target Audience

The target audience includes FHWA, State Department of Transportation (State DOT), Metropolitan Planning Organization (MPO), and other agency staff who expend or administer Federal-aid funds, including planning, engineering, and fiscal staff.

Training Level: Basic

Fee: 2016: $25 Per Person; 2017: N/A

Length: 2.5 HOURS (CEU: 0 UNITS)

Class Size: MINIMUM: 1; MAXIMUM: 1

NHI Customer Service: (877) 558-6873 • nhicustomerservice@dot.gov

Course Number

FHWA-NHI-151049

Course Title

FHWA Planning and Research Grants: Audits - WEB-BASED

NHI 151049 is a 2-hour WBT that covers basic audit requirements and reviews OMB Circular A-133. The course provides examples of when a single audit is required and explains what triggers a single audit.

This Web-based training (WBT) course is one of a series designed as an introduction to FHWA planning and research grant administration. The series includes four independent WBTs that cover the history of FHWA planning and research grants, the Common Grant Rule, Cost Principles, and Audits:

FHWA-NHI-151046--FHWA Planning and Research Grants: History, Sources, and Regulations;

FHWA-NHI-151047--FHWA Planning and Research Grants: Common Grant Rule;

FHWA-NHI-151048--FHWA Planning and Research Grants: Cost Principles; and

FHWA-NHI-151049--FHWA Planning and Research Grants: Audits.

These WBTs are designed to complement FHWA-NHI-151021 Administration of FHWA Planning and Research Grants, a 2-day Instructor-led training course. They are not intended to replace NHI 151021.

Outcomes

Upon completion of the course, participants will be able to:

- Describe basic audit requirements
- Identify single audit roles and responsibilities

Target Audience

The target audience includes FHWA, State Department of Transportation (State DOT), Metropolitan Planning Organization (MPO), and other agency staff who expend or administer Federal-aid funds, including planning, engineering, and fiscal staff.

Training Level: Basic

Fee: 2016: $25 Per Person; 2017: N/A

Length: 2 HOURS (CEU: 0 UNITS)

Class Size: MINIMUM: 1; MAXIMUM: 1

NHI Customer Service: (877) 558-6873 • nhicustomerservice@dot.gov

Course Number

FHWA-NHI-151050

Course Title

Traffic Monitoring Programs: Guidance and Procedures

Participants are encouraged to bring their own copy of the FHWA Traffic Monitoring Guide 2013 and a calculator. The training room must be large enough to allow for group exercises, as well as room to display local traffic data collection equipment.

Additionally, the FHWA Office of Highway Policy Information offers a complimentary presentation of the Travel Monitoring Analysis System (TMAS) in conjunction with this training course. Please contact Steven Jessberger (Steven.Jessberger@dot.gov) for more information.

Developed in conjunction with the 5th revision of the FHWA Traffic Monitoring Guide (TMG 2013), this new course replaces NHI 151018 and offers guidance on how to manage a successful traffic monitoring program. The training begins with an overview of Federal traffic monitoring regulations and a presentation of the host State's traffic monitoring program. Subsequent lessons introduce federal guidance, best practices, and recommended procedures for developing a data collection framework for traffic volume, speed, classification, weight, and non-motorized programs. The course also incorporates related traffic monitoring elements of transportation management and operations, traffic data needs and uses, traffic data submittal requirements, and relevant traffic monitoring research. The critical importance of quality data collection is emphasized to support project planning, programming, design, and maintenance decisions-- all of which affect the Nation's transportation network.

Outcomes

Upon completion of the course, participants will be able to:

- Describe the appropriate use of the TMG
- Describe the TMG procedures for obtaining traffic monitoring data for Federal and State programs
- Explain how to apply traffic monitoring data to answer specific questions on Federal and State issues
- Explain traffic data reporting requirements
- Explain the value of cooperative and multi-disciplinary approaches to traffic monitoring programs

Target Audience

This Instructor-led training (ILT) course is designed for transportation professionals involved in traffic monitoring programs. Primarily intended for FHWA and State DOT staff, this training is also relevant to regional and local government staff, as well as others whose roles include development and/or oversight of traffic monitoring programs. There are no course pre-requisites or assumed pre-training competencies.

Training Level: Basic

Fee: 2016: $450 Per Person; 2017: N/A

Length: 2 DAYS (CEU: 1.4 UNITS)

Class Size: MINIMUM: 20; MAXIMUM: 30

NHI Customer Service: (877) 558-6873 • nhicustomerservice@dot.gov

COURSE NUMBER

FHWA-NHI-151051

WBT

COURSE TITLE

Highway Program Funding: An Overview WBT

This self-paced web-based training provides an overview of the Federal-aid Highway Program, focusing on various aspects of highway program funding unique to the Federal Highway Administration (FHWA). Topics include: Federal and State roles in the Federal-aid Highway Program, the Highway Trust Fund, the phases of the Federal-Aid Highway funding lifecycle, and the current highway authorization law, the Moving Ahead for Progress in the 21st Century Act (MAP-21).

OUTCOMES

Upon completion of the course, participants will be able to:

- Describe the Federal-aid Highway Program and the Federal and State roles in relation to it
- Explain the purpose of the Highway Trust Fund and identify its major sources of revenue
- Identify the major phases of the lifecycle of Federal-aid highway funding
- List characteristics of the current highway authorization law, the Moving Ahead for Progress in the 21st Century Act (MAP-21)

TARGET AUDIENCE

This training is intended for Federal, State, regional and local government employees; Congressional staff; consultants; and others interested in the process by which Congress authorizes the Federal-aid Highway program and the FHWA distributes Federal-aid highway funding.

TRAINING LEVEL: Basic

FEE: 2016: $25 Per Person; 2017: N/A

LENGTH: 1.5 HOURS (CEU: 0 UNITS)

CLASS SIZE: MINIMUM: 1; MAXIMUM: 1

NHI Customer Service: (877) 558-6873 • nhicustomerservice@dot.gov

Course Number

FHWA-NHI-151056

Course Title

Highway Performance Monitoring System (HPMS): Concepts, Data Collection & Reporting Requirements

The National Highway Institute (NHI) workshop titled, Highway Performance Monitoring System (HPMS): Concepts, Data Collection & Reporting Requirements, is a two-day workshop intended to provide advanced, in-depth, hands-on understanding of data collection and reporting requirements for HPMS. The workshop is designed to cover:

HPMS Program Background

The HPMS 2010+ Data Model

HPMS Data Collection and Reporting Requirements

Statistical Sampling Requirements; and

The HPMS Submittal Process

Outcomes

Upon completion of the course, participants will be able to:

- Upon completion of the workshop, participants will be able to:
- Describe the Scope of HPMS
- Describe the Background of HPMS
- Describe the structure of the HPMS Data Model, in terms of the various catalogs and datasets that comprise the model
- Describe the various HPMS datasets
- Differentiate between the datasets that are to be developed/submitted by the States, and the datasets that will be developed/ maintained by FHWA
- Explain how geo-referencing is performed in HPMS for analysis and reporting purposes
- Describe the structure of the Sections and Sample Panel Identification datasets
- Explain the relationship between the Sections and Sample Panel Identification datasets and how these are used for sampling purposes
- Interpret the data collection, coding, and reporting requirements for the Sections dataset
- Describe the Sampling Framework that is used within the context of the Highway Performance Monitoring System (HPMS)
- Discuss the way in which AADT Volume Groups and Precision Levels are used for sampling purposes in HPMS
- Explain the Sample Size Estimation procedure and how it is used in HPMS
- Discuss the importance of Sample Adequacy and Sample Maintenance in HPMS
- Describe the steps involved in the annual submittal of the various HPMS datasets

Target Audience

This two‐day workshop is intended for State DOT HPMS staff, including: Staff responsible for data collection, processing, analysis, and production of the annual HPMS submittal.While primarily targeted for those responsible for the assembling the annual HPMS submittal, others who can benefit from this training are:GIS staff responsible for developing/providing HPMS‐related dataTraffic data providersPavement data providersRoad Inventory data providersMPO Staff who provide HPMS‐related data to State DOTsLocal agency staff that provides HPMS‐related data to their State DOTThis workshop is designed for those individuals seeking to obtain an understanding or expand their basic knowledge of the annual data collection and reporting requirements for HPMS. The material covered in this workshop is primarily based on requirements which were a product of the 2010+ Reassessment.

TRAINING LEVEL: Basic

FEE: 2016: $425 Per Person; 2017: N/A

LENGTH: 2 DAYS (CEU: 0 UNITS)

CLASS SIZE: MINIMUM: 20; MAXIMUM: 30

NHI Customer Service: (877) 558-6873 • nhicustomerservice@dot.gov

COURSE NUMBER

FHWA-NHI-152054

COURSE TITLE

Introduction to Urban Travel Demand Forecasting

Through classroom lectures and interactive workshops, this introductory course covers the traditional four-step modeling process of trip generation, trip distribution, mode choice, and trip assignment. The course includes presentations on land use inputs, network and zone structures, time of day factoring, and reasonableness checking.

In order to ensure that participants have a basic overview of travel demand forecasting, each registered participant will receive a Self-Instructional CD--entitled Introduction to Travel Forecasting--in advance of a scheduled session. To ensure that these CDs are shipped, we request that the Host provide the instructor coordinator with names and mailing addresses of their registrants. Participants are expected to complete the CD in advance of the session

A half day computer lab exercise is included to reinforce the concepts presented in the classroom. The hosting organization is responsible for providing MS Windows microcomputers with color graphics, color monitors, and at least 10 megabytes of hard disk space. There should be no more than two participants per computer station.

Prerequisites: Computer experience and an understanding of college-level algebra. Participants must bring scientific calculators to the session.

OUTCOMES

Upon completion of the course, participants will be able to:

- Describe the role of travel forecasting within transportation planning
- Explain the principles of the four-step model: trip generation, trip distribution, mode choice, and trip assignment
- Demonstrate how input data is used in each step of the four-step model
- Identify reasonableness checks for model inputs, outputs, and equations
- Interpret the outputs from each step

TARGET AUDIENCE

Federal, State, local planners, and engineers, and consultants who wish to gain a better understanding of the principles and applications of travel demand forecasting models.

TRAINING LEVEL: Intermediate

FEE: 2016: $800 Per Person; 2017: N/A

LENGTH: 4 DAYS (CEU: 2.4 UNITS)

CLASS SIZE: MINIMUM: 20; MAXIMUM: 30

NHI Customer Service: (877) 558-6873 • nhicustomerservice@dot.gov

COURSE NUMBER

FHWA-NHI-152072

COURSE TITLE

Highway Program Funding

Please note the 152072 course title has changed to more accurately reflect the curriculum materials.

This instructor-led training provides an overview of the Federal-aid Highway Program, focusing on various aspects of highway program funding unique to the Federal Highway Administration (FHWA). Topics include: the operation of the Highway Trust Fund and its significance to the funding level of the Federal-aid Highway Program; the content and policy implications of authorizing and appropriating legislation; the FHWA apportionment process; discussion of obligation limitation, allocations, deductions, earmarking, and transferability; and the effect of policy and budget considerations on the use of Federal-aid funds. The course has been updated to complement the new Federal-aid authorization bill.

OUTCOMES

Upon completion of the course, participants will be able to:

- Describe the flow of Federal highway funding from authorization to outlay
- Explain authorization, appropriation, apportionment, allocation, and obligation limitation
- Discuss the impact contract authority and obligation limitation have on the use of Federal funds
- Explain how the Federal budgetary process applies to the Federal-Aid Highway Program
- Describe the significance of the Highway Trust Fund to the funding levels for the Federal-Aid Highway Program

TARGET AUDIENCE

This training is intended for Federal, State, regional and local government employees; Congressional staff; consultants; and others interested in the process by which Congress authorizes the Federal-aid Highway program and the FHWA distributes Federal-aid highway funding. NHI encourages a mix of participants at each session.

TRAINING LEVEL: Basic

FEE: 2016: $425 Per Person; 2017: N/A

LENGTH: 1.5 DAYS (CEU: .9 UNITS)

CLASS SIZE: MINIMUM: 20; MAXIMUM: 40

NHI Customer Service: (877) 558-6873 • nhicustomerservice@dot.gov

TRANSPORTATION PLANNING

COURSE NUMBER

FHWA-NHI-152072A

COURSE TITLE

Highway Program Funding- Executive Session

An overview of the Federal-Aid Highway Program, focusing on various aspects of highway program financing unique to the Federal Highway Administration (FHWA). Topics include: the operation of the Highway Trust Fund and its significance to the funding level of the Federal-Aid Highway Program; the content and policy implication of authorizing and appropriating legislation; the FHWA apportionment process; discussion of obligation limitation, allocations, deductions, earmarking, and transferability; and the effect of policy and budget considerations on the use of Federal-Aid funds. This course has been updated to complement the new Federal-Aid authorization bill.

OUTCOMES

Upon completion of the course, participants will be able to:

- Describe the flow of Federal funding from authorization to reimbursement
- Explain authorization, appropriation, apportionment, allocation, and obligation limitation
- Discuss the impact contract authority and obligation limitation have on the use of Federal funds
- Explain how the Federal budgetary process applies to the Federal-aid Highway Program
- Describe the significance of the Highway Trust Fund to the funding levels for the Federal-aid Highway Program

TARGET AUDIENCE

Executives, Consultants and Senior Managers-- who work for and with governmental agencies-- and seek a broad understanding of the framework for Federal-aid Highway Financing.

TRAINING LEVEL: Intermediate

FEE: 2016: $375 Per Person; 2017: N/A

LENGTH: 5 HOURS (CEU: 0 UNITS)

CLASS SIZE: MINIMUM: 20; MAXIMUM: 30

NHI Customer Service: (877) 558-6873 • nhicustomerservice@dot.gov

Course Number

FHWA-NHI-231027

Course Title

Funds Management for FHWA Employees - WEB-BASED

The course specifically focuses on the Principles of Appropriations Law as codified in the GAO "Red Book" Volume 1 and 2 (Chapters 6-8, 10 and 11) and it's Application to the unique requirements of the HTF. Course content covers the following topics:

Provide information on the historical perspective, life cycle of an appropriation, constitutional basis, definitions, and effect of decisions of the Comptroller General.

The legal framework: basic concepts about appropriations, relationships between appropriations, Congressional intent, authorization acts, appropriation acts, supplemental and deficiency appropriations, and apportionment and allotment.

Agency regulations and administrative control: agency regulations and interpretations, agency discretion.

Availability of appropriations, purpose: principles/concepts, "Necessary Expense" doctrine, specific purpose authorities and limitations, attendance at meetings, attorney's fees, compensation restrictions, entertainment, recreation, morale, welfare, food, fines and penalties, municipal services, gifts and awards, guard services, rewards, lobbying, membership fees, personal expenses, state and local taxes, telephone services, and a group study of Comptroller General decisions relating to the above material.

Availability of appropriations, time: general principles, "Bona Fide Needs" rule, concepts, prior and future year needs, replacing and modifying contracts, exceptions, advance payments considerations, dispositions of appropriations balances, close of fiscal year, and a group study of the Comptroller General decisions related to above.

Availability of appropriations, amount: earmarking, avoiding Anti-Deficiency

Act violations, supplemental appropriations, augmentation of appropriations, and a group study, and a group study of Comptroller General decisions related to the above.

Obligation of appropriations: nature of obligations, recording obligations and reporting, de-obligation, (tied to 23 USC Sec 106(3)) and contingent liabilities.

Operating under continuing resolutions: rate for operations, project or activities, and relationship to other legislation.

Liability and relief of accountable officers: general principles, physical loss or deficiency, illegal or improper payment, relief, and a group study of Comptroller decisions relate to the above.

Grants and Cooperative Agreements, guaranteed and insured loans, to include discussion of Transportation Innovative Finance vehicles such as State Infrastructure Banks (SIBs), Grant Anticipated Revenue Vehicles (GARVEEs) Section 129 Loans and Transportation Infrastructure Finance and Innovation Act projects, overview of GAO principles of appropriations, law, claims against the United States, debt collection, payment of judgments, and review and general discussion of recent Comptroller decisions.

Outcomes

Upon completion of the course, participants will be able to:

- Explain how resources are requested and approved.
- Describe the different types of appropriations available
- Describe each type of appropriation
- Explain the use and significance of each type of appropriation
- Discuss the general guidelines for controlling the use of federal resources.
- Explain the limitations and latitudes on the use of the federal resources.

Target Audience

The target audience includes Financial Managers and Specialists, Division Administrators and Assistant Division Administrators, Program and Project Managers, and Supervisors and Team Leaders who are drawn from a cross section of these occupations. The target audience is primarily employees involved with the HTF.

TRAINING LEVEL: Basic

FEE: 2016: $0 Per Person; 2017: N/A

LENGTH: 6 HOURS (CEU: 0 UNITS)

CLASS SIZE: MINIMUM: 1; MAXIMUM: 1

NHI Customer Service: (877) 558-6873 • nhicustomerservice@dot.gov

Course Number

FHWA-NHI-231028

Course Title

Using the AASHTO Audit Guide for the Procurement and Administration of A/E Contracts

Using the AASHTO Audit Guide for the Procurement and Administration of A/E Contracts course is a one-day introductory course of interest to a wide variety of practitioners whose jobs require that they work with Architectural and Engineering (A/E) contracts. The course incorporates small- and large group discussions, case study activities, and both a scored and unscored assessment to reinforce learning.

The course begins with an overview of government contracting for A/E services and the related roles and responsibilities. Participants learn about the A/E Project Cycle and discuss cost components common to A/E contracts.

Next, participants learn about important regulations and standards applicable to the administration of A/E contracts and the role of each. Key cost principles are covered so that participants can learn to distinguish between direct and indirect costs and to differentiate between the concepts of allowability, allocability, and reasonableness.

The importance of internal controls is emphasized as participants are taught to recognize risk factors and indicators of control deficiencies. In a discussion of key areas of costs, participants learn to use the AASHTO Uniform Audit & Accounting Guide to better understand directly associated costs and whether specific indirect costs are allowable. A case study helps participants to practice the application of these principles.

After an overview of A/E firm audits and related roles and responsibilities, participants review a sample cost proposal and related contract wording in order to begin linking audit information, cost proposals, and contracts. The course ends with a discussion of cognizance and the risk management framework followed by a review of select tools and resources that support the administration of A/E contracts.

Outcomes

Upon completion of the course, participants will be able to:

- Explain the Federal and State laws, regulations, policies and procedures that relate to the procurement and administration of A/E contracts.
- Explain how to use audit information in the procurement and administration of A/E contracts.
- Identify and discuss concepts of direct and indirect, allowable and unallowable costs in A/E contracts.
- Locate selected tools and resources to assist in the procurement and administration of A/E contracts.

Target Audience

This course is particularly suited for practitioners associated with procurement, audit, and the administration of A/E contracts.

Training Level: Basic

Fee: 2016: $350 Per Person; 2017: N/A

Length: 9 HOURS (CEU: .9 UNITS)

Class Size: MINIMUM: 20; MAXIMUM: 30

NHI Customer Service: (877) 558-6873 • nhicustomerservice@dot.gov

COURSE NUMBER

FHWA-NHI-231029

COURSE TITLE

Using the AASHTO Audit Guide for the Development of A/E Consultant Indirect Cost Rates

This two-day advanced course is of interest to a wide variety of practitioners who want to be able to apply the AASHTO Audit Guide in the development and administration of A/E design consultant direct and indirect costs and rates. This course is written for both prime A/E consultants and subconsultants.

OUTCOMES

Upon completion of the course, participants will be able to:

- Employ appropriate requirements, concepts, and tools necessary to develop and apply indirect cost rates to A/E contracts.
- Describe the required components of compliant internal controls.
- Prepare an appropriate analysis necessary to demonstrate the reasonableness of compensation.
- Interpret and apply Federal and State laws, regulations, policies and procedures.
- Explain various components of the external oversight framework including ethics, dispute resolution, and the FHWA function.
- Compare and distinguish between contract types and implications on account costing and billing.

TARGET AUDIENCE

This course is intended for those who perform one or more of the following roles: o Performing indirect cost rate audits for A/E Design firmso Ensuring compliance with the AASHTO Audit Guideo Administering contracts or subcontracts and procuring services o Managing contracts or subcontractso Ensuring compliance of contracts or subcontractso Providing oversight of local contracts or subcontractso Building and reviewing cost proposalso Approving the payment of A/E design consultant invoiceso Auditing indirect cost and contract proposalso Closing out and performing final reconciliations of contracts o Designing and enforcing internal control systemso Reviewing RFPs and contracts for government projectso Sell A/E design services to State DOTs

TRAINING LEVEL: Intermediate

FEE: 2016: $740 Per Person; 2017: N/A

LENGTH: 14 HOURS (CEU: 1.4 UNITS)

CLASS SIZE: MINIMUM: 20; MAXIMUM: 30

NHI Customer Service: (877) 558-6873 • nhicustomerservice@dot.gov

COURSE NUMBER

FHWA-NHI-231030

COURSE TITLE

Using the AASHTO Audit Guide for the Auditing and Oversight of A/E Consultant Indirect Cost Rates

Note NHI-231029 AASHTO Uniform Audit and Accounting Guide Part 1 is a pre-requisite for this course. Participants that have not successfully completed NHI-231029 will be turned away.

This two-day advanced course is of interest to A/E design firms; State DOT and local government auditors; CPAs; and FHWA, State DOT, and A/E design firm financial and/or consultant services management who perform the audit or audit compliance review function in accordance with the AASHTO Uniform Audit & Accounting Guide (AASHTO Audit Guide). The course focuses primarily on audit requirements and procedures designed to develop reasonable assurance that indirect cost rates are developed in accordance with applicable Federal regulations and guidance. The course incorporates small- and large-group discussions, document reviews, case study activities, un-scored self-assessments, and a scored final assessment to reinforce learning.

OUTCOMES

Upon completion of the course, participants will be able to:

- Perform audit functions related to the planning, performance, or oversight of A/E consultant indirect cost rate audits.
- Determine and attest to A/E consultant compliance with applicable guidance and/or requirements.
- Discuss how State DOTs will use the CPA Workpaper Review Program (AASHTO Audit Guide Appendix A) to evaluate audits performed by CPAs.
- Identify and apply appropriate audit tools and techniques as specified in the AASHTO Audit Guide.
- Describe the components of a complete audit report and how to evaluate the report presentation.
- Describe various components of the State DOT's oversight and risk management framework.
- Describe at a high level the FHWA's roles and responsibilities in its stewardship and oversight of Federal-Aid funds related to procurement of A/E design services and administration of related agreements.

TARGET AUDIENCE

This course is primarily for those who perform one or more of the following functions:o Perform indirect cost rate audits for A/E design firmso Ensure consistency with the AASHTO Audit Guideo Ensure compliance of contracts or subcontractso Provide oversight of local agency contracts or subcontractso Review cost proposalso Audit indirect cost and contract proposalso Close out and perform final reconciliation of contractso Design and enforce internal control systemso Review RFPs and contracts for government projectso Evaluate the effectiveness of the State DOT oversight and risk management framework

TRAINING LEVEL: Accomplished

FEE: 2016: $740 Per Person; 2017: N/A

LENGTH: 14 HOURS (CEU: 1.4 UNITS)

CLASS SIZE: MINIMUM: 20; MAXIMUM: 30

NHI Customer Service: (877) 558-6873 • nhicustomerservice@dot.gov

COURSE NUMBER

FHWA-NHI-134061

COURSE TITLE

Construction Program Management and Inspection

The Federal Highway Administration's (FHWA) responsibilities for construction project and program oversight has changed considerably throughout the years. Today, the FHWA field engineers are typically involved in a diverse array of issues that were not common in construction projects of decades past. Changes in legislation, declines in staffing resources and expertise, and increased complexity of the Federal-aid construction program have all had an impact on how the FHWA conducts construction program management and oversight. Today's FHWA field engineers must have a more focused and programmatic approach in fulfilling construction stewardship and oversight responsibilities.

This 2-day training workshop highlights the FHWA roles and resources to assist the State in delivering a quality construction program. The training will assist the FHWA field engineers in maintaining and improving technical competence and in selecting a balanced program of construction management techniques.

The workshop uses the "Construction Program Management and Inspection Guide" as instructional material. While the workshop is focused primarily at FHWA's staff and FHWA oversight activities, participation by State partners and other relevant entities is highly encouraged to further educate and train Federal Aide partners to "act on FHWA's behalf in line with the Divisions/State DOT Stewardship Agreement.

OUTCOMES

Upon completion of the course, participants will be able to:

- Manage and oversee Federal-aid construction programs.

TARGET AUDIENCE

This training is targeted at FHWA Division field engineers and State agencies, and will provide staff with the background and knowledge they need for managing and overseeing their Federal-aid construction programs. The training is geared towards the new FHWA generalist employee but is also intended as a refresher for the veteran FHWA engineer.

TRAINING LEVEL: Basic

FEE: 2016: $335 Per Person; 2017: N/A

LENGTH: 2 DAYS (CEU: 0 UNITS)

CLASS SIZE: MINIMUM: 15; MAXIMUM: 40

NHI Customer Service: (877) 558-6873 • nhicustomerservice@dot.gov

COURSE NUMBER

FHWA-NHI-134064

COURSE TITLE

Transportation Construction Quality Assurance (1.5-Day)

The Federal Highway Administration (FHWA) identified the need for transportation construction and materials personnel to increase their knowledge of the fundamentals of effective transportation construction Quality Assurance (QA). This course was developed to ensure that agency, contractor, producer, and consultant personnel responsible for interpreting and applying quality assurance specifications in transportation construction are properly qualified. The course will utilize a Quality Assurance Reference Manual, adapted from the current NETTCP manual.

This one and a half-day version of the course covers Chapters 1 through 6 of the course materials and will be available to, and appropriate for, all audiences including management level personnel. The content covered in this first day includes how quality assurance is featured in a transportation construction quality assurance program, quality assurance program elements, the evolution of quality assurance specifications, measuring quality, and the roles and responsibilities of both contractor and agency personnel.

OUTCOMES

Upon completion of the course, participants will be able to:

- Consistently apply fundamental Quality Assurance concepts, terminology, and definitions
- Differentiate QA specifications from other specifications
- Explain each of the six core elements of a QA program and how each is essential to successful implementation of Quality Assurance
- Describe the respective roles and responsibilities of the project decision makers (Contractor QC and Agency Acceptance personnel) and how their interaction contributes to construction quality

TARGET AUDIENCE

This is an intermediate-level course for personnel who are implementing QA specifications on construction projects. Necessary background knowledge for participants is 3-5 years minimum in transportation construction specifications inspections. The suggested list of personnel that may consider attending, if they have the requisite background knowledge are Contractor/Consultant Personnel (QC managers/QC Plan Administrators, Senior Production Facility QC Technician/Inspectors, Senior QC Laboratory Personnel, and Senior Field QC Technicians/Inspectors) and Agency Personnel (Project Managers/Resident Engineers, Senior Production Facility Acceptance Technicians/Inspectors, Senior Acceptance Laboratory Personnel, and Senior Field Acceptance Technicians/Inspectors.

TRAINING LEVEL: Intermediate

FEE: 2016: $250 Per Person; 2017: $250 Per Person

LENGTH: 1.5 DAYS (CEU: 1.1 UNITS)

CLASS SIZE: MINIMUM: 20; MAXIMUM: 30

NHI Customer Service: (877) 558-6873 • nhicustomerservice@dot.gov

Course Number

FHWA-NHI-134064A

Course Title

Transportation Construction Quality Assurance (3-Day)

The Federal Highway Administration (FHWA) identified the need for transportation construction and materials personnel to increase their knowledge of the fundamentals of effective transportation construction Quality Assurance (QA). This course was developed to ensure that agency, contractor, producer, and consultant personnel responsible for interpreting and applying quality assurance specifications in transportation construction are properly qualified. The course will utilize a Quality Assurance Reference Manual, adapted from the current NETTCP manual.

This three-day version of the course covers Chapters 1 through 10 of the course materials and will be available to, and appropriate for, production, laboratory, and field QC and Acceptance technicians and inspectors. This version contains mathematical terms and principles used in QA sampling, testing, and decision-making. The content also includes how quality assurance is featured in a transportation construction quality assurance program, quality assurance program elements, the evolution of quality assurance specifications, measuring quality, and the roles and responsibilities of both contractor and agency personnel.

Outcomes

Upon completion of the course, participants will be able to:

- Consistently apply fundamental Quality Assurance concepts, terminology, and definitions
- Differentiate QA specifications from other specifications
- Explain each of the six core elements of a QA program and how each is essential to successful implementation of Quality Assurance
- Describe the respective roles and responsibilities of the project decision makers (Contractor QC and Agency Acceptance personnel) and how their interaction contributes to construction quality
- Apply the mathematical concepts of variability, statistical distribution, and sampling protocols to measure construction quality
- Describe the primary components of inspection, properly document the results of inspection, and utilize inspection data to quantify quality of workmanship

Target Audience

This is an intermediate-level course for personnel who are implementing QA specifications on construction projects. Necessary background knowledge for participants: 3-5 years minimum in transportation construction specifications inspections, basic statistical knowledge/training, some usage of tools necessary to the Quality Assurance process (contractor test results). The suggested list of personnel that may consider attending, if they have the requisite background knowledge are Contractor/Consultant Personnel (QC managers/QC Plan Administrators, Senior Production Facility QC Technician/Inspectors, Senior QC Laboratory Personnel, and Senior Field QC Technicians/Inspectors) and Agency Personnel (Project Managers/Resident Engineers, Senior Production Facility Acceptance Technicians/Inspectors, Senior Acceptance Laboratory Personnel, andSenior Field Acceptance Technicians/Inspectors).

Training Level: Intermediate

Fee: 2016: $350 Per Person; 2017: $350 Per Person

Length: 3 DAYS (CEU: 1.8 UNITS)

Class Size: MINIMUM: 20; MAXIMUM: 30

NHI Customer Service: (877) 558-6873 • nhicustomerservice@dot.gov

Course Number

FHWA-NHI-134069

Course Title

Ethics Awareness for the Transportation Industry - WEB-BASED

This training is provided by the Transportation Curriculum Coordination Council (TCCC) in partnership with NHI to provide good practices for ethical behavior of transportation employees. The training was prepared by State DOT personnel for State DOT personnel. This course is primarily intended for inspectors and technicians.

The training contains good practices from various agencies. The topics of discussion in this training are: conflict of interest, safety, fraud, falsification of documentation, reporting ethical concerns, gifts and favors, fairness, personal use of agency property, and consequences.

Not all State agencies' codes of conduct are the same but they all demand similar ethical behavior of their employees. Be sure to access to your agency's codes or check with your supervisor for more information specific to your organization. Each State agency/company has their own work rules, which the viewer needs to review and follow.

NHI is hosting this and other TCCC Web-based developments to serve a critical need for training. We need your feedback to determine whether we should continue posting other Web-based trainings like this one. Please take the time to complete the evaluation form provided at the end of the training, or email NHIMarketing@dot.gov with your feedback.

Outcomes

Upon completion of the course, participants will be able to:

- Describe agency expectations on ethics
- Give an example of a current code of conduct policy
- Recognize and practice good ethics as an employee in the transportation industry
- Explain the consequences when rules and regulations are not followed

Target Audience

This training is designed for Level I and Level II State and local public agency personnel and their industry counterparts involved in the construction, maintenance and testing process for highways and structures. Level I or Entry refers to employees/ trainees with little to no experience in the subject area and perform his/her activities under direct supervision. Level II or Intermediate refers to employees that understand and demonstrate skills in one or more areas of the entry level and perform specific tasks under general supervision.

Training Level: Basic

Fee: 2016: $25 Per Person; 2017: $25 Per Person

Length: 1 HOURS (CEU: 0 UNITS)

Class Size: MINIMUM: 1; MAXIMUM: 1

NHI Customer Service: (877) 558-6873 • nhicustomerservice@dot.gov

Course Number

FHWA-NHI-134070

Course Title

SpecRisk Quality Assurance Specification Development and Validation Course - WEB-BASED

This course will provide an introduction to statistical analysis and the development of statistically valid quality assurance specifications, introducing general guidelines established and put forth by the Federal Government and FHWA policy. The course also provides participants with an introduction to SpecRisk, the resource that is necessary to successfully develop statistically valid specifications. The course is designed and delivered to motivate members of the target audience to use SpecRisk software to develop their specifications. Although the course demonstrates basic functions of the software, it is not intended to be an in-depth training on how to use SpecRisk.

This course requires a solid foundation in basic statistics. Completion of FHWA-NHI 134042, or equivalent training, is also recommended. NHI 134042 trains participants to identify the importance of organizing data and how to plot frequency histograms. It explains how a sample relates to the population, the relationship between single and multiple samples, and the use of random stratified sampling tables. This knowledge provides an excellent foundation for this course.

Outcomes

Upon completion of the course, participants will be able to:

- Recognize key concepts to develop an effective, statistically valid Quality Assurance (QA) specification.
- Make an informed selection among available options when developing an acceptance plan.
- Develop QA specifications in alignment with best practices, Federal regulations, and FHWA policy.
- Apply SpecRisk software to understand risks and develop statistically valid specifications.

Target Audience

Personnel involved in specification development: Federal, State, and local highway agency engineers and technicians in materials, construction, and research. The training is also appropriate for industry personnel that are involved in reviewing and providing input to the specification development process.

Training Level: Basic

Fee: 2016: $50 Per Person; 2017: $50 Per Person

Length: 8 HOURS (CEU: 0 UNITS)

Class Size: MINIMUM: 1; MAXIMUM: 1

NHI Customer Service: (877) 558-6873 • nhicustomerservice@dot.gov

Course Number

FHWA-NHI-136065

Course Title

Risk Management

During planning for a transportation improvement, thorough consideration of potential risks can help reduce or eliminate them, ensuring a smoother project life cycle. In the same way, accounting for risk in program management and strategic planning can help transportation agencies minimize or avoid unexpected costs or delays. Recent legislation reflects the growing importance of understanding risk in government programs. Congress included requirements for risk management in the latest highway authorization, the Moving Ahead for Progress in the 21st Century Act (MAP-21).

This course provides an understanding of risk management concepts and processes, including terminology, benefits of use, risk management planning, and a framework for implementation. The course presents a cyclical risk management framework through classroom training and an independent study assignment, focusing on gathering information; identifying risk; analyzing and prioritizing risk events; documenting risk; identifying risk response strategies; incorporating response strategies into a plan; and monitoring, evaluating, and adjusting strategies. The 1-hour independent study assignment is completed by participants before attending the 2-day classroom session.

Participants use tools and methods from each step of the framework in a series of exercises that provide realistic, job-relevant practice in applying the risk management process. In order to maximize the impact of the training and increase the likelihood of participants' mastery of the risk management process, the agency can select active agency issues (project, program, or network) to be used during the exercises. In addition, the agency can provide problem statements and pre-select the teams for the exercises. Contact the NHI Training Program Manager for further information on making such arrangements.

Outcomes

Upon completion of the course, participants will be able to:

- Explain the overall organizational context, framework, and importance of risk management to others.
- Follow a consistent process for managing risk.
- Utilize standard risk terminology, tools, and methods.
- Apply the risk management process to a project or program within established time-constraints.
- Develop effective plans for internal and external communication and consultation about risk and risk-related issues.

Target Audience

The target audience for this course includes Federal, State and local highway employees who are responsible for directing and managing all aspects of highway related programs and projects such as planning, environment, project development, design, construction, operations, maintenance, and finance. Audience experience, background, knowledge, skills and abilities are expected to vary widely. No previous experience with risk management is required.

Training Level: Basic

Fee: 2016: $375 Per Person; 2017: N/A

Length: 2 DAYS (CEU: 1.2 UNITS)

Class Size: MINIMUM: 20; MAXIMUM: 30

NHI Customer Service: (877) 558-6873 • nhicustomerservice@dot.gov

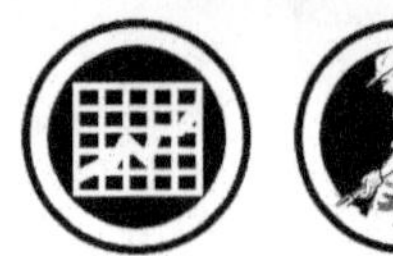

COURSE NUMBER

FHWA-NHI-136106

COURSE TITLE

Introduction to Transportation Asset Management

Asset management principles are becoming increasingly important to help agencies manage their assets as they face fewer available resources, higher expectations for customer service, and increased demand for more transparency in the decision process. In an asset management environment, investment decisions are linked to targeted performance levels that have been established based on current and expected asset conditions. Trade-offs between investments in different types of assets and different investment priorities can be assessed because of the availability of reliable data and a clear set of performance metrics that the agency hopes to achieve. As a result, agencies are better able to use their funding effectively and to defend their need for additional resources.

The Transportation Asset Management course introduces a strategic approach to managing physical transportation infrastructure. This 1-day course covers the principles of asset management and introduces the five core questions every agency should be able to answer about its assets. The course contains modules on the following topics: asset management principles, performance measures, long-term financial planning, and risk assessment. Through a series of workshops, the course material introduces the participants to the application of asset management principles in real life situations.

This course can be delivered with the addition of a half-day workshop. During the workshop, participants review self-assessments to identify agency gaps between the desired and actual application of asset management principles. The workshop also includes recommendations for advancing the implementation of asset management practices within the agency. Refer to 136106A "Introduction to Transportation Asset Management with Workshop" on the NHI Web site for additional information.

This course is the first in a series and previously cataloged under 131106. The other courses in this series are 136106B "Development of a Transportation Asset Management Plan" and 136106C "Introduction to Transportation Asset Management Plans", which is Web-based. See the NHI Web site for additional information on each of these courses.

OUTCOMES

Upon completion of the course, participants will be able to:

- Champion the use of asset management principles and concepts within the organization.
- Define their role in supporting the agency's asset management efforts.
- Describe the maturity of the agency's asset management program.

TARGET AUDIENCE

This training is intended for senior-level and mid-level managers from State departments of transportation and other transportation agencies, who typically have the responsibility for decision-making in one or more areas addressed by transportation asset management. Participants should represent a number of organizational units, including (but not limited to) planning, engineering (e.g., facility management, design, construction), capital programming, maintenance and operations, financial management, traffic and safety engineering, system operation and management, and information technology. The course is also intended for individuals who manage or provide critical information to senior managers, or who have direct responsibility for meeting specific transportation system performance or program delivery targets.

TRAINING LEVEL: Basic

FEE: 2016: $450 Per Person; 2017: N/A

LENGTH: 1 DAYS (CEU: 0 UNITS)

CLASS SIZE: MINIMUM: 20; MAXIMUM: 30

NHI Customer Service: (877) 558-6873 • nhicustomerservice@dot.gov

COURSE NUMBER

FHWA-NHI-136106A

COURSE TITLE

Introduction to Transportation Asset Management with Workshop

Asset management principles are becoming increasingly important to help agencies manage their assets as they face fewer available resources, higher expectations for customer service, and increased demand for more transparency in the decision process. In an asset management environment, investment decisions are linked to targeted performance levels that have been established based on current and expected asset conditions. Trade-offs between investments in different types of assets and different investment priorities can be assessed because of the availability of reliable data and a clear set of performance metrics that the agency hopes to achieve. As a result, agencies are better able to use their funding effectively and to defend their need for additional resources.

The Transportation Asset Management course introduces a strategic approach to managing physical transportation infrastructure. This 1.5-day course covers the principles of asset management and introduces the five core questions every agency should be able to answer about its assets. The course contains modules on the following topics: asset management principles, performance measures, long-term financial planning, risk assessment, and implementation. Through a series of workshops, the course material introduces the participants to the application of asset management principles in real life situations. It also provides a review of a self-assessment that can be used to identify agency gaps between the desired and actual use of these principles. The course concludes with recommendations for advancing the implementation of asset management practices within the agency.

This course can be delivered without the half-day workshop with the approval of the FHWA Technical Lead. See 136106 "Introduction to Transportation Asset Management" on the NHI Web site for further information.

This course is the first in a series and previously cataloged under 131106A. It serves as a prerequisite for 136106B "Development of a Transportation Asset Management Plan." The other course in the series is 136106C "Introduction to Transportation Asset Management Plans," which is Web-based. See the NHI Web site for additional information on each of these courses.

OUTCOMES

Upon completion of the course, participants will be able to:

- Champion the use of asset management principles and concepts within the organization.
- Define their role in supporting the agency's asset management efforts.
- Describe the maturity of the agency's asset management program.
- Draft a plan for using existing resources to advance the agency's use of asset management principles.

TARGET AUDIENCE

This training is designed for senior-level and mid-level managers from State departments of transportation and other transportation agencies, who typically have the responsibility for decision-making in one or more areas addressed by transportation asset management. Participants should represent a number of organizational units, including (but not limited to) planning, engineering (e.g., facility management, design, construction), capital programming, maintenance and operations, financial management, traffic and safety engineering, system operation and management, and information technology. The course is also intended for individuals who manage or provide critical information to senior managers, or who have direct responsibility for meeting specific transportation system performance or program delivery targets.

TRAINING LEVEL: Basic

FEE: 2016: $575 Per Person; 2017: N/A

LENGTH: 1.5 DAYS (CEU: 0 UNITS)

CLASS SIZE: MINIMUM: 20; MAXIMUM: 30

NHI Customer Service: (877) 558-6873 • nhicustomerservice@dot.gov

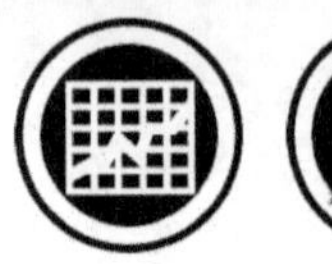

Course Number

FHWA-NHI-136106B

Course Title

Developing a Transportation Asset Management Plan

"Developing a Transportation Asset Management Plan" is blend of online and classroom training that introduces the role of the Transportation Asset Management Plan (TAMP) as a planning, communication, and accountability tool. This course also provides the information necessary for an agency to develop or enhance a TAMP. This course is the second in the series of courses on transportation asset management and previously cataloged under 131106B.

It is required that participants take the Web-based training 136106C and attend 136106 or 136106A prior to attending this course, or have a solid background in transportation asset management principles and planning.

The course focuses on three primary components to the TAMP, including strategic performance measurement, risk assessment, and financial management. It also provides templates and guidelines for use in developing a TAMP.

The course is 1.5-days in length. Prior to attending the first day of class, participants must complete the online, 1-hour training module NHI-136106C "Introduction to a Transportation Asset Management Plan." Information on how participants access this training is provided to the host by the instructor once the session is confirmed.

Outcomes

Upon completion of the course, participants will be able to:

- Describe the role of a Transportation Asset Management Plan in a transportation agency.
- Identify strategies for incorporating risk into investment decisions.
- Explain how to determine whether an agency is making sustainable, long-term investments in its assets.
- Develop a Transportation Asset Management Plan that matches the amount of data and the sophistication of the analysis tools available.

Target Audience

The course is intended for senior-level and mid-level managers from State departments of transportation and other transportation agencies, who have the responsibility for decision-making in one or more areas addressed by transportation asset management. Course participants should represent a broad range of organizational units, such as (but not limited to) planning, engineering (facility management, design, and construction), capital programming, maintenance and operations, financial management, traffic and safety engineering, system operation and management, and information technology. If the agency has an Asset Management Steering Committee, its members would benefit from this course. In addition, individuals who manage individual assets or provide critical information to senior managers, or who have direct responsibility for meeting specific transportation system performance or program delivery targets, are also excellent candidates for attending the course.

Training Level: Intermediate

Fee: 2016: $575 Per Person; 2017: N/A

Length: 1.5 DAYS (CEU: 0 UNITS)

Class Size: MINIMUM: 20; MAXIMUM: 30

NHI Customer Service: (877) 558-6873 • nhicustomerservice@dot.gov

Course Number

FHWA-NHI-136106C

Course Title

Introduction to a Transportation Asset Management Plan (WEB-BASED)

This training is a prerequisite of another NHI training and is offered at no cost.

A Transportation Asset Management Plan (TAMP) can be considered a business plan that builds the case for making infrastructure investments and explains how resources will be used. This course, "Introduction to a Transportation Asset Management Plan," is a 1-hour, Web-based training (WBT) that introduces the content and organization of a TAMP and the typical TAMP development process. This course was previously cataloged under 131106C

This training includes the following topics.

- The use of a TAMP in transportation agencies
- The typical content of a TAMP (including a comparison with requirements in MAP-21)
- Key components, including performance projections and the financial summary
- Examples of TAMPs at various levels of maturity
- Existing and anticipated use of a TAMP in state highway agencies
- The expected involvement of agency personnel in developing and updating a TAMP

This training includes audio clips from leaders in state highway agencies that convey the anticipated benefits from the development of a TAMP and the way they expect to use their TAMP. In addition, the WBT highlights the use of existing documentation to develop the TAMP and plans for enhancing the content of future TAMPs.

This training serves as a prerequisite for NHI-136106B "Developing a Transportation Asset Management Plan", which describes the role of a TAMP in a transportation agency and explores in some detail three important components: strategic performance management, risk assessment and management, and financial management.

Outcomes

Upon completion of the course, participants will be able to:

- Describe the role of a TAMP as a communication tool with internal and external stakeholders.
- List the typical content of a TAMP.
- Identify several sources of information that will contribute to the development of a TAMP.

Target Audience

The course is intended for senior-level and mid-level managers from State departments of transportation and other transportation agencies, who typically have the responsibility for decision-making in one or more areas addressed by transportation asset management. Participants should represent a number of organizational units, including (but not limited to) planning, engineering (e.g., facility management, design, construction), capital programming, maintenance and operations, financial management, traffic and safety engineering, system operation and management, and information technology. The course is also intended for individuals who manage or provide critical information to senior managers, or who have direct responsibility for meeting specific transportation system performance or program delivery targets.

Training Level: Intermediate

Fee: 2016: $0 Per Person; 2017: N/A

Length: 1 HOURS (CEU: 0 UNITS)

Class Size: MINIMUM: 0; MAXIMUM: 0

NHI Customer Service: (877) 558-6873 • nhicustomerservice@dot.gov

COURSE NUMBER

FHWA-NHI-136113

COURSE TITLE

Transportation Asset Management Overview - WEB-BASED

This training was developed by the Transportation Curriculum Coordination Council (TCCC) in partnership with AASHTO and NHI. This training explains the basics of asset management and why asset management is important. After you complete this training, you'll have new terms, and new ways of thinking about what you're already doing. More importantly, you'll understand why it's so important to be strategic and systematic when you're responsible for managing huge numbers of assets.

This training contains the following lessons:

Lesson 1: What is Transportation Asset Management? This lesson will explain the concept of asset management; give examples of how asset management is used in the planning process; and explain how current asset management practices have been impacted by past transportation needs.

Lesson 2: Asset Management Principles and Practices. This lesson lists the categories of activity that inform spending decisions; explain how policy goals and objectives impact asset management; relate planning and programming to managing assets; describe how asset management principles apply to program delivery; explain why system monitoring is necessary; and explain how quality data and analysis impact asset management.

OUTCOMES

Upon completion of the course, participants will be able to:

- Explain what transportation asset management is and why it is important
- Describe the asset management principles and practices used to make informed spending decisions

TARGET AUDIENCE

This training was developed by the Transportation Curriculum Coordination Council (TCCC) in partnership with AASHTO, NHI, and is recommended for TCCC levels II through IV.

TRAINING LEVEL: Basic

FEE: 2016: $25 Per Person; 2017: N/A

LENGTH: 2 HOURS (CEU: 0 UNITS)

CLASS SIZE: MINIMUM: 1; MAXIMUM: 1

NHI Customer Service: (877) 558-6873 • nhicustomerservice@dot.gov

Course Number

FHWA-NHI-310108

Course Title

Federal Lands 101

During these times of economic expansion and growth, there are dramatic workforce changes taking place. With the passage of TEA-21, the program for Federal Lands Highway (FLH) nearly doubled and there is the prospect that it will again increase under pending transportation reauthorization. Due to this change, coupled with the increasing demand by our partners and customers for more technical assistance, FLH needs to develop the knowledge of their new/mid-career hires in the area of FLH operations and regulations.

Therefore, the overall course goal is to provide FLH employees with an overview of how FLH operates in order to administer programs, deliver projects, develop and transfer technology, and provide external training.

Outcomes

Upon completion of the course, participants will be able to:

- Identify the role and authorities of FLH within the FHWA and its interactions with Federal-aid divisions
- Describe unique aspects of FLH customers and programs
- Describe how FLH delivers projects
- Describe how FLH conducts business, including processes and resources

Target Audience

New hires to FLH in all positions and grades and Federal-aid (particularly new employees) and Federal Lands Management Agency employees upon request.

Training Level: Basic

Fee: 2016: $400 Per Person; 2017: N/A

Length: 2 DAYS (CEU: 1.2 UNITS)

Class Size: MINIMUM: 20; MAXIMUM: 30

NHI Customer Service: (877) 558-6873 • nhicustomerservice@dot.gov

Course Number

FHWA-NHI-310109

Course Title

Federal-Aid 101 (FHWA Employee Session)

During this time of economic expansion and growth, there are dramatic workforce changes taking place. Given the increasing demand by our partners and customers for more technical assistance, FHWA needs to develop the knowledge of their new/mid-career hires in the area of the Federal-aid processes and regulations.

Therefore, the overall course goal is to provide FHWA employees, particularly mid-career hires, with an overview of the key elements of the Federal-Aid Highway Program. Specifically, this course focuses on general requirements and laws that govern the Federal-Aid Highway Program, processes and procedures followed in the project development, and identifying flexibility inherent in the Federal-Aid Program.

Outcomes

Upon completion of the course, participants will be able to:

- Identify the elements and project milestones of the Federal-Aid Highway program.
- Describe the financial aspects and requirements of the Federal-Aid Highway program.
- Describe how the Federal-Aid Highway program fits with other laws (23 U.S.C and other laws that affect the Federal-Aid Highway program).
- Identify the requirements for using Federal-Aid Highway funding.
- Identify how FHWA initiatives such as civil rights, safety, and innovative financing impact the Federal-Aid Highway program.
- Explain the risk-based stewardship and oversight approach,
- Identify key responsibilities or elements of the risk-based stewardship and oversight approach.
- Identify the flexibility inherent in the Federal-Aid Highway program.

Target Audience

New/mid career hires from all disciplines (i.e., planners, engineers, environmental specialists, financial specialists or managers).

Training Level: Intermediate

Fee: 2016: $530 Per Person; 2017: N/A

Length: 3 DAYS (CEU: 2.1 UNITS)

Class Size: MINIMUM: 20; MAXIMUM: 30

NHI Customer Service: (877) 558-6873 • nhicustomerservice@dot.gov

COURSE NUMBER

FHWA-NHI-310110

COURSE TITLE

Federal-Aid Highways - 101 (State Version)

During this time of economic expansion and growth, there are dramatic workforce changes taking place. Given the increasing demand by our partners and customers, it is critical to develop the knowledge of State DOT employees in the area of the Federal-aid highway development processes and regulations.

Therefore, the overall course goal is to provide participants with an overview of the key elements of the Federal-Aid Highway Program. Specifically, this course focuses on general requirements and laws that govern the Federal-Aid Highway Program; processes and procedures followed in the entire project development process, including financing, planning, environment, right of way, highway and bridge design, construction, operations/ITS, maintenance, and technology; and identifying flexibility inherent in the Federal-Aid Program.

OUTCOMES

Upon completion of the course, participants will be able to:

- Identify the key elements of the overall highway project development process
- Identify the elements and requirements of the Federal-Aid Highway Program and the associated Federal/State relationships
- Develop a flowchart of the project development process from the initial planning concept through the environmental and right-of-way processes, on through design, construction, and opening to traffic
- Identify the roles of safety, intelligent transportation systems, operations, research, and development in the Federal-aid process
- Identify the need for public involvement early in the process, opportunities for application of the principles of environmental justice/civil rights, context sensitive solutions, etc.
- Learn the fundamentals of Federal-aid financing, including several innovative financing techniques that will maximize the use of Federal-aid funds
- Develop a network of professionals that can be contacted for help
- Discuss how the Federal-aid laws and regulations relate to other laws (i.e., NEPA, Uniform Act, the Davis Bacon Act, OMB Circular A-87, 49 CFR Part 18 (Common Rule)) and the application of FHWA regulations, policies, technical guidance, etc.

TARGET AUDIENCE

State and local government employees and private-sector participants interested in the process by which the Federal-Aid Highway Program is carried out.

TRAINING LEVEL: Intermediate

FEE: 2016: $400 Per Person; 2017: N/A

LENGTH: 2 DAYS (CEU: 1.2 UNITS)

CLASS SIZE: MINIMUM: 20; MAXIMUM: 30

NHI Customer Service: (877) 558-6873 • nhicustomerservice@dot.gov

COURSE NUMBER

FHWA-NHI-310115

COURSE TITLE

Introducing Highway Federal-Aid - WEB-BASED Standard Version

This training is undergoing revisions and this version is offered to you at no cost.

Based upon customer feedback, NHI has summarized portions of FHWA-NHI-310109 Federal-Aid 101 into this self-paced, Web-based format. We encourage everyone to take advantage of the opportunity to attend the full, Instructor-led training. However, for those constrained by time or travel money, this WBT provides a good short-term option.

The overall goal of this training is to provide an overview of the key elements of the Federal-Aid Highway Program. Specifically, this training focuses on the general requirements and laws that govern the Federal-Aid Highway Program, the processes and procedures of project development, and the identification of inherent flexibility in the Federal-Aid Program. NHI is continuously expanding our Web-based training offerings and would love to hear what you think about this training. When you complete it, please take the time to fill out the online course evaluation form provided. We've also prepared an accessible, 508-compliant version, see FHWA-NHI-310115W for more information.

OUTCOMES

Upon completion of the course, participants will be able to:

- Identify the key elements of the overall Highway Federal-Aid project development process
- Explain the FHWA civil rights programs (i.e., Title VI, Disadvantaged Business Enterprise (DBE), EEO Contract Compliance, Title VII, Americans with Disabilities Act (ADA), Indian Outreach) and their relationship to the Federal-Aid Highway Program and the Federal/State relationships
- Identify where environmental justice is included in all aspects of project planning, development, and construction
- Develop a flowchart of the project development process from the initial planning concept through the environmental and right-of-way processes, on to construction and opening to traffic
- Identify the roles of safety, intelligent transportation systems, operations, research, and development in the Federal-Aid process
- Identify timing and use of public involvement throughout the Highway Federal-Aid project development process
- Explain the similarities and relations among the Federal-Aid laws and regulations and other laws (i.e., NEPA, Uniform Act, the Davis Bacon Act, OMB Circular A-87, 49 CFR Part 18 (Common Rule))

TARGET AUDIENCE

New/mid career hires from all disciplines (i.e., planners, engineers, environmental specialists, financial specialists or managers) or FHWA employees who took the FHWA-NHI-310109 Federal Aid 101 Instructor-Led Training before 2005.

TRAINING LEVEL: Basic

FEE: 2016: $0 Per Person; 2017: N/A

LENGTH: 8 HOURS (CEU: .6 UNITS)

CLASS SIZE: MINIMUM: 1; MAXIMUM: 1

NHI Customer Service: (877) 558-6873 • nhicustomerservice@dot.gov

Course Number

FHWA-NHI-310119

Course Title

Writing Effective Program Review Reports: Moving People to Action

The ability to provide clear communication is vital to the business of FHWA and good writing skills are a key element in the communication process. FHWA uses program reviews as tools to fulfill its stewardship and oversight responsibilities, manage program risk, and identify process improvements for the Federal-aid program. Each year, FHWA conducts approximately 200 program reviews. The product of these reviews is usually a review that details the observations and recommendations of the review team in an effort to improve a process or product. The review's effectiveness is largely determined by how well the review is communicated to the target audience.

The goal of this course is to improve the writing skills of FHWA's employees. Improved writing skills should lead to higher quality review reviews, which in turn should increase FHWA's ability to motivate the reading audience to act upon the review's recommendations. Action on the part of the reader will ultimately lead to improved effectiveness in delivering FHWA programs by reducing costs, accelerating project delivery, and improving stewardship and oversight. Throughout this course, you will learn that effective writing is more than proper punctuation and using spell-check. It's learning how to write for your audience, the busy reader. You will also learn writing skills that will aid in motivating your readers to action.

Outcomes

Upon completion of the course, participants will be able to:

- write an executive summary that informs the audience about potential problems and persuades them to act on your recommendations or solution;
- write recommendations that motivate the audience to take corrective action;
- discuss usefulness and readability;
- describe how review content is generated by questions;
- develop and answer review objectives;
- evaluate the logical link of review objectives, observations, and recommendations;
- focus on the relevant elements of an observation finding to create convincing support;
- use the deductive message-first structure throughout reviews;
- design/organize reviews to benefit the busy reader;
- control paragraph unity (one main topic) and coherence (flow);
- avoid information overload within sentences;
- control common sentence problems; and
- develop objective criteria for writing and reviewing reviews.

Target Audience

This course is primarily intended for FHWA personnel who are responsible for writing program reviews. It is anticipated that participants may not have in-depth writing background. More knowledgeable persons may be expected to attend and will add to the overall effectiveness of the training through their active participation.

Training Level: Intermediate

Fee: 2016: $380 Per Person; 2017: N/A

Length: 1.5 DAYS (CEU: 0 UNITS)

Class Size: MINIMUM: 22; MAXIMUM: 30

NHI Customer Service: (877) 558-6873 • nhicustomerservice@dot.gov

Course Number

FHWA-NHI-310120

Course Title

Conducting Effective Program Reviews

The Conducting Effective Program Reviews, provides participants with an introduction and/or review of the best practices and tools involved in the planning and conducting of effective program reviews.

To accomplish FHWA's Stewardship Mission, units at every level and in every program area need the expertise to plan, design and carry out, often jointly with partners, reviews to ensure that operational processes are consistent with established standards and expectations, performing at the most effective and efficient level, and that best practices are captured and made available to units at all levels.

Building on FHWA experience and expertise gained through Program Reviews, Process Reviews, and Continuous Process Improvement Reviews, an improved workshop, tailored to the unit's needs is now being offered.

The Workshop consists of assistance in the form of consultation, training and hands on assistance in the methodology and tools for conducting successful reviews.

Outcomes

Upon completion of the course, participants will be able to:

- Write a review Objective and create a Team Charter
- Develop a Review Workplan, including the steps to collect data
- Select data analysis tools to be used during a program review
- Define the key parts of an observation
- Link the parts of an observation to an objective and recommendation
- Develop a program review report and conduct effective close-out meetings

Target Audience

The target audience for Conducting Effective Program Reviews includes FHWA staff who participate in and/or lead program or process reviews. As such, the FHWA staff will be primarily the division offices, but may include staff from FHWA headquarters, State DOTs, or the Resource Centers (RC).

Training Level: Basic

Fee: 2016: $400 Per Person; 2017: N/A

Length: 2 DAYS (CEU: 1.2 UNITS)

Class Size: MINIMUM: 20; MAXIMUM: 30

NHI Customer Service: (877) 558-6873 • nhicustomerservice@dot.gov

COURSE NUMBER

FHWA-NHI-310122

COURSE TITLE

Introduction to Data Analysis

In this age of information, FHWA's business requires skill to analyze, evaluate, and present data and information so that others understand, gain knowledge and have a greater ability to make decisions. Our agency has a unique perspective to help identify Federal Highway program improvements by analyzing the data available either in-depth and across many core highway programs including civil rights, finance, environment, planning, design, construction, operations, and safety. This perspective coupled with the Agency's stewardship and oversight role provides opportunities to interact with decision makers to improve the quality of the highway program. We use Program Reviews as a tool to manage program risk and assist our partners in identifying program improvements, and it is important to have the tools and techniques to analyze highway program data and be able to present it as easily understood information in our Program Review reports.

By taking this course tailored to the FHWA business context, participants will be better able to make data driven decisions in a performance measurement operating environment. FHWA staff will be better able to present information in Program Review Reports based on sound data analysis by applying the skills learned in data collection, analysis, sampling, and statistical analysis. As the agency continues to prepare for Performance Management, our agency staff would benefit from this suite of training modules that delve deeper into the above-referenced data oriented areas.

The overall goal of this course is to improve FHWA's ability to apply techniques and tools when evaluating and presenting data and information used to develop recommendations and solutions in the program improvement and performance measurement context. So much information is accessible and can be easily stored electronically. There will continue to be a need for skills and refresher training to do data analysis and evaluation. To make the best use of available data it is important to have fundamental statistical knowledge and be able to use tools when analyzing data. The data analysis skills gained by taking this course are expected to be applied during the conduct of program and process reviews and while implementing performance management strategies.

OUTCOMES

Upon completion of the course, participants will be able to:

- Identify useful analysis models such as root cause analysis, six sigma model, and Plan-Do-Check-Act.
- Evaluate data and information that support recommendations for program improvements.
- Demonstrate an understanding of the tools and techniques for graphically presenting data and information so that others understand, gain knowledge, and have a greater ability to make decisions.

TARGET AUDIENCE

This course is appropriate for any individual that needs to apply techniques and tools for evaluating and presenting data and information.

TRAINING LEVEL: Basic

FEE: 2016: $400 Per Person; 2017: N/A

LENGTH: 1.5 DAYS (CEU: 0 UNITS)

CLASS SIZE: MINIMUM: 20; MAXIMUM: 30

NHI Customer Service: (877) 558-6873 • nhicustomerservice@dot.gov

COURSE NUMBER

FHWA-NHI-310123

COURSE TITLE

FHWA Basic Contracting Officers Representative (COR) Training

Contracting Officer's Representatives (COR) are integral to the acquisition process and perform critical acquisition functions; FHWA relies on CORs to help the Contracting Officer (CO) monitor work conducted under contracts in order to meet the Agency mission. Because of this important role, FHWA needs to develop the knowledge of their new and mid-career hires in the area of acquisition management.

The overall course goal is to address the essential core competencies, outlined by the Office of Federal Procurement Policy (OFPP), required for CORs to effectively monitor Federal Government contracts. The class is tailored to meet the specific needs of FHWA CORs with examples and content directed to common contract types and issues faced by FHWA and Federal Lands Programs.

Participants who successfully complete the course will earn 40 Continuous Learning Points in support of a Level II FAC-COR certification.

OUTCOMES

Upon completion of the course, participants will be able to:

- Explain the duties and responsibilities of the Contracting Officer's Representative (COR)
- Discuss COR best practices
- Define key acquisition terminology
- Associate the importance of professional business skills with effectively monitoring the work under the contract
- Determine the elements of contract monitoring appropriate for a given contract
- Describe the process leading up to contract award
- Appropriately respond to legal and ethical issues that may arise

TARGET AUDIENCE

New/mid-career hires who anticipate being appointed as a COR, current CORs who desire refresher/updated COR training. Agreement Officer's Technical Representatives, persons monitoring task orders under an Indefinite Delivery/ Indefinite Quantity contract, and anyone desiring Basic COR Training.

TRAINING LEVEL: Basic

FEE: 2016: $675 Per Person; 2017: N/A

LENGTH: 5 DAYS (CEU: 0 UNITS)

CLASS SIZE: MINIMUM: 20; MAXIMUM: 30

NHI Customer Service: (877) 558-6873 • nhicustomerservice@dot.gov

Course Number

FHWA-NHI-310124

Course Title

Highway Research 101: Administering the FHWA Highway Research Program

In advancing Federal highway research goals, collaboration between FHWA, grant recipients, and sub-recipients is critical. The Highway Research 101: Administering the FHWA Highway Research Program Web-based Training (WBT) is intended to highlight the responsibilities of FHWA Division Office staff members responsible for research oversight and to acquaint them with the key aspects of regulation and practice that satisfy the agency's responsibility, as well as expose them to FHWA R&T priorities and programs to help them advance agency goals.

Implementation of RD&T programs is highly contextual, as is implementation of the overall federally assisted, State-administered programs. Those considered among the best are developed and executed to meet the unique priorities and needs of each FHWA partner. Thus, the emphasis of this course is not to communicate the one best way to administer programs using specific professional disciplines. Instead, it communicates the basics of sound project and program management, ranging from practices that lay a framework for optimizing return on investment and provide for accountability to stimulating innovation and improvements to the state of the practice. Formal case studies are available in this course to illustrate the concepts.

Outcomes

Upon completion of the course, participants will be able to:

- Define FHWA's Research Development and Technology (RD&T) policy
- Explain the Research Program Management Process
- Describe how to administer the requirements for SP&R Subpart B work programs
- Explain how to determine what costs are eligible
- Define a peer exchange program
- Identify the RD&T Coordinator's role in determining state highway problems and RD&T needs
- Identify how national programs and organizations impact/complement SP&R Part B

Target Audience

The target audience for this course is the staff person deployed in each FHWA Division Office to carry out research oversight. Responsibility for the research portion of SP&R is normally only one of several functional programs administered by this individual. This course is applicable to FHWA research coordinators and other FHWA staff who need training and knowledge to administer the research portion of the SP&R program and support the development and execution of State research programs.

Training Level: Basic

Fee: 2016: $0 Per Person; 2017: N/A

Length: 4 DAYS (CEU: .4 UNITS)

Class Size: MINIMUM: 0; MAXIMUM: 0

NHI Customer Service: (877) 558-6873 • nhicustomerservice@dot.gov

Course Number

FHWA-NHI-310125

Course Title

Risk-Based Stewardship and Oversight (Federal Version)

This Instructor-led Training (ILT) course will expand participants' understanding of the risk-based processes, roles and responsibilities that Federal Highway Administration (FHWA) Division and Program Office personnel are using to help optimize the effective and efficient delivery of the Federal-aid Highway Program (FAHP), and to help ensure its compliance with Federal laws and regulations. This approach, as it is being implemented, is commonly known as the Risk-based Stewardship and Oversight (RBSO) approach and builds upon the risk management foundation in the agency's strategic and performance planning processes. The RBSO model is designed to identify risk-based S&O actions and initiatives at both the national and Division levels for both programs and projects.

Outcomes

Upon completion of the course, participants will be able to:

- Describe the FHWA vision and rationale for using Risk-Based Stewardship and Oversight (RBSO), built on four core principles, to optimize the successful delivery of the Federal Highway Program and ensure compliance with Federal law and regulations
- Explain how RBSO integrates strategic and performance planning to allocate limited resources in order to achieve stewardship and oversight objectives
- Explain how FHWA/State DOT Stewardship and Oversight (S&O) Agreements are used to ensure respective S&O responsibilities and expectations are set
- Demonstrate how RBSO integrates the FHWA risk management process to identify program and project risks, and develop the appropriate S&O response strategies to effectively manage those risks
- Explain how program involvement optimizes successful program and project delivery and helps ensure compliance with Federal requirements
- Explain how project involvement optimizes successful program and project delivery and helps ensure compliance with Federal requirements
- Demonstrate the RBSO tools FHWA uses to provide a reasonable level of assurance of both project and program compliance, while also informing other S&O strategies and actions
- Demonstrate how Divisions use risk-based project level S&O activities to: (1) manage project level risks, and (2) provide value-added stewardship to help optimize successful project and program delivery
- Demonstrate how the various RBSO tools work together to optimize the successful delivery of the Federal Highway Program

Target Audience

Since every member of FHWA is either directly or indirectly engaged in carrying out the agency's role of stewardship and oversight on a routine basis, the target audience for this course includes FHWA personnel at all levels and in all disciplines, in both Division and Program Offices. State DOT management officials, and other DOT staff involved in the delivery of the FAHP, would also benefit from taking this course.

Training Level: Basic

Fee: 2016: $1000 Per Person; 2017: $1000 Per Person

Length: 2 DAYS (CEU: 1.2 UNITS)

Class Size: MINIMUM: 20; MAXIMUM: 30

NHI Customer Service: (877) 558-6873 • nhicustomerservice@dot.gov

Course Number

FHWA-NHI-361031

Course Title

DBE/ACDBE Certification Training

On November 3, 2014, the Department of Transportation issued a final rule amending its disadvantaged business enterprise program at 49 CFR Part 26. This final rule contains amendments to various certification provisions that are not reflected in the 9-module DBE/ACDBE Certification Training series (FHWA-NHI-361031). We are working to update the 9 modules to reflect the changes. In the meantime, we recommend that upon completion of training you watch a recorded presentation of the final rule amendments available at the Departmental Office of Civil Rights website here: http://www.civilrights.dot.gov/disadvantaged-business-enterprise

-- This training is provided to you at no cost by the Office of the Secretary of Transportation (OST) --

Gain the skills necessary to perform a full review and analysis of Disadvantaged Business Enterprise (DBE) and Airport Concession Disadvantaged Business Enterprise (ACDBE) certification eligibility. The course is delivered through 12 hours of web based training consisting of 9 critical module segments. This training helps ensure that all persons responsible for determining whether or not a firm qualifies as a DBE or ACDBE, as well as those who have general DBE/ACDBE program responsibilities, are knowledgeable concerning all requirements for eligibility, and that the interpretation and application of requirements are consistent throughout the country.

Outcomes

Upon completion of the course, participants will be able to:

- Identify and understand the historical foundation of the DBE/ACDBE program, its objectives, and the overall program operation
- Identify basic certification eligibility requirements according to the regulation 49 CFR Part 26
- Assess whether applicant firms and existing DBE/ACDBEs meet the small business size requirements of the regulation
- Assess ownership/control requirements according to the regulation
- Determine how applicant owners can make an individual showing of social and economic disadvantage according to 49 CFR Part 26 and Appendix E
- Assess whether firm owners meet the economic disadvantage requirements of the regulation
- Perform on-site reviews and collect necessary data
- Properly deny applicant firms entry into the program or remove existing firms' DBE/ACDBE certification
- Properly apply the interstate certification provisions of the regulation
- Understand fraud and fraud prevention strategies applicable to the DBE/ACDBE program
- Identify and understand DBE/ACDBE certification requirements

Target Audience

All persons responsible for determining whether a firm qualifies as a DBE or ACDBE should take this training, including certifiers and DBE Liaison Officers. Certifiers are required to be knowledgeable concerning all requirements for eligibility and that the interpretation and application of the regulatory requirements are applied consistently nationwide. Ensuring that individuals processing DBE certifications apply the same measure of scrutiny and subjectivity is integral to maintaining the integrity of the program.

TRAINING LEVEL: Basic

FEE: 2016: $0 Per Person; 2017: N/A

LENGTH: 12 HOURS (CEU: 0 UNITS)

CLASS SIZE: MINIMUM: 0; MAXIMUM: 0

NHI Customer Service: (877) 558-6873 • nhicustomerservice@dot.gov

Course Number

FHWA-NHI-133078

Course Title

Access Management, Location and Design

This course covers the complex technical issues that underlie effective access management practices on streets and highways and provides the technical rationale for proper signal spacing, driveway spacing and design, the application and design of auxiliary lanes. "Before" and "after" case studies illustrate the impacts of projects to improve traffic safety and operations. In addition, the course addresses the issues involved in developing and administering an effective access management program. The course references the state-of-the-practice as presented in the Transportation Research Board's Access Management Manual, the latest edition of AASHTO's A Policy on Geometric Design of Highways and Streets (Green Book), and pertinent NCHRP reports. In summary, this training provides a lasting reference and specific applications of techniques and practices that will enable transportation engineering and planning personnel to implement successful access management strategies and programs. All participants will receive the class notebook and a copy of the TRB Access Management Manual.

Outcomes

Upon completion of the course, participants will be able to:

- Discuss the impact of access on highway safety and operations
- Choose access management techniques to mitigate challenges
- Identify practices needed for implementing access management programs

Target Audience

This course targets transportation and planning professionals involved in traffic operations, roadway design, the planning of circulation systems, and land development. Specifically, the course is designed for those individuals directly involved in implementing access management solutions in their jurisdictions, as it focuses heavily on resources and solutions to reduce the impact of access points on traffic flow.

Training Level: Basic

Fee: 2016: $875 Per Person; 2017: N/A

Length: 3 DAYS (CEU: 1.8 UNITS)

Class Size: MINIMUM: 20; MAXIMUM: 30

NHI Customer Service: (877) 558-6873 • nhicustomerservice@dot.gov

Course Number

FHWA-NHI-133116

Course Title

Maintenance of Traffic for Technicians - WEB BASED

The Maintenance of Traffic for Technicians Web-based training presents information about the placement of, field maintenance required for, and inspection of traffic control devices. In addition, drafting work zone traffic control plans and flaggering are discussed.

We've broken this training into five modules:

1. General Terms and Procedures
2. Traffic Channelizing and Control Devices
3. Traffic Control Zones
4. Flagger Operations
5. Traffic Control Zone Operations

Outcomes

Upon completion of the course, participants will be able to:

- Identify the correct placement of work zone traffic control devices
- Perform field maintenance of work zone traffic control devices
- Inspect placement or operational functions of work zone traffic control devices
- Generate work zone traffic control plans
- Explain the basics of flagging

Target Audience

This training is designed for all persons with duties that include: Direct responsibility for placement of work zone traffic control devices; Direct responsibility for field maintenance of work zone traffic control devices; Inspection of the placement or operational function of work zone traffic control devices; and Drafting or electronic generation of work zone traffic control plans. The target audience could be geographically dispersed, in need of immediate training or information, or not have access to travel funds.

Training Level: Basic

Fee: 2016: $50 Per Person; 2017: N/A

Length: 5 HOURS (CEU: 0 UNITS)

Class Size: MINIMUM: 1; MAXIMUM: 1

NHI Customer Service: (877) 558-6873 • nhicustomerservice@dot.gov

WBT

Course Number

FHWA-NHI-133117

Course Title

Maintenance of Traffic for Supervisors - WEB BASED

The Maintenance of Traffic for Supervisors Web-based training presents information about the placement of, field maintenance required for, and inspection of traffic control devices. In addition, drafting work zone traffic control plans and flagging are discussed. This training focuses on the design of a traffic control plan, and how and why one needs to operate and implement traffic control in the work zone.

We've broken this training into five modules:

1. Fundamental Principles of Temporary Traffic Control Zones
2. Temporary Traffic Control Devices
3. Traffic Control Zones
4. Transportation Management Plans
5. Flagger Operations

Outcomes

Upon completion of the course, participants will be able to:

- Describe how to create clear, organized traffic control plans
- Identify acceptable temporary traffic control devices
- Determine good and bad flagging techniques

Target Audience

This training is designed for personnel with responsibility or authority to decide on the specific maintenance of traffic requirements to be implemented. These positions include engineers responsible for work zone traffic control development and work site traffic supervisors. The target audience could be geographically dispersed, in need of immediate training or information, or not have access to travel funds.

Training Level: Basic

Fee: 2016: $50 Per Person; 2017: N/A

Length: 5 HOURS (CEU: 0 UNITS)

Class Size: MINIMUM: 1; MAXIMUM: 1

NHI Customer Service: (877) 558-6873 • nhicustomerservice@dot.gov

COURSE NUMBER

FHWA-NHI-134107

COURSE TITLE

Recognizing Roadside Weeds (Southeastern States) - WEB-BASED

This training was prepared by the Transportation Curriculum Coordination Council (TCCC) in partnership with NHI and has been designed for someone learning the first steps in the vegetation management. However, it does not go into the education of weed prevention. This training is recommended for the Transportation Curriculum Coordination Council levels I, and II. This course is primarily intended for inspectors and technicians.

The first step in determining an appropriate weed control strategy is to identify the weed plant. There are numerous different plants growing along many roadsides that can be considered weeds. This is a basic course in the area of weed identification. Most weeds are territorial to different climates and regions, therefore, making it difficult to identify nationally weeds that are dealt with by different State DOT's. This training does focus on southeastern states and is organized in alphabetical order of the weeds that will be covered.

For more information on how stop the migration of weeds contact your State Vegetation Management Program.

OUTCOMES

Upon completion of the course, participants will be able to:

- Understand the definition of a weed
- Describe the reasons for weed control
- Identify several of the most common weeds

TARGET AUDIENCE

This course is designed for entry level individuals working in vegetation management.

TRAINING LEVEL: Basic

FEE: 2016: $25 Per Person; 2017: $25 Per Person

LENGTH: 1 HOURS (CEU: 0 UNITS)

CLASS SIZE: MINIMUM: 1; MAXIMUM: 1

NHI Customer Service: (877) 558-6873 • nhicustomerservice@dot.gov

Updated Training

COURSE NUMBER

FHWA-NHI-142045

COURSE TITLE

Pedestrian Facility Design

To emphasize the importance of planning for pedestrians, the course focuses on case examples involving corridor and intersection design issues. Participants are engaged through lecture, discussion, video demonstrations of problem areas in corridors and intersections, small group problem identification, and the development of design alternatives. This training was developed to provide information and application opportunities to those involved in the design of pedestrian facilities. The Americans with Disabilities Act (ADA) requires newly constructed and altered sidewalks to be accessible and usable by people with disabilities, and accessibility improvements need to be implemented for existing facilities.

OUTCOMES

Upon completion of the course, participants will be able to:

- List the characteristics of pedestrians and motorized traffic that influence pedestrian facility design
- Apply the concepts of universal design and applicable design reference material to redesigning an existing location and/or designing a new location that meets the needs of motorized and nonmotorized users
- Given a case example, identify potential conflicts between pedestrians and other traffic and propose design options that improve access and safety
- Given a case example, analyze the network for improvement options to meet the needs of pedestrian and other traffic

TARGET AUDIENCE

Engineers with planning, design, construction, or maintenance responsibilities; pedestrian and bicycle specialists, disability and orientation specialists, transportation planners, architects, landscape architects, as well as decisionmakers at the project planning level.

TRAINING LEVEL: Intermediate

FEE: 2016: $700 Per Person; 2017: N/A

LENGTH: 1.5 DAYS (CEU: .9 UNITS)

CLASS SIZE: MINIMUM: 20; MAXIMUM: 30

NHI Customer Service: (877) 558-6873 • nhicustomerservice@dot.gov

Course Number

FHWA-NHI-142046

Course Title

Bicycle Facility Design

This training will assist planners and designers in learning how to apply the existing standards and how to deal with other technical issues involved. The availability of Federal, State, and local transportation funding for bicycle facilities that serve transportation and recreational users is resulting in a dramatic increase in the number of bicycling (and shared use) facilities being planned and built. Although there are no Federal design standards for bicycle facilities, the AASHTO Guide for the Development of Bicycle Facilities, or a modification thereof, serves as a design guide. As with most guides, the AASHTO guide cannot address every possible scenario so designers often need to apply engineering judgment where specific information is not provided. The training fee includes a copy of the AASHTO Guide for the Development of Bicycle Facilities.

Outcomes

Upon completion of the course, participants will be able to:

- List the needs of bicyclists as transportation facility users
- Identify common roadway and traffic conditions that affect bicyclists
- Describe the characteristics of a roadway and a shared-use path that are designed to accommodate bicyclists
- List the benefits to the transportation system of accommodating bicyclists with different abilities
- Recognize opportunities to accommodate bicyclists during the planning, design, construction, and operational phases of a project

Target Audience

Federal, State, or local engineers with planning, design, construction, or maintenance responsibilities; bicycle specialists, transportation planners, landscape architects, as well as decisionmakers at the project planning level.

Training Level: Accomplished

Fee: 2016: $700 Per Person; 2017: N/A

Length: 1.5 DAYS (CEU: 1 UNITS)

Class Size: MINIMUM: 20; MAXIMUM: 30

NHI Customer Service: (877) 558-6873 • nhicustomerservice@dot.gov

Course Number

FHWA-NHI-380005

Course Title

Railroad-Highway Grade Crossing Improvement Program

The training provides information on rail-highway crossings, grade crossing components, including program/project development and administration. Workshops will provide the participants a chance to make hands-on applications of the training material, which include such topics as historical background, railroad-highway intersection definition and components, collection and maintenance of data, assessment of crossing safety and operations, identification and selection of alternate improvements, program and project development and implementation, maintenance, and other topics (i.e., private crossings, operation lifesaver).

Outcomes

Upon completion of the course, participants will be able to:

- Describe Active and Passive Devices used in connection with at-grade crossings
- Identify techniques and engineering principles used for at-grade crossings
- Appraise existing at-grade crossings
- Develop alternate methods to improve railroad-highway grade crossings

Target Audience

Federal, State, and local transportation agencies responsible for the design, construction, and/or maintenance of railroad-highway crossings. State and local traffic engineers responsible for highway-railroad grade crossing safety.

Training Level: Accomplished

Fee: 2016: $520 Per Person; 2017: N/A

Length: 2 DAYS (CEU: 1.2 UNITS)

Class Size: MINIMUM: 20; MAXIMUM: 30

NHI Customer Service: (877) 558-6873 • nhicustomerservice@dot.gov

Course Number

FHWA-NHI-380032A

Course Title

Roadside Safety Design (3-Day)

This course provides an overview of the AASHTO Roadside Design Guide. At the end of the course, you will be able to apply the clear zone concept to all classes of roadways; recognize unsafe roadside design features and elements and make appropriate changes; identify the need for a traffic barrier; and apply other highway hardware core competencies.

This course is intended for experienced safety and design engineers.

Outcomes

Upon completion of the course, participants will be able to:

- Apply the clear zone concept to all classes of roadway
- Warrant roadside and median barriers
- Design roadside barriers
- Select the most appropriate end treatment
- Select the most appropriate safety hardware
- Correctly locate safety hardware
- Describe the elements of economic analysis

Target Audience

Experienced Federal, State, and local highway engineers involved in the formulation and/or application of policies and standards relating to the design of safe roadside hardware.

Training Level: Accomplished

Fee: 2016: $800 Per Person; 2017: N/A

Length: 3 DAYS (CEU: 1.8 UNITS)

Class Size: MINIMUM: 20; MAXIMUM: 30

NHI Customer Service: (877) 558-6873 • nhicustomerservice@dot.gov

COURSE NUMBER

FHWA-NHI-380034

COURSE TITLE

Design, Construction, and Maintenance of Highway Safety Appurtenances and Features (1-Day)

The course has been developed for a 3-day course presentation but can also be structured into a 1- or 2-day training course. The sponsoring agency will be able to choose the modules for presentation that will best meet its needs. The course covers the design, construction, and maintenance of highway safety appurtenances and features. It covers the purpose and performance requirements of state-of-the-art highway safety features, such as breakaway sign supports, breakaway utility poles, traffic barriers, impact attenuators, traversable terrain, and hardware features such as drainage inlets. The course describes how these features function, what can go wrong, and how to recognize and correct improper installations.

OUTCOMES

Upon completion of the course, participants will be able to:

- Identify advantages and disadvantages of different types of longitudinal barriers and crash cushions
- Identify National Cooperative Highway Research Program 350 tested safety appurtenances
- Identify application of highway safety appurtenances, why they are used, when and where they should be used, and what is necessary to ensure their function
- Design the placement of, and determine the need for, longitudinal barriers
- Use required installation, construction, and maintenance procedures for proprietary longitudinal barriers, terminals, transitions, crash cushions, bridge railings, and sign supports
- Recognize substandard or potentially hazardous highway appurtenances and features
- Develop alternatives to eliminate, correct, or mitigate unsatisfactory operational characteristics of existing safety devices

TARGET AUDIENCE

Highway engineers, including local personnel involved in the design, construction, or maintenance of highway safety appurtenances and features. This course is suitable for all local, State, and Federal employees that are involved with the installation and repair of highway appurtenances.

TRAINING LEVEL: Accomplished

FEE: 2016: $330 Per Person; 2017: N/A

LENGTH: 1 DAYS (CEU: 0 UNITS)

CLASS SIZE: MINIMUM: 20; MAXIMUM: 30

NHI Customer Service: (877) 558-6873 • nhicustomerservice@dot.gov

Course Number

FHWA-NHI-380034A

Course Title

Design, Construction, and Maintenance of Highway Safety Appurtenances and Features (2-Day)

The course has been developed for a 3-day course presentation but can also be structured into a 1- or 2-day training course. The sponsoring agency will be able to choose the modules for presentation that will best meet its needs. The course covers the design, construction, and maintenance of highway safety appurtenances and features. It covers the purpose and performance requirements of state-of-the-art highway safety features, such as breakaway sign supports, breakaway utility poles, traffic barriers, impact attenuators, traversable terrain, and hardware features such as drainage inlets. The course describes how these features function, what can go wrong, and how to recognize and correct improper installations.

Outcomes

Upon completion of the course, participants will be able to:

- Identify advantages and disadvantages of different types of longitudinal barriers and crash cushions
- Identify National Cooperative Highway Research Program 350 tested safety appurtenances
- Identify application of highway safety appurtenances, why they are used, when and where they should be used, and what is necessary to ensure their function
- Design the placement of, and determine the need for, longitudinal barriers
- Use required installation, construction, and maintenance procedures for proprietary longitudinal barriers, terminals, transitions, crash cushions, bridge railings, and sign supports
- Recognize substandard or potentially hazardous highway appurtenances and features
- Develop alternatives to eliminate, correct, or mitigate unsatisfactory operational characteristics of existing safety devices

Target Audience

Highway engineers, including local personnel involved in the design, construction, or maintenance of highway safety appurtenances and features. This course is suitable for all local, State, and Federal employees that are involved with the installation and repair of highway appurtenances.

Training Level: Accomplished

Fee: 2016: $650 Per Person; 2017: N/A

Length: 2 DAYS (CEU: 0 UNITS)

Class Size: MINIMUM: 20; MAXIMUM: 30

NHI Customer Service: (877) 558-6873 • nhicustomerservice@dot.gov

Course Number

FHWA-NHI-380034B

Course Title

Design, Construction, and Maintenance of Highway Safety Appurtenances and Features (3-Day)

The course has been developed for a 3-day course presentation but can also be structured into a 1- or 2-day training course. The sponsoring agency will be able to choose the modules for presentation that will best meet its needs. The course covers the design, construction, and maintenance of highway safety appurtenances and features. It covers the purpose and performance requirements of state-of-the-art highway safety features, such as breakaway sign supports, breakaway utility poles, traffic barriers, impact attenuators, traversable terrain, and hardware features such as drainage inlets. The course describes how these features function, what can go wrong, and how to recognize and correct improper installations.

Outcomes

Upon completion of the course, participants will be able to:

- Identify advantages and disadvantages of different types of longitudinal barriers and crash cushions
- Identify National Cooperative Highway Research Program 350 tested safety appurtenances
- Identify application of highway safety appurtenances, why they are used, when and where they should be used, and what is necessary to ensure their function
- Design the placement of, and determine the need for, longitudinal barriers
- Use required installation, construction, and maintenance procedures for proprietary longitudinal barriers, terminals, transitions, crash cushions, bridge railings, and sign supports
- Recognize substandard or potentially hazardous highway appurtenances and features
- Develop alternatives to eliminate, correct, or mitigate unsatisfactory operational characteristics of existing safety devices

Target Audience

Highway engineers, including local personnel involved in the design, construction, or maintenance of highway safety appurtenances and features. This course is suitable for all local, State, and Federal employees that are involved with the installation and repair of highway appurtenances.

Training Level: Accomplished

Fee: 2016: $800 Per Person; 2017: N/A

Length: 3 DAYS (CEU: 0 UNITS)

Class Size: MINIMUM: 20; MAXIMUM: 30

NHI Customer Service: (877) 558-6873 • nhicustomerservice@dot.gov

Course Number

FHWA-NHI-380069

Course Title

Road Safety Audits/Assessments

Performing effective road safety audits/assessments, (RSAs), improves safety and demonstrates to the public an agency's dedication to crash reduction. An RSA is a formal safety performance examination of an existing or future road or intersection by an independent audit team. The RSA training provides practical information on how to conduct an RSA, select a location, and build an independent, multi-disciplinary team. The costs, time, benefits, and common myths and concerns surrounding RSAs will be discussed. Participants learn how to improve transportation safety by applying a new proactive approach. Emphasis is placed on using low cost safety improvements as well as understanding the interaction between the highway and all road users.

The training includes hands-on application of the training materials, which includes information on each stage of a road safety audit and easy-to-use-prompt lists. A copy of "FHWA Road Safety Audit Guidelines" is provided.

Outcomes

Upon completion of the course, participants will be able to:

- Express the road safety audit process terminology
- Perform a simple road safety audit, as a member of a team
- Assess the benefits of a road safety audit on a local or statewide basis

Target Audience

Personnel who are likely to serve on a road safety audit team including Federal, State, local transportation personnel, first responders and consultants who conduct highway safety studies should also attend.

Training Level: Accomplished

Fee: 2016: $400 Per Person; 2017: N/A

Length: 2 DAYS (CEU: 1.2 UNITS)

Class Size: MINIMUM: 20; MAXIMUM: 30

NHI Customer Service: (877) 558-6873 • nhicustomerservice@dot.gov

Course Number

FHWA-NHI-380070

Course Title

Highway Safety Manual Practitioners Guide for Geometric Design Features

This course includes both 2-lane and multi-lane highways and provides a proven methodology for the safety performance of geometric design decisions in a like manner to that of predicting capacity and level of service based upon large scale definitive research. The crash prediction models for total crashes and cross-section related crashes based upon lane width, shoulder width, roadside hazard, traffic volume (exposure) and other characteristics are presented. Examples of safety performance prediction are presented for highway segments and intersections.

Discussion of research and the interactive effects of lane and shoulder widths, hazard rating, and access density (driveways) on safety performance are presented. Each student receives a copy of the "Safety Effects of Highway Design Features" manual.

IMPORTANT: Participants should bring a scientific notation calculator as the course involves calculating decimal value to decimal power for crash prediction values.

Outcomes

Upon completion of the course, participants will be able to:

- Recognize the safety effects of geometric design features
- Predict the safety performance of geometric design features
- Compare alternative designs based upon an assessment of the safety effects of geometric design features

Target Audience

State and local highway engineers and consultants involved in the design of both two-lane rural and/or multilane highways.

Training Level: Accomplished

Fee: 2016: $400 Per Person; 2017: N/A

Length: 2 DAYS (CEU: 1.2 UNITS)

Class Size: MINIMUM: 20; MAXIMUM: 30

NHI Customer Service: (877) 558-6873 • nhicustomerservice@dot.gov

COURSE NUMBER

FHWA-NHI-380070A

COURSE TITLE

Highway Safety Manual Practitioners Guide for Two-Lane Rural Highways

This course provides a proven methodology for the safety performance of geometric design decisions in a like manner to that of predicting capacity and level of service based upon large scale definitive research. The crash prediction models for total crashes and cross-section related crashes based upon lane width, shoulder width, roadside hazard, traffic volume (exposure) and other characteristics are presented. Examples of safety performance prediction are presented for highway segments and intersections.

Discussion of research and the interactive effects of lane and shoulder widths, hazard rating, and access density (driveways) on safety performance are presented. Each student receives a copy of the "Safety Effects of Highway Design Features for Two-Lane Rural Highways" manual.

IMPORTANT: Participants should bring a scientific notation calculator as the course involves calculating decimal value to decimal power for crash prediction values.

OUTCOMES

Upon completion of the course, participants will be able to:

- Recognize the safety effects of geometric design features
- Predict the safety performance of geometric design features
- Compare alternative designs based upon an assessment of the safety effects of geometric design features

TARGET AUDIENCE

State and local highway engineers and consultants involved in the design of two-lane rural highways.

TRAINING LEVEL: Accomplished

FEE: 2016: $300 Per Person; 2017: N/A

LENGTH: 1 DAYS (CEU: .6 UNITS)

CLASS SIZE: MINIMUM: 20; MAXIMUM: 30

NHI Customer Service: (877) 558-6873 • nhicustomerservice@dot.gov

Course Number

FHWA-NHI-380070B

Course Title

Highway Safety Manual Practitioners Guide for Multilane Highways

This course provides proven methodology for the safety performance of geometric design decisions for multilane highways in a like manner to that of predicting capacity and level of service based upon large scale definitive research. The crash prediction models for total crashes based upon lane width, shoulder width, roadside hazard, traffic volume (exposure) and other characteristics are presented. Examples of safety performance prediction are presented for highway segments and intersections.

Discussion of research and the interactive effects on safety performance for median width and barriers, of access (driveways) and side streets and intersection turning lanes are presented. Each student receives a copy of the "Safety Effects of Highway Design Features" manual.

IMPORTANT: Participants should bring a scientific notation calculator as the course involves calculating decimal value to decimal power for crash prediction values.

Outcomes

Upon completion of the course, participants will be able to:

- Recognize the safety effects of geometric design features
- Predict the safety performance of geometric design features
- Compare alternative designs based upon an assessment of the safety effects of geometric design features

Target Audience

State and local highway engineers and consultants involved in the design of multilane highways.

Training Level: Accomplished

Fee: 2016: $300 Per Person; 2017: N/A

Length: 1 DAYS (CEU: .6 UNITS)

Class Size: MINIMUM: 20; MAXIMUM: 30

NHI Customer Service: (877) 558-6873 • nhicustomerservice@dot.gov

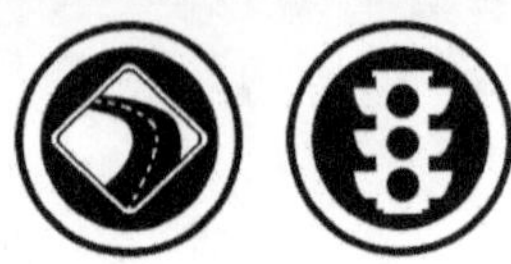

COURSE NUMBER

FHWA-NHI-380071

COURSE TITLE

Interactive Highway Safety Design Model

This course instructs highway design project managers, planners, designers, and traffic and safety reviewers in the application of the Interactive Highway Safety Design Model (IHSDM) software and provides guidance on interpretation of the output.

IHSDM is a suite of software tools to evaluate safety of two-lane rural highways. The software, developed for FHWA, was released in 2003 after several years of research and development to provide state-of-the-art techniques for safety analysis. IHSDM contains five tools that can be used to apply the most recent safety analysis techniques in a relatively straightforward and automated manner. For more information about IHSDM, go to http://www.tfhrc.gov/safety/ihsdm/ihsdm.htm.

Participants gain hands-on experience with the software. Therefore, the training facility must be equipped with computers. There should be no more than two participants per computer. Minimum system specifications for the computers are as follows: Operating System - Microsoft Vista, Windows XP or Windows 2000 Professional; HTML Browser - Microsoft Internet Explorer, Netscape Navigator, or Foxfire; Spreadsheet Program, Microsoft Excel or equivalent; Hardware - At least 450 MHz Pentium III (or equivalent) CPU, 256 MB RAM or greater desirable, 800x600 high colors (16 bit) display; and 300 MB free disk space

OUTCOMES

Upon completion of the course, participants will be able to:

- Describe key capabilities and limitations of IHSDM
- Evaluate a two-lane rural highway using IHSDM
- Recognize when and how IHSDM can be used in the project development process

TARGET AUDIENCE

Highway design project managers, planners, designers, and traffic and safety reviewers with at least one or two years of experience with highway design, preferably two-lane rural highway design.

TRAINING LEVEL: Accomplished

FEE: 2016: $400 Per Person; 2017: N/A

LENGTH: 2 DAYS (CEU: 0 UNITS)

CLASS SIZE: MINIMUM: 20; MAXIMUM: 30

NHI Customer Service: (877) 558-6873 • nhicustomerservice@dot.gov

Course Number

FHWA-NHI-380073

Course Title

Fundamentals of Planning, Design and Approval of Interchange Improvements to the Interstate System

This course provides participants with key knowledge of freeway systems and interchange types, FHWA policy on justification for interchange access approval, and applications of technical knowledge and policy understanding to interchange project decisions. Topics covered in this course include service and system interchange types, 8-point interchange justification process, interchange study and selection process, fundamentals of freeway system operations and planning, urban freeway diagnosis, geometric design considerations, and technical and documentation procedures.

Outcomes

Upon completion of the course, participants will be able to:

- Compare and contrast design and operational attributes of different of freeway interchange types
- Interpret and apply the elements of the FHWA Policy for approving Interstate access
- Describe and apply principles of good freeway systems and interchange design
- Describe the application of design exceptions to interchange project decisions
- Describe the content of an appropriate safety and operational analysis to support an access request
- Compare alternative designs based upon an assessment of appropriate measures of effectiveness (MOEs)
- Apply an interchange design study procedure

Target Audience

The target audience for the course includes traffic engineers and transportation professionals with one to five years of working experience.

Training Level: Accomplished

Fee: 2016: $500 Per Person; 2017: N/A

Length: 3 DAYS (CEU: 1.8 UNITS)

Class Size: MINIMUM: 20; MAXIMUM: 30

NHI Customer Service: (877) 558-6873 • nhicustomerservice@dot.gov

Course Number

FHWA-NHI-380075

Course Title

New Approaches to Highway Safety Analysis

The primary purpose of this course is to help attendees gain an understanding of the Highway Safety Improvement Program (HSIP) process, safety engineering principles and human factors issues related to traffic and road safety. It also provides the participant with an explanation of the latest methods for identifying collision causes and selecting cost-effective safety improvements. Finally, this course will serve as a prerequisite for those who will be utilizing SafetyAnalyst, a set of software tools currently under development that are designed to assist State and local agencies to improve the decisionmaking process in implementing safety improvement projects.

Outcomes

Upon completion of the course, participants will be able to:

- Describe the components of the Highway Safety Improvement Program (HSIP)
- Explain safety engineering principles relevant to planning for highway safety improvement measures specific to three types of crashes ¿ roadway departures, intersection-related, and pedestrian
- Describe the relevance and impact of human factors in the planning of highway safety improvement measures for three types of crashes ¿ roadway departures, intersection-related, and pedestrian
- Determine strategies for the selection of cost-effective highway safety improvement measures for three types of crashes ¿ roadway departures, intersection-related, and pedestrian

Target Audience

This course is intended primarily for State DOT staff involved with the Highway Safety Improvement Program, and for FHWA safety specialists. These specialists include engineers, planners, and technicians.

Training Level: Accomplished

Fee: 2016: $910 Per Person; 2017: N/A

Length: 3 DAYS (CEU: 0 UNITS)

Class Size: MINIMUM: 20; MAXIMUM: 30

NHI Customer Service: (877) 558-6873 • nhicustomerservice@dot.gov

COURSE NUMBER

FHWA-NHI-380076

COURSE TITLE

Low-Cost Safety Improvements Workshop

This course provides a comprehensive presentation of low-cost, ready-to-use improvements that enhance the safety of highways. The course covers a synthesis of countermeasures and their associated crash reduction factors as identified in the "AASHTO Strategic Highway Safety Plan -- NCHRP 500 Guidebooks." Countermeasures for specific areas of highway safety, including roadside hazards; signing, markings, and lighting; traffic control devices; intersections; traffic signals; and railroad grade crossings are discussed. The course also introduces recent low-cost safety improvements that have been developed by States and local engineers. Through exercises, participants learn how to analyze highway safety situations and apply appropriate countermeasures to those situations.

OUTCOMES

Upon completion of the course, participants will be able to:

- Identify appropriate engineering countermeasures from crash patterns
- Recognize deficiencies in operation/design and select appropriate countermeasures for roadside hazards
- Recognize deficiencies in safety performance of signing, markings, and lighting, and elect appropriate countermeasures
- Recognize deficiencies in operation/design of intersections and select appropriate countermeasures
- Recognize deficiencies in operation/design of traffic signals and select appropriate countermeasures
- Recognize deficiencies in operation/design of railroad grade crossings and select appropriate countermeasures
- Illustrate new and innovative low-cost safety improvement measures developed by State DOTs

TARGET AUDIENCE

Federal, State, and local transportation, traffic and safety engineers, and planners involved in reducing crashes.

TRAINING LEVEL: Accomplished

FEE: 2016: $300 Per Person; 2017: N/A

LENGTH: 1 DAYS (CEU: 0 UNITS)

CLASS SIZE: MINIMUM: 20; MAXIMUM: 35

NHI Customer Service: (877) 558-6873 • nhicustomerservice@dot.gov

Course Number

FHWA-NHI-380077

Course Title

Intersection Safety Workshop

Beginning with an introduction to intersection and crash characteristics, this course provides information on ready-to-use, direct-application safety measures for rural unsignalized and signalized intersections. Participants are presented with a synthesis of countermeasures and their associated crash reduction factors as identified in the "AASHTO Strategic Highway Safety Plan - NCHRP 500 Guidebooks." The course focuses on the application of these countermeasures and design and safety operations best practices for substantive improvements to intersection safety. During the course, participants have the opportunity to present intersection safety situations that they are currently facing and discuss appropriate countermeasures and best practices to address those situations.

Outcomes

Upon completion of the course, participants will be able to:

- Apply models (equations) to predict the number of crashes for an intersection based upon traffic volumes
- Identify high crash intersections and recognize appropriate engineering countermeasures
- Identify crash reduction factors/crash modification factors associated with countermeasures
- Describe safety performance of intersection geometric design features and the models to quantify the safety effect
- List regulatory, warning, and guide signing and markings countermeasures and associated safety benefits
- List highway lighting countermeasures and associated safety benefits
- List traffic signal countermeasures and associated safety benefits

Target Audience

Federal, State, and local transportation traffic and safety engineers, and planners involved in reducing intersection crashes.

Training Level: Accomplished

Fee: 2016: $320 Per Person; 2017: N/A

Length: 1 DAYS (CEU: 0 UNITS)

Class Size: MINIMUM: 20; MAXIMUM: 30

NHI Customer Service: (877) 558-6873 • nhicustomerservice@dot.gov

Course Number

FHWA-NHI-380078

Course Title

Signalized Intersection Guidebook Workshop

This course provides an overview of the "Signalized Intersections: Informational Guide FHWA-HRT-04-091." The guide is a comprehensive document containing methods for evaluating the safety and operations of signalized intersections and tools to remedy deficiencies. It takes a holistic approach to signalized intersections and considers the safety and operational implications of a particular treatment on all system users, including motorists, pedestrians, bicyclists, and transit users. Using the guide, participants learn to make insightful intersection assessments, understand the tradeoffs of potential improvement measures, and apply guidebook measures and best practices to reduce the incidence of intersection crashes.

Outcomes

Upon completion of the course, participants will be able to:

- Recognize and apply fundamentals of signalized intersections in terms of user needs, geometric design, traffic design, and illumination
- Describe signalized intersection project process, safety analysis methods, and operational analysis methods
- Describe the more than 100 signalized intersection treatments and their advantages and disadvantages

Target Audience

Federal, State, and local transportation, traffic and safety engineers, and planners involved in planning, designing, operating, and remedying crash problems for signalized intersections.

Training Level: Intermediate

Fee: 2016: $330 Per Person; 2017: N/A

Length: 1 DAYS (CEU: .6 UNITS)

Class Size: MINIMUM: 20; MAXIMUM: 30

NHI Customer Service: (877) 558-6873 • nhicustomerservice@dot.gov

Course Number

FHWA-NHI-380083

Course Title

Low-Cost Safety Improvements - WEB-BASED

This course helps to equip the target audience with the knowledge and skills needed to analyze crash data, identify crash patterns, and select appropriate "low cost" countermeasures. Upon completion of this course, participants will be able to identify appropriate (i.e., cost effective) engineering countermeasures by using the Six-Step Crash Mitigation Process (CMP).

The course uses a combination of web-conferences and self-paced materials that aid in application to current safety projects. You will need access to both a telephone and internet connection to participate in the live web sessions.

Outcomes

Upon completion of the course, participants will be able to:

- Identify appropriate engineering countermeasures from crash patterns.
- Select appropriate countermeasures for:
- Roadside hazards based upon deficiencies in operations or design.
- Deficiencies in signage, roadway markings, and lighting.
- Deficiencies in operation/design of highway intersections.

Target Audience

This course is intended for individuals responsible for identifying, recommending, selecting, installing and/or maintaining appropriate low cost countermeasures to help reduce the number of crashes.

Training Level: Basic

Fee: 2016: $270 Per Person; 2017: N/A

Length: 10 HOURS (CEU: 1 UNITS)

Class Size: MINIMUM: 10; MAXIMUM: 45

NHI Customer Service: (877) 558-6873 • nhicustomerservice@dot.gov

Course Number

FHWA-NHI-380085

Course Title

Guardrail Installation Training

This course provides instruction in the principles and practices of guardrail installation and performance. Instruction focuses on the Length of Need of barriers (including a field expedient procedure) but also includes instruction on guardrail transitions and guardrail end treatments. Participants will evaluate existing installations for proper performance characteristics.

Outcomes

Upon completion of the course, participants will be able to:

- Explain the "Roadside Safety" problem and the warrants for barrier.
- Explain how barrier systems operate.
- Describe the installation principles necessary for proper barrier operation.
- Describe the installation principles necessary for proper terminal operation.
- Inspect barrier systems for proper installation and operation.

Target Audience

Due to the amount of material in this one day course, the pace is best suited for the experienced guardrail installer or inspector. Others that may benefit from the course include construction and maintenance engineers.

Training Level: Accomplished

Fee: 2016: $300 Per Person; 2017: N/A

Length: 1 DAYS (CEU: .6 UNITS)

Class Size: MINIMUM: 20; MAXIMUM: 30

NHI Customer Service: (877) 558-6873 • nhicustomerservice@dot.gov

COURSE NUMBER

FHWA-NHI-380089

COURSE TITLE

Designing for Pedestrian Safety

The Designing for Pedestrian Safety course is intended to help state and local transportation engineering professionals address pedestrian safety issues through design and engineering solutions. The training course includes a field exercise in the application of the principles, concepts, and strategies covered in the course. Also the participants will share and prioritize potential policies, programs, and strategies.

OUTCOMES

Upon completion of the course, participants will be able to:

- Describe the influence of planning factors: land use, street connectivity, access management, site design, and level of service.
- Describe how pedestrians should be considered and provided for during the planning, design, work zone, maintenance, and operations phases.
- Describe how human behavior affects the interaction between pedestrians and drivers
- Identify good practices and effective solutions to enhance pedestrian safety and accessibility.

TARGET AUDIENCE

This course is intended primarily for state DOT staff involved with the Highway Safety Improvement Program, and for FHWA Safety Specialists. These specialists shall include: Engineers, planners, traffic safety and enforcement professionals, public health and injury prevention professionals, and decision-makers who have the responsibility of improving pedestrian safety at the state or local level.

TRAINING LEVEL: Basic

FEE: 2016: $450 Per Person; 2017: N/A

LENGTH: 2 DAYS (CEU: 1.5 UNITS)

CLASS SIZE: MINIMUM: 20; MAXIMUM: 30

NHI Customer Service: (877) 558-6873 • nhicustomerservice@dot.gov

Course Number

FHWA-NHI-380090

Course Title

Developing a Pedestrian Safety Action Plan

The Developing a Pedestrian Safety Action Plan course is designed to help state and local officials learn "HOW TO" address pedestrian safety issues in the development of a pedestrian safety action plan, program, and activities tailored to their community. It is also intended to assist agencies in the further enhancement of their existing pedestrian safety plan, programs, and activities, including involving partners and stakeholders, collecting and analyzing data and information, prioritizing issues and concerns, selecting and implementing an optimal combination of education, enforcement, engineering strategies. The training course includes a field exercise in the application of the principles, concepts, and strategies covered in the course. Also the participants will share and prioritize potential policies, programs, and strategies.

Outcomes

Upon completion of the course, participants will be able to:

- Develop and implement a Pedestrian Safety Action Plan addressing your specific issues, problems, needs and resources
- Describe how pedestrians should be considered and provided for during the planning, design, work zone, maintenance, and operations phases.
- Describe how human behavior affects the interaction between pedestrians and drivers
- Identify good practices and effective solutions to enhance pedestrian safety and accessibility.

Target Audience

This course is intended primarily for state DOT staff involved with the Highway Safety Improvement Program, and for FHWA Safety Specialists. These specialists shall include: Engineers, planners, traffic safety and enforcement professionals, public health and injury prevention professionals, and decision-makers who have the responsibility of improving pedestrian safety at the state or local level.

Training Level: Basic

Fee: 2016: $450 Per Person; 2017: N/A

Length: 2 DAYS (CEU: 1.2 UNITS)

Class Size: MINIMUM: 20; MAXIMUM: 30

NHI Customer Service: (877) 558-6873 • nhicustomerservice@dot.gov

HIGHWAY SAFETY

Course Number

FHWA-NHI-380091

Course Title

Planning and Designing for Pedestrian Safety

The Planning and Designing for Pedestrian Safety is a combination of the information from the 2-day "Developing a Pedestrian Safety Action Plan" (NHI-380089) and 2-day "Designing for Pedestrian Safety" (NHI-380090) course. This comprehensive course is designed to help state and local officials learn "HOW TO" address pedestrian safety issues in the development of a pedestrian safety action plan, and specific programs and activities tailored to their community. It is also intended to assist agencies in the further enhancement of their existing pedestrian safety plan, programs, and activities, including involving partners and stakeholders, collecting and analyzing data and information, prioritizing issues and concerns, selecting and implementing an optimal combination of education, enforcement, engineering strategies. This course goes into more detail on engineering strategies than the "Developing a Pedestrian Safety Action Plan" (NHI-380089) course. This course includes two field exercises in the application of the principles, concepts, and strategies covered in the course. Also the participants will share and prioritize potential policies, programs, and strategies.

Outcomes

Upon completion of the course, participants will be able to:

- Describe the role that planning and street design play in pedestrian safety.
- Demonstrate how pedestrians should be considered and provided for during the planning, design, work zone maintenance, and operations phases of the pedestrian safety action plan.
- Describe how human behavior issues related to pedestrians and drivers interacting safely and common pedestrian crash types.
- Identify good practices and effective solutions to enhance pedestrian safety and accessibility.
- Explain the significance of land-use, street connectivity, and site design in helping to make a safer pedestrian environment.
- Recognize human behavior issues related to pedestrians and drivers interacting safely and common pedestrian crash types.
- Collect and analyze data in a meaningful way to identify safety deficiencies and priorities for improvement.
- Employ commonly used and effective pedestrian crash countermeasures
- Effectively involve stakeholders to create publicly supported and trusted policies, programs, and projects.

Target Audience

Engineers, planners, traffic safety and enforcement professionals, public health and injury prevention professionals, and decision-makers who have the responsibility of improving pedestrian safety at the state or local level.

Training Level: Basic

Fee: 2016: $530 Per Person; 2017: N/A

Length: 3 DAYS (CEU: 1.8 UNITS)

Class Size: MINIMUM: 20; MAXIMUM: 30

NHI Customer Service: (877) 558-6873 • nhicustomerservice@dot.gov

Course Number

FHWA-NHI-380093

Course Title

Application of Crash Modification Factors (CMF)

This course focuses on the application of Crash Modification Factors to select countermeasures. The course covers the project development cycle (starting from network screening and site selection for safety review), diagnostics of safety concerns, cost-benefit evaluation, and countermeasure selection.

This course combines a web-conference and a self-paced lesson that aids in application to your current projects. You will need access to both a telephone and internet connection to participate in the live web sessions.

Outcomes

Upon completion of the course, participants will be able to:

- Explain how Crash Modification Factors are used to estimate the safety effects of highway improvements.
- Apply Crash Modification Factors to compare and select highway safety improvements.

Target Audience

This course is intended for individuals that have the responsibility for identifying, recommending, selecting, installing, and maintaining appropriate countermeasures to help reduce the number of crashes.

Training Level: Basic

Fee: 2016: $150 Per Person; 2017: N/A

Length: 3 HOURS (CEU: .3 UNITS)

Class Size: MINIMUM: 10; MAXIMUM: 45

NHI Customer Service: (877) 558-6873 • nhicustomerservice@dot.gov

COURSE NUMBER

FHWA-NHI-380094

COURSE TITLE

Science of Crash Modification Factors

This course provides participants with the knowledge and skills needed to critically assess the quality of Crash Modification Factors (CMFs). The course covers concepts underlying the measurement of safety and the development of CMFs, key statistical issues that affect the development of quality CMFs, key methodological issues that affect the development of quality CMFs, and the general and methodological issues and statistical thresholds used to recognize quality CMFs.

This course combines self-paced material that will orient you to CMFs and what constitutes a quality CMF followed by a web-conference that will help you evaluate CMFs found in the CMF Clearinghouse. You will need access to both a telephone and internet connection to participate in the live web session.

OUTCOMES

Upon completion of the course, participants will be able to:

- Explain the concepts of Crash Modification Factors (CMFs) and the measurement of safety.
- List and describe important statistical issues that affect safety research.
- Describe and compare three methodologies for evaluating the safety effect of a countermeasure.
- Select the most appropriate CMF for a given application.

TARGET AUDIENCE

The professionals who would be most interested in completing this course are those who are responsible for identifying, recommending, selecting, and installing appropriate countermeasures to help reduce the number of crashes.

TRAINING LEVEL: Intermediate

FEE: 2016: $150 Per Person; 2017: N/A

LENGTH: 3 HOURS (CEU: .3 UNITS)

CLASS SIZE: MINIMUM: 10; MAXIMUM: 45

NHI Customer Service: (877) 558-6873 • nhicustomerservice@dot.gov

Course Number

FHWA-NHI-380095

Course Title

Geometric Design: Applying Flexibility and Risk Management

Highway designers often face complex trade-offs when developing projects. A "quality" design may be thought of as satisfying the needs of a wide variety of users while balancing the often competing interests of cost, safety, mobility, social and environmental impacts. Applying flexibility and risk management in highway design requires more than simply assembling geometric elements from the available tables, charts and equations of design criteria. This course provides participants with knowledge of the functional basis of critical design criteria to enable informed decisions when applying engineering judgment and flexibility. The course exercises and case studies provide practical applications of current knowledge from research and operational experience of human factors and safety effects for various design elements.

Outcomes

Upon completion of the course, participants will be able to:

- Define the relationship among design criteria, design guidelines and design standards
- Describe the concepts of design speed, target speed, posted speed and operating speed
- Describe the FHWA Policy for Design Standards and Design Exceptions
- List the 13 controlling geometric design criteria that require a formal written design exception from FHWA
- Evaluate the safety effects and qualitative risk of proposed design exceptions
- Evaluate the effectiveness and appropriateness of mitigation strategies for design exceptions
- Describe the relationship between safety and key geometric features of highway alignment and cross section
- Describe the applicability of a human-centered approach to geometric design considerations

Target Audience

This course is targeted toward engineers that are involved in applying engineering judgment in the selection of design criteria and in the assessment of design exceptions. It is most practical for practicing engineers and highway decision makers from state highway agencies, local agencies, design consultants, and FHWA field offices.

Training Level: Accomplished

Fee: 2016: $400 Per Person; 2017: N/A

Length: 2 DAYS (CEU: 1.2 UNITS)

Class Size: MINIMUM: 20; MAXIMUM: 30

NHI Customer Service: (877) 558-6873 • nhicustomerservice@dot.gov

Course Number

FHWA-NHI-380096

Course Title

Modern Roundabouts: Intersections Designed for Safety

The modern roundabout is a proven strategy for improving the safety and operations of intersections. The physical characteristics of a well-designed modern roundabout reduce the frequency and severity of intersection crashes for all users including pedestrians and bicyclists. This course highlights the benefits of modern roundabouts and gives participants the fundamental knowledge needed to plan and consider applying roundabout intersection projects in their area. This course is an introductory level course with a blend of technical and non-technical planning, design and operations considerations.

Outcomes

Upon completion of the course, participants will be able to:

- Distinguish a modern roundabout from other types of circular intersections
- Describe the safety advantages of roundabouts
- Describe the operational advantages roundabouts provide
- Identify what type of locations roundabouts may be appropriate
- Describe strategies to overcome common barriers to implementation of roundabouts, such as negative public perceptions
- Describe the key considerations when planning an area's first roundabout
- Apply basic traffic operational models and capacity calculations for roundabouts
- Describe key geometric design principles of a modern roundabout
- Apply signing and marking suggested practices
- Apply design strategies for pedestrians and bicyclists

Target Audience

Transportation professionals with at least one year of working experience

Training Level: Basic

Fee: 2016: $320 Per Person; 2017: N/A

Length: 1 DAYS (CEU: 0 UNITS)

Class Size: MINIMUM: 20; MAXIMUM: 30

NHI Customer Service: (877) 558-6873 • nhicustomerservice@dot.gov

Course Number

FHWA-NHI-380097

Course Title

An Overview of the Railroad-Highway Grade Crossing Improvement Program

A highway-rail grade crossing is the intersection of two transportation modes: railroads and highways. Many crossings are "at-grade" which significantly increases incidents of often-fatal crashes between trains and motor vehicles. Over the past 30 years, there have been substantial reductions in crashes and fatalities at highway-rail grade crossings due to the efforts of federal, state, and local governments, the railroads, and non-profit organizations such as Operation Lifesaver, Inc. Nonetheless, crashes still occur.

The goal of this one-day training course is to provide attendees with the knowledge and tools needed to plan, implement, and evaluate safety improvements to highway-rail grade crossings. The course presents:

An overview of the regulations, responsibilities, and funding mechanisms that apply to today's Highway-Rail Grade Crossing (HRGX) program.

The steps involved in planning, implementing, and evaluating highway-rail grade crossing improvement projects.

Outcomes

Upon completion of the course, participants will be able to:

- Identify the highway-rail grade crossing program components and processes and the regulations that apply to the program.
- Identify highway-rail grade crossing improvement work that is required as part of highway improvement projects under other federal-aid programs.
- Describe the purpose and benefits of assessing highway-rail grade crossing safety and operations.
- Explain considerations for implementing and maintaining a grade crossing improvement project.
- Identify techniques and tools for improving highway-rail grade crossing safety and operations.

Target Audience

The target audience for this training course includes: State DOT personnel involved in highway-rail grade crossings; Public project engineers from railroad industries; Transportation consultants; FHWA safety engineers; MPO/City/county DOT personnel; and FRA crossing managers.

Training Level: Basic

Fee: 2016: $300 Per Person; 2017: N/A

Length: 1 DAYS (CEU: .6 UNITS)

Class Size: MINIMUM: 20; MAXIMUM: 30

NHI Customer Service: (877) 558-6873 • nhicustomerservice@dot.gov

Course Number

FHWA-NHI-380100

WCT

Course Title

Using IHSDM

NHI delivers Web-conference Training to you!

The IHSDM course is a training that gives participants the opportunity to use the IHSDM software tools to evaluate and analyze highway designs.

The delivery format consists of 4 live Web Conference Trainings (WCT), which participants are required to attend. In between Web-conferences, participants must complete self-paced assignments.

The Interactive Highway Safety Design Model (IHSDM) is a suite of software analysis tools used to evaluate the safety and operational effects of geometric design decisions on highways.

IHSDM is a decision-support tool, which provides estimates of a highway design's expected safety and operational performance and checks existing or proposed highway designs against relevant design policy values. Results of the IHSDM support decisionmaking in the highway design process. Intended users include highway project managers, designers, and traffic and safety reviewers in State and local highway agencies and in engineering consulting firms.

The IHSDM, which supports the Data-Driven Safety Analysis initiative that is part of Federal Highway Administration's (FHWA's) Every Day Counts 3 efforts, includes six evaluation modules (Crash Prediction, Design Consistency, Intersection Review, Policy Review, Traffic Analysis, and Driver/Vehicle). This Web site summarizes the capabilities and applications of the IHSDM evaluation modules, and provides a library of the research reports documenting their development.

The IHSDM - HSM Predictive Method 2015 Release (version 11.0.1, October 2015) may be downloaded free of charge at http://www.ihsdm.org. The new version includes major enhancements to the Policy Review Module, which was expanded to include policy checks for rural multilane highways.

www.fhwa.dot.gov/research/tfhrc/projects/safety/comprehensive/ihsdm/index.cfm

Outcomes

Upon completion of the course, participants will be able to:

- Explain the scope and uses for the IHSDM tool.
- Input rural highway data to IHSDM.
- Explain the purpose of each of the six IHSDM modules.
- Demonstrate the workflow for each IHSDM module.
- Interpret and apply data from IHSDM reports and graphs to make rural highways safer.

Target Audience

The Using IHSDM Course is designed for personnel working on highway design projects who will be directly interacting with the IHSDM software tools or applying the data generated by them. The IHSDM course benefits highway design project managers, planners, designers, safety engineers, and other personnel responsible for reviewing operations and safety on rural highways. Participants should have general familiarity with highway design elements and terminology.

Training Level: Intermediate

Fee: 2016: $280 Per Person; 2017: N/A

Length: 12 HOURS (CEU: 1.2 UNITS)

Class Size: MINIMUM: 10; MAXIMUM: 45

NHI Customer Service: (877) 558-6873 • nhicustomerservice@dot.gov

COURSE NUMBER

FHWA-NHI-380105

COURSE TITLE

Highway Safety Manual Practitioners Guide for Intersections

The new Highway Safety Manual is the state of the art "toolbox" for the "science of safety" for the analysis and prediction of crash frequency for highways and streets. The HSM reflects the evolution in safety analysis from descriptive methods to quantitative, predictive analyses.

The Highway Safety Manual (HSM) provides analytical tools and techniques for quantifying the potential effects on crashes as a result of decisions made in planning, design, operations, and maintenance. A universal objective is to reduce the number and severity of crashes within the limits of available resources, science, and technology, while meeting legislatively mandated priorities. The information in the HSM is provided to assist agencies in their effort to integrate safety into their decision-making processes. The HSM is intended to be a resource document that is used nationwide to help transportation professionals conduct safety analyses in a technically sound and consistent manner thereby improving decisions made based on safety performance.

This course introduces practitioners at the state, county, metropolitan planning organization (MPO), or local level to the new techniques and knowledge in the HSM. The users and professionals described above include, but are not limited to transportation planners, highway designers, traffic engineers, and other transportation professionals who make discretionary road planning, design and operational decisions.

OUTCOMES

Upon completion of the course, participants will be able to:

- Recognize the Highway Safety Manual purpose, structure, and benefits
- Describe and apply Safety Performance Functions and Crash Modification Factors to analyze and predict crash frequency performance of highways, streets, and intersections

TARGET AUDIENCE

The course is intended practitioners at the state, county, metropolitan planning organization (MPO), or local level.

TRAINING LEVEL: Basic

FEE: 2016: $300 Per Person; 2017: N/A

LENGTH: 1 DAYS (CEU: 0 UNITS)

CLASS SIZE: MINIMUM: 20; MAXIMUM: 30

NHI Customer Service: (877) 558-6873 • nhicustomerservice@dot.gov

COURSE NUMBER

FHWA-NHI-380106

COURSE TITLE

Highway Safety Manual Online Overview

Implementation of the HSM requires an understanding of the Science of Safety which supports the quantitative methodologies presented in the manual. This course is an overview of the HSM structure, concepts and principles.

The free selection format of the course allows the student to select modules and concepts of interest in the order preferable to their:

learning style

time availability

and previous knowledge level.

It includes an introduction of terminology, examples of the Roadway Safety Management Process (Part B) and Predictive Methods (Part C), explains the relationship of Crash Modification Factors (CMFs) to decision making and quantitative safety analysis, and human factors. FHWA will continue to develop courses, products and services to meet the needs of the HSM implementation community.

OUTCOMES

Upon completion of the course, participants will be able to:

- Identify the parts of HSM and what they are used for.
- Explain the overall concepts and principles promoted in HS for safety decision making.
- Recognize the benefits of using a quantitative safety analysis in various stages of the transportation project development process.

TARGET AUDIENCE

This course is for all interested students. It is an introductory course intended to provide a broad, base level understanding of HSM.

TRAINING LEVEL: Basic

FEE: 2016: $0 Per Person; 2017: N/A

LENGTH: 12 HOURS (CEU: 0 UNITS)

CLASS SIZE: MINIMUM: 1; MAXIMUM: 1

NHI Customer Service: (877) 558-6873 • nhicustomerservice@dot.gov

Course Number

FHWA-NHI-380108

Course Title

Maintenance of Drainage Features for Safety - WEB-BASED

The purpose of this training is to highlight common roadway drainage problems that can cause an unsafe condition and suggest inspection methods and corrective action. Maintaining roadway drainage is important for safety and for ensuring the long life of the roadway by preventing erosion of the roadway, saturation of the subbase, and damage to roadway structures. The training is broken into two modules:

Module 1: Effects of Drainage describes common roadway safety hazards and how to recognize drainage problems.

Module 2: Safe Drainage Features and Work Zones covers solutions to common roadway safety issues and work zone safety.

This training is not intended to be a design guide. Participants may want to contact their State Local Technical Assistance Program (LTAP) for more details on drainage design.

Outcomes

Upon completion of the course, participants will be able to:

- Identify problems created by ponding and standing water on the roadway
- Describe safety issues related to ditches and side slopes
- Describe how drainage features can become safety hazards
- Identify methods for identifying drainage problems
- Recall conditions to look for during field inspections
- Explain how to fix or prevent common roadway side slope problems
- Describe work zone safety procedures

Target Audience

This training is intended to help local road agency maintenance workers understand the importance of maintaining and upgrading drainage features on their road system to avoid an unsafe condition.

Training Level: Basic

Fee: 2016: $25 Per Person; 2017: N/A

Length: 1 HOURS (CEU: 0 UNITS)

Class Size: MINIMUM: 1; MAXIMUM: 1

NHI Customer Service: (877) 558-6873 • nhicustomerservice@dot.gov

COURSE NUMBER

FHWA-NHI-380109

COURSE TITLE

Alternative Intersections and Interchanges

Transportation professionals are continually challenged with finding improved ways for satisfying the mobility needs of an increasing population. Highway intersections pose particular challenges with regard to safety and mobility as traffic volumes and congestion levels continue to increase. As a result, drivers, pedestrians, and bicyclists experience longer delays and greater exposure to safety risks. Today's traffic and safety problems are becoming increasingly more complex, and conventional intersections and interchange designs are sometimes found to be insufficient to mitigate transportation problems. Consequently, many engineers are investigating and implementing innovative treatments in an attempt to alleviate these issues.

This course provides participants with an overview of various non-traditional intersection concepts that may offer advantages compared to conventional at-grade intersections and grade-separated interchanges. The training presents the salient geometric, operational, and safety features associated with the alternative design concepts, and will illustrate how intersections are selected using an analysis tool. It also will identify potential advantages and disadvantages of each design.

OUTCOMES

Upon completion of the course, participants will be able to:

- Describe key design and operation features of the six non-traditional intersections and interchanges: 1. Displaced Left-Turn Intersections; 2. Median U-turn Intersection; 3. Restricted Crossing U-Turn Intersection; 4. Quadrant Roadway Intersection; 5. Double Crossover Diamond Interchange (Diverging Diamond); 6. Displaced Left Turn Diamond Interchange
- List the advantages and disadvantages of their use
- Describe where they are best suited for existing and planned conditions
- Identify resources to acquire additional information on these designs and their implementations

TARGET AUDIENCE

Federal, State, and local transportation traffic and safety engineers, and planners involved in improving the performance of intersections.

TRAINING LEVEL: Intermediate

FEE: 2016: $300 Per Person; 2017: N/A

LENGTH: 1 DAYS (CEU: .6 UNITS)

CLASS SIZE: MINIMUM: 20; MAXIMUM: 30

NHI Customer Service: (877) 558-6873 • nhicustomerservice@dot.gov

Course Number

FHWA-NHI-380110

Course Title

Highway Safety Improvement Program Overview - WEB BASED

This course is intended to provide you with a basic understanding of the purpose of the Highway Safety Improvement Program (HSIP) and relationship of HSIP programs, background on data collection and quality measures, and an overview of the HSIP processes for planning, implementation and evaluation.

This training course provides a basic understanding of the purpose of the HSIP and relationship of HSIP programs, as well as a basic understanding of the HSIP processes for planning, implementation, and evaluation. Since data is the foundation of the HSIP, the course provides an overview of safety data including safety data collection and management methods, safety data sources, data quality measures, and methods for overcoming data challenges.

A primary challenge in bringing highway safety professionals, traffic and safety engineers, and transportation planners together is a lack of understanding of each area's responsibilities and a common language. The course provides a basic understanding of how the HSIP works; encourages managers to make employees knowledgeable about the program; and begins to establish a common language among HSIP practitioners. The HSIP Overview Course can help overcome the barriers to cross-discipline collaboration.

NHI hosts the HSIP Overview Course and four other Web-based HSIP-related training courses: 380113 Strategic Highway Safety Plan (SHSP) Development, 380114 SHSP Implementation, 380111 HSIP Project Identification, and 380112 HSIP Project Evaluation.

Outcomes

Upon completion of the course, participants will be able to:

- Describe HSIP program structure;
- Recognize HSIP legislative/regulatory requirements;
- Identify potential safety data issues and method for overcoming data challenges; and
- Explain the processes involved in HSIP planning, implementation and evaluation.

Target Audience

This training course is designed for a wide range of transportation professionals from transportation agency leadership to new practitioners in the transportation safety field, HSIP managers and SHSP partners.

Training Level: Basic

Fee: 2016: $25 Per Person; 2017: N/A

Length: 4 DAYS (CEU: 0 UNITS)

Class Size: MINIMUM: 0; MAXIMUM: 0

NHI Customer Service: (877) 558-6873 • nhicustomerservice@dot.gov

Course Number

FHWA-NHI-380111

Course Title

Highway Safety Improvement Program (HSIP) Project Identification

The Highway Safety Improvement Program (HSIP) Project Identification Course provides participants with the necessary background and tools needed to identify projects for the HSIP. Background knowledge critical to understanding project identification includes data analysis concepts such as regression-to-the-mean. Tools essential to the network screening process includes approaches that support systemic safety improvements and those aimed at identifying particular sites with potential for safety improvement. The course will allow participants to choose between different network screening methods by distinguishing between the data needs, strengths and weaknesses of the different approaches; identify and evaluate different countermeasures; prioritize projects based on measures of economic effectiveness; and identify potential funding sources and strategies.

Responsibilities:

You will be expected to complete ten online lessons and two facilitated Web conferences. It is recommended that you complete the modules in sequential order. You must complete all ten online lessons and participate in the two Web conferences to obtain your certificate. By passing the online test at the end of the course, you can also receive Continuing Education Units (CEUs) for the course. All participants will need their own computer with internet connection and a telephone line to participate in the Web conferences.

Outcomes

Upon completion of the course, participants will be able to:

- Identify the need for data driven decision making in the HSIP project identification process;
- Identify data sources to use in HSIP project identification;
- Recognize fundamentals of data analysis for HSIP project identification;
- Describe the HSIP project identification process;
- Differentiate between systemic and site specific network screening processes;
- Recognize strengths and limitations of various approaches to network screening;
- Recognize the countermeasure identification process; and
- Identify methods for prioritizing countermeasures and projects for implementation.

Target Audience

The intended audience for this course is planners and engineers who conduct technical analysis to support HSIP project identification; professionals developing emphasis areas for the SHSP; and data analysts responsible for identifying sites with potential for improvement and locations for systemic improvements.

Training Level: Basic

Fee: 2016: $300 Per Person; 2017: N/A

Length: 8 HOURS (CEU: 0 UNITS)

Class Size: MINIMUM: 15; MAXIMUM: 30

NHI Customer Service: (877) 558-6873 • nhicustomerservice@dot.gov

COURSE NUMBER

FHWA-NHI-380112

COURSE TITLE

Highway Safety Improvement Program (HSIP) Project Evaluation

The Highway Safety Improvement Program (HSIP) Project Evaluation Course provides the necessary fundamentals to perform project evaluation. The course presents a description of safety effectiveness evaluation, an overview of fundamentals needed to perform safety effectiveness evaluation, and information about why safety effectiveness evaluation is important to a Highway Safety Improvement Program. Examples of project evaluation methodologies that account for regression-to-the-mean are discussed and you will be given an opportunity to calculate simple observational before-after studies, observational before-after studies with Empirical Bayes adjustment, and observational before-after studies using comparison groups.

This course can be a stand-alone course for professionals requiring an in-depth knowledge of project evaluation methods or part of a series of courses for professionals performing analysis for the HSIP process. Professionals performing analysis for the HSIP process are encouraged to complete the HSIP Overview and HSIP Project Identification courses prior to enrolling in this course.

Responsibilities:

You will be expected to complete six online lessons and two facilitated web conferences. The modules should be taken in order. Self-paced Modules 1 through 5 must be completed prior to web conference Modules 5.1 and 5.2. Module 6 is the final course module and is self-paced. You must complete all six of the online lessons and participate in the two Web conferences to obtain your certificate. By passing the online test at the end of the course, you can also receive Continuing Education Units (CEUs) for the course. All participants will need their own computer with internet connection and a telephone line to participate in the Web conference.

OUTCOMES

Upon completion of the course, participants will be able to:

- Identify the role of project evaluation in the HSIP;
- Recognize data needs of each project evaluation methodology;
- Conduct project evaluation using each methodology;
- Describe how project evaluation supports the development of crash modification factors; and
- Explain how project evaluation results can benefit the planning process.

TARGET AUDIENCE

The intended audience for this course is planners and engineers who evaluate the safety impact of projects on crash frequency and severity and those conducting technical analysis to support HSIP project and program evaluation.

TRAINING LEVEL: Basic

FEE: 2016: $190 Per Person; 2017: N/A

LENGTH: 5 HOURS (CEU: .5 UNITS)

CLASS SIZE: MINIMUM: 15; MAXIMUM: 30

NHI Customer Service: (877) 558-6873 • nhicustomerservice@dot.gov

Course Number

FHWA-NHI-380113

WBT

Course Title

Strategic Highway Safety Plan Development

This course provides applications for States presently in the implementation stage or for those in the process of updating their SHSP. This course will also benefit regional and local agencies who are considering or in the process of developing their first regional safety plan, or updating their existing plan.

The course is designed to appeal to experienced SHSP stakeholders and those that are new to the process.

This training course provides a basic understanding of the Strategic Highway Safety Plan (SHSP) development processes. The course will benefit States presently in the implementation stage or those in the process of updating their SHSP, along with regional and local agencies that are developing or updating a regional safety plan. The intended audience for SHSP Development encompasses the many federal, state and local stakeholders which partner on state SHSPs but will be especially useful for individuals who are new to the SHSP.

The SHSP Development Course contains relevant information for all SHSP stakeholders. Many states have updated or are in the process of updating their SHSPs, and a refresher course may be helpful to the oversight committees, emphasis area team members, or as training for new stakeholders. Metropolitan Planning Organizations (MPO), counties, and communities who are encouraged by the state departments of transportation (DOT) to participate in SHSP implementation by developing local safety plans related to the SHSP will find this course instructive. New hires are continually joining the workforce, which creates a demand for a basic tutorial on the background, history, contents, development, and maintenance of the SHSP.

NHI hosts the SHSP Development Course and four other Web-based Highway Safety Improvement Program (HSIP)-related training courses: HSIP Overview, SHSP Implementation, HSIP Project Identification, and HSIP Project Evaluation.

Outcomes

Upon completion of the course, participants will be able to:

- Identify the purpose and benefits of Strategic Highway Safety Plans (SHSPs);
- Recognize SHSP legislative/regulatory requirements;
- Identify the SHSP development process;
- Recognize the importance of data in SHSP development and implementation;
- Recognize the importance of collaboration and leadership in the SHSP development process; and
- Identify the purpose of problem identification, monitoring, and evaluation.

Target Audience

The target audience for this course encompasses a wide range of safety stakeholders involved in SHSP efforts. Stakeholders may include State departments of transportation safety engineers/specialists, transportation planning and safety professionals representing metropolitan planning organizations, local safety and planning organizations/agencies; highway safety offices; motor carrier safety offices; law enforcement agencies; EMS offices and first responders; ; nonprofit and private sector partners; others involved in transportation safety; and representatives from Federal agencies (FHWA, NHTSA, FMCSA, FTA).

TRAINING LEVEL: Basic

FEE: 2016: $25 Per Person; 2017: N/A

LENGTH: 4 DAYS (CEU: 0 UNITS)

CLASS SIZE: MINIMUM: 15; MAXIMUM: 30

NHI Customer Service: (877) 558-6873 • nhicustomerservice@dot.gov

Course Number

FHWA-NHI-380114

Course Title

Strategic Highway Safety Plan Implementation

The Strategic Highway Safety Plan (SHSP) Implementation Course provides strategies and examples of SHSP implementation processes. It draws from the SHSP Implementation Process Model (IPM) and other sources to assist safety practitioners in managing SHSP implementation. The course recognizes the highly variable nature from one jurisdiction to another, both within and among the states which implement their SHSP according to available opportunities and resources.

While this course is primarily based on ideal practice, each State implements their SHSP according to the available opportunities and resources. Models, such as the SHSP IPM, are representations of ideal processes, and all parts of the model may not work or be necessary for all States. Regardless, this course includes take-aways for everyone. Each course participant should use the pieces that work best for their State.

This training course provides strategies and examples of Strategic Highway Safety Plan (SHSP) implementation processes that will help safety partners manage their state's SHSPs. The intended audience for the SHSP Implementation Course encompasses a wide range of safety stakeholders involved in SHSP management and implementation efforts at all levels (e.g., local, regional, state, and Federal) including: engineers (e.g., safety, traffic, design, operations, maintenance, and management); transportation planners; safety practitioners; law enforcement officers and managers; emergency responders; and nonprofit and private sector partners.

The course recognizes States implement their SHSP according to available opportunities and resources. Models, such as the SHSP Implementation Process Model (IPM) presented in this course, are representations of ideal processes, and all parts of the model may not work or be necessary for all States. The model presents the ideal framework to help states assess, compare, and adjust their own SHSP implementation efforts.

Responsibilities:

You will be expected to complete three online lessons and three facilitated Web conferences. It is recommended that Modules 1 and 2 be completed prior to any other modules. Module 6 can be taken at anytime following the first two, but should be completed prior to taking Modules 7 and 8. You must complete all eight of the online lessons and participate in the Web conferences to obtain your certificate. By passing the online test at the end of the course, you can also receive Continuing Education Units (CEUs) for the course. All participants will need their own computer with internet connection and a telephone line to participate in the Web conference.

NHI hosts SHSP Implementation and four other Web-based Highway Safety Improvement Program (HSIP)-related training courses HSIP Overview, SHSP Development, HSIP Project Identification, and HSIP Project Evaluation.

To register go to NHI Web site at www.nhi.fhwa.dot.gov and search for course number 380114

Outcomes

Upon completion of the course, participants will be able to:

- Identify the "Essential Eight" discussed in the SHSP Implementation Process Model (leadership; communication; collaboration; data collection & analysis; Emphasis Area action plans; SHSP integration into other transportation plans; marketing; and monitoring, evaluation & feedback);
- Recognize the relationship of SHSP to other transportation plans including, the Long Range Transportation Plan, the State Transportation Improvement Program (S/TIP), the HSIP; the Highway Safety Plan (HSP), and the Commercial Vehicle Safety Plan (CVSP) ; and
- Identify effective strategies for implementing the SHSP.

Target Audience

The target audience for this course encompasses the wide range of safety stakeholders involved in SHSP management and implementation efforts at all levels (e.g., local, regional, state, and Federal) including: engineers (e.g., safety, traffic, design, operations, maintenance, and management); transportation planners; safety practitioners; highway safety office personnel; law enforcement executives and officers; EMS office personnel and emergency responders; motor carrier safety office personnel; and nonprofit and private sector partners; others involved in transportation safety; and Federal representatives (FHWA, NHTSA, FMCSA, and FTA).

TRAINING LEVEL: Basic

FEE: 2016: $260 Per Person; 2017: N/A

LENGTH: 4 HOURS (CEU: 0 UNITS)

CLASS SIZE: MINIMUM: 15; MAXIMUM: 30

NHI Customer Service: (877) 558-6873 • nhicustomerservice@dot.gov

Course Number

FHWA-NHI-380117

Course Title

Combating Roadway Departures

This course provides participants with some tools for addressing roadway departure crashes. Topics covered in this course include a discussion of engineering countermeasures as well as implementation strategies.

Outcomes

Upon completion of the course, participants will be able to:

- Describe the Roadway Departure crash problem
- Discuss countermeasures to:
- - Reduce potential for leaving the roadway
- - Reduce potential for a crash if a vehicle does leave the roadway
- - Minimize severity if a crash does occur
- Compare methods for deploying countermeasures

Target Audience

The target audience for the course includes Federal, State and local highway engineers, consulting highway design engineers, and maintenance workers. This training program is intended for individuals that have the responsibility for identifying, recommending, selecting, installing and/or maintaining appropriate countermeasures to help improve highway safety.

Training Level: Basic

Fee: 2016: $310 Per Person; 2017: N/A

Length: 1 DAYS (CEU: 0 UNITS)

Class Size: MINIMUM: 20; MAXIMUM: 30

NHI Customer Service: (877) 558-6873 • nhicustomerservice@dot.gov

COURSE NUMBER

FHWA-NHI-380118

COURSE TITLE

Integrating Geometric Design & Traffic Control for Improved Safety

This course provides an overview of the inter-relationship of geometric design and traffic control device applications. The primary focus of the course concerns interchange areas where lane elimination, lane configurations, and traffic control devices on freeways and expressways may present challenges for both designers and motorists. This course addresses lane balance effects, degree of control (markings) practices, arrows (signs and markings) usage, advance vehicle positioning, short auxiliary lanes, and geometric design influences on signing and marking. This course includes discussion and guidance for meeting driver expectations and the human factors associated with roadway geometry and the application, selection, and placement of traffic control devices. Participants engage in group exercises to strengthen and apply the principles covered in the workshop,

OUTCOMES

Upon completion of the course, participants will be able to:

- o Identify Human Factors concepts
- o Compare and evaluate lane configuration designs and methods of lane elimination
- o Explain the role of TCDs (signs and pavement markings)
- o Identify the basic signing and marking concepts, types and purposes from the MUTCD
- o Describe the flexibility and interdependence of geometric and traffic control design .

TARGET AUDIENCE

Engineers, engineering practitioners, technologists, and engineering assistants involved in freeway and expressway design, construction, and operations including Sections such as Roadway Design, Traffic Engineering, District personnel with responsible charge of plan review of TCDs (striping, signing, other markings), plan preparation, development/revision of standards for the same and Consultant Management staff, as well as consultants performing work on such projects and/or related duties.

TRAINING LEVEL: Intermediate

FEE: 2016: $310 Per Person; 2017: N/A

LENGTH: 1 DAYS (CEU: .6 UNITS)

CLASS SIZE: MINIMUM: 20; MAXIMUM: 30

NHI Customer Service: (877) 558-6873 • nhicustomerservice@dot.gov

Course Number

FHWA-NHI-380120

Course Title

Introducing Human Factors in Roadway Design and Operations

The primary purpose of this course, Implementing Human Factors into Roadway Design: A Workshop on How to Use and Apply the Human Factors Guidelines (HFG) for Road Systems, is to help attendees gain an understanding of the HFG and how they apply to road system design and operational decisions. This course will provide an overview of human factors as they relate to the roadway environment. The course will describe why it is necessary to incorporate human factors in the design and operation of roadways as a complement to existing standards and manuals for roadway design and operation. Finally, the course offers a review of specific guidelines, as well as scenario-based case studies that allow attendees to apply the HFG to real roadway situations.

Outcomes

Upon completion of the course, participants will be able to:

- Describe basic human characteristics relevant to being a road user.
- List ways in which the vehicle, road user, and roadway elements interact to influence operations and safety outcomes.
- Identify how individual characteristics impact a road user's experience of the road environment.
- Describe the HFG and list its intended usage.
- Describe how the HFG relates to reference sources such as the HSM, MUTCD, and AASHTO's Policy on Geometric Design of Highways and Streets.
- Select and apply specific HFG guidelines for roadway location or design engineering elements to common scenarios.
- Select and apply specific HFG guidelines for traffic engineering elements to common scenarios.
- Analyze case studies, identify critical human factors issues associated with these case studies, and select applicable guidance from the HFG.

Target Audience

The primary audience for the HFG course is composed of the following:Engineers (state departments of transportation (DOT), metropolitan planning organizations (MPO), counties, local municipalities, and consultants to the public agencies)Safety EngineersTraffic EngineersDesign EngineersSafety (non-engineers) Professionals (state DOTs, MPOs, counties, local municipalities, and consultants to the public agencies)Planners (state DOTs, MPOs, counties, local municipalities, and consultants to the public agencies)

Training Level: Basic

Fee: 2016: $500 Per Person; 2017: N/A

Length: 2 DAYS (CEU: 1.2 UNITS)

Class Size: MINIMUM: 20; MAXIMUM: 30

NHI Customer Service: (877) 558-6873 • nhicustomerservice@dot.gov

Course Number

FHWA-NHI-381002

Course Title

Safe Use of Hand and Power Operated Tools - WEB-BASED

This training was prepared by the Transportation Curriculum Coordination Council (TCCC) in partnership with NHI and has been designed for someone learning the basic steps in tool safety. This training is recommended for the Transportation Curriculum Coordination Council levels I and II. This course is primarily intended for inspectors and technicians.

Hand and power tools are a common part of our everyday lives and are present in nearly every industry. These tools help us to easily perform tasks that otherwise would be difficult or impossible. However, these simple tools can be hazardous and have the potential for causing severe injuries when used or maintained improperly. Special attention toward hand and power tool safety is necessary in order to reduce or eliminate these hazards

In the process of removing or avoiding the hazards, workers must learn to recognize the hazards associated with the different types of tools and the safety precautions necessary to prevent those hazards.

In this training we'll discuss the proper use and maintenance of hand tools and a variety of power tools. This is a basic course in the safe use of hand and power operated tools. However, it does not go into regulatory compliance or manufacturer's instructions.

For more information on hand and power operated tool safety, contact your State safety office or the manufacturer.

Outcomes

Upon completion of the course, participants will be able to:

- Describe how to properly and safely use a hand tool
- Describe how to properly and safely use a power tool
- List five types of power tools
- List the five general safety rules for power tools

Target Audience

This course is designed for any individuals wanting to learn more about hand and power tool safety.

Training Level: Basic

Fee: 2016: $25 Per Person; 2017: N/A

Length: 1 HOURS (CEU: 0 UNITS)

Class Size: MINIMUM: 1; MAXIMUM: 1

NHI Customer Service: (877) 558-6873 • nhicustomerservice@dot.gov

Course Number

FHWA-NHI-381004

Course Title

CDL Series - General Knowledge - WEB-BASED

This training was prepared by the Transportation Curriculum Coordination Council (TCCC) in partnership with NHI has been designed for someone interested in commercial driver's license (CDL) general knowledge. This training is recommended for the Transportation Curriculum Coordination Council levels I and II or anyone interested in obtaining a CDL. This course is primarily intended for inspectors and technicians.

This training contains the general knowledge and safe driving information that all commercial drivers should know. It is broken into three modules:

Module 1 reviews vehicle control, shifting gears, seeing the road, communicating, speed control, and space management.

Module 2 covers night driving, driving in cold and hot weather, mountain driving, and railroad crossings.

Module 3 discusses seeing hazards, driving and road emergencies, staying alert and fit to drive, and transporting hazards.

This general knowledge training does not have specific information on air brakes or pre-trip inspection. You may complete other training in the CDL series to learn more about them.

For more information on the CDL examination and requirements that apply to your State, contact your State license agencies.

Outcomes

Upon completion of the course, participants will be able to:

- Describe the procedures in controlling your vehicle and shifting gears
- Define the steps to seeing the road in various situations
- Recognize the importance of signaling and communicating your presence
- Identify the important components of speed control and space management
- Describe the proper ways to drive at night
- Identify the correct practices for driving in cold weather and hot weather
- Describe the procedures for driving on a mountain
- Recognize the proper way to cross a railroad
- Describe the procedures in responding to driving emergencies and emergencies on the road
- Identify the guidelines to staying alert and fit to drive
- Define the proper way to transport hazardous materials

Target Audience

This course is designed for any individuals wanting to learn more about commercial driver's license (CDL) general information.

Training Level: Basic

Fee: 2016: $25 Per Person; 2017: N/A

Length: 3 HOURS (CEU: 0 UNITS)

Class Size: MINIMUM: 1; MAXIMUM: 1

NHI Customer Service: (877) 558-6873 • nhicustomerservice@dot.gov

WBT

COURSE NUMBER

FHWA-NHI-381005

COURSE TITLE

CDL Series - Air Brakes - WEB-BASED

This training was prepared by the Transportation Curriculum Coordination Council (TCCC) in partnership with NHI has been designed for someone interested in commercial driver's license (CDL) air brake systems. This training is recommended for the Transportation Curriculum Coordination Council levels I and II or anyone interested in obtaining a CDL. This course is primarily intended for technicians.

In this training we'll discuss the parts of an air brake system, dual air brake systems, how to inspect your air brake system, and how to effectively use your air brake system. If you want to drive a truck or bus with air brakes, or pull a trailer with air brakes, you'll need to take a test on this material.

This training contains information on air brakes system that all commercial drivers should know. It is broken into two modules:

Module 1 consists of air brake system parts and dual air brakes systems.

Module 2 consists of inspecting air brakes and using air brakes.

This air brakes training does not have specific information on general knowledge or pre-trip inspection. You may complete other training in the CDL series to learn more about them.

For more information on the CDL examination and requirements that apply to your State, contact your State license agencies.

OUTCOMES

Upon completion of the course, participants will be able to:

- Identify the important parts of the air brake system
- Define dual air brakes
- Recognize key elements in the air flow process of the dual air brake system
- Identify the important components of air brakes inspection
- Recognize the proper ways to use air brakes

TARGET AUDIENCE

This course is designed for any individuals wanting to learn more about commercial driver's license (CDL) air brake systems.

TRAINING LEVEL: Basic

FEE: 2016: $25 Per Person; 2017: N/A

LENGTH: 1.5 HOURS (CEU: 0 UNITS)

CLASS SIZE: MINIMUM: 1; MAXIMUM: 1

NHI Customer Service: (877) 558-6873 • nhicustomerservice@dot.gov

COURSE NUMBER

FHWA-NHI-381006

COURSE TITLE

CDL Series - Pre-Trip Inspection - WEB-BASED

This training was prepared by the Transportation Curriculum Coordination Council (TCCC) in partnership with NHI has been designed for someone interested in commercial driver's license (CDL) pre-trip inspection. This training is recommended for the Transportation Curriculum Coordination Council levels I and II or anyone interested in obtaining a CDL. This course is primarily intended for inspectors and technicians.

This is a basic course in the area of pre-trip inspection. This training covers different parts of the vehicle you would check before a trip. We'll approach the different parts of the vehicle in the order that we would in a standard pre-trip inspection. It is broken into two modules:

Module 1 covers front of the vehicle; engine compartment; engine start and cab check; steering; and suspension. Module 2 reviews brakes; wheels; side of vehicle; back of vehicle; and trailer.

This pre-trip inspection training does not have specific information on air brakes or general knowledge. You may complete other training in the CDL series to learn more about them.

For more information on the CDL examination and requirements that apply to your State, contact your State license agencies.

OUTCOMES

Upon completion of the course, participants will be able to:

- Describe the inspection items in front of the vehicle and engine compartment
- Identify the important steps in the cab check and air brake check
- Describe the important components of steering inspection
- Define the parts of the front and rear suspension
- Identify the components that are checked in front and rear brakes and front and rear wheels
- Describe the inspection steps for the side and back of the vehicle
- Identify the inspection items for parts of the trailer

TARGET AUDIENCE

This course is designed for any individuals wanting to learn more about commercial driver's license (CDL) pre-trip inspection.

TRAINING LEVEL: Basic

FEE: 2016: $25 Per Person; 2017: N/A

LENGTH: 2 HOURS (CEU: 0 UNITS)

CLASS SIZE: MINIMUM: 1; MAXIMUM: 1

NHI Customer Service: (877) 558-6873 • nhicustomerservice@dot.gov

COURSE NUMBER

FHWA-NHI-381002

COURSE TITLE

Safe Use of Hand and Power Operated Tools - WEB-BASED

This training was prepared by the Transportation Curriculum Coordination Council (TCCC) in partnership with NHI and has been designed for someone learning the basic steps in tool safety. This training is recommended for the Transportation Curriculum Coordination Council levels I and II. This course is primarily intended for inspectors and technicians.

Hand and power tools are a common part of our everyday lives and are present in nearly every industry. These tools help us to easily perform tasks that otherwise would be difficult or impossible. However, these simple tools can be hazardous and have the potential for causing severe injuries when used or maintained improperly. Special attention toward hand and power tool safety is necessary in order to reduce or eliminate these hazards

In the process of removing or avoiding the hazards, workers must learn to recognize the hazards associated with the different types of tools and the safety precautions necessary to prevent those hazards.

In this training we'll discuss the proper use and maintenance of hand tools and a variety of power tools. This is a basic course in the safe use of hand and power operated tools. However, it does not go into regulatory compliance or manufacturer's instructions.

For more information on hand and power operated tool safety, contact your State safety office or the manufacturer.

OUTCOMES

Upon completion of the course, participants will be able to:

- Describe how to properly and safely use a hand tool
- Describe how to properly and safely use a power tool
- List five types of power tools
- List the five general safety rules for power tools

TARGET AUDIENCE

This course is designed for any individuals wanting to learn more about hand and power tool safety.

TRAINING LEVEL: Basic

FEE: 2016: $25 Per Person; 2017: N/A

LENGTH: 1 HOURS (CEU: 0 UNITS)

CLASS SIZE: MINIMUM: 1; MAXIMUM: 1

NHI Customer Service: (877) 558-6873 • nhicustomerservice@dot.gov

WBT

COURSE NUMBER

FHWA-NHI-381004

COURSE TITLE

CDL Series - General Knowledge - WEB-BASED

This training was prepared by the Transportation Curriculum Coordination Council (TCCC) in partnership with NHI has been designed for someone interested in commercial driver's license (CDL) general knowledge. This training is recommended for the Transportation Curriculum Coordination Council levels I and II or anyone interested in obtaining a CDL. This course is primarily intended for inspectors and technicians.

This training contains the general knowledge and safe driving information that all commercial drivers should know. It is broken into three modules:

Module 1 reviews vehicle control, shifting gears, seeing the road, communicating, speed control, and space management.

Module 2 covers night driving, driving in cold and hot weather, mountain driving, and railroad crossings.

Module 3 discusses seeing hazards, driving and road emergencies, staying alert and fit to drive, and transporting hazards.

This general knowledge training does not have specific information on air brakes or pre-trip inspection. You may complete other training in the CDL series to learn more about them.

For more information on the CDL examination and requirements that apply to your State, contact your State license agencies.

OUTCOMES

Upon completion of the course, participants will be able to:

- Describe the procedures in controlling your vehicle and shifting gears
- Define the steps to seeing the road in various situations
- Recognize the importance of signaling and communicating your presence
- Identify the important components of speed control and space management
- Describe the proper ways to drive at night
- Identify the correct practices for driving in cold weather and hot weather
- Describe the procedures for driving on a mountain
- Recognize the proper way to cross a railroad
- Describe the procedures in responding to driving emergencies and emergencies on the road
- Identify the guidelines to staying alert and fit to drive
- Define the proper way to transport hazardous materials

TARGET AUDIENCE

This course is designed for any individuals wanting to learn more about commercial driver's license (CDL) general information.

TRAINING LEVEL: Basic

FEE: 2016: $25 Per Person; 2017: N/A

LENGTH: 3 HOURS (CEU: 0 UNITS)

CLASS SIZE: MINIMUM: 1; MAXIMUM: 1

NHI Customer Service: (877) 558-6873 • nhicustomerservice@dot.gov

Course Number

FHWA-NHI-381005

WBT

Course Title

CDL Series - Air Brakes - WEB-BASED

This training was prepared by the Transportation Curriculum Coordination Council (TCCC) in partnership with NHI has been designed for someone interested in commercial driver's license (CDL) air brake systems. This training is recommended for the Transportation Curriculum Coordination Council levels I and II or anyone interested in obtaining a CDL. This course is primarily intended for technicians.

In this training we'll discuss the parts of an air brake system, dual air brake systems, how to inspect your air brake system, and how to effectively use your air brake system. If you want to drive a truck or bus with air brakes, or pull a trailer with air brakes, you'll need to take a test on this material.

This training contains information on air brakes system that all commercial drivers should know. It is broken into two modules:

Module 1 consists of air brake system parts and dual air brakes systems.

Module 2 consists of inspecting air brakes and using air brakes.

This air brakes training does not have specific information on general knowledge or pre-trip inspection. You may complete other training in the CDL series to learn more about them.

For more information on the CDL examination and requirements that apply to your State, contact your State license agencies.

Outcomes

Upon completion of the course, participants will be able to:

- Identify the important parts of the air brake system
- Define dual air brakes
- Recognize key elements in the air flow process of the dual air brake system
- Identify the important components of air brakes inspection
- Recognize the proper ways to use air brakes

Target Audience

This course is designed for any individuals wanting to learn more about commercial driver's license (CDL) air brake systems.

Training Level: Basic

Fee: 2016: $25 Per Person; 2017: N/A

Length: 1.5 HOURS (CEU: 0 UNITS)

Class Size: MINIMUM: 1; MAXIMUM: 1

NHI Customer Service: (877) 558-6873 • nhicustomerservice@dot.gov

Course Number

FHWA-NHI-381006

Course Title

CDL Series - Pre-Trip Inspection - WEB-BASED

This training was prepared by the Transportation Curriculum Coordination Council (TCCC) in partnership with NHI has been designed for someone interested in commercial driver's license (CDL) pre-trip inspection. This training is recommended for the Transportation Curriculum Coordination Council levels I and II or anyone interested in obtaining a CDL. This course is primarily intended for inspectors and technicians.

This is a basic course in the area of pre-trip inspection. This training covers different parts of the vehicle you would check before a trip. We'll approach the different parts of the vehicle in the order that we would in a standard pre-trip inspection. It is broken into two modules:

Module 1 covers front of the vehicle; engine compartment; engine start and cab check; steering; and suspension. Module 2 reviews brakes; wheels; side of vehicle; back of vehicle; and trailer.

This pre-trip inspection training does not have specific information on air brakes or general knowledge. You may complete other training in the CDL series to learn more about them.

For more information on the CDL examination and requirements that apply to your State, contact your State license agencies.

Outcomes

Upon completion of the course, participants will be able to:

- Describe the inspection items in front of the vehicle and engine compartment
- Identify the important steps in the cab check and air brake check
- Describe the important components of steering inspection
- Define the parts of the front and rear suspension
- Identify the components that are checked in front and rear brakes and front and rear wheels
- Describe the inspection steps for the side and back of the vehicle
- Identify the inspection items for parts of the trailer

Target Audience

This course is designed for any individuals wanting to learn more about commercial driver's license (CDL) pre-trip inspection.

Training Level: Basic

Fee: 2016: $25 Per Person; 2017: N/A

Length: 2 HOURS (CEU: 0 UNITS)

Class Size: MINIMUM: 1; MAXIMUM: 1

NHI Customer Service: (877) 558-6873 • nhicustomerservice@dot.gov

Course Number

FHWA-NHI-381008

Course Title

Job Hazard Analysis - WEB-BASED

This course reviews what a job hazard analysis is and why it should be performed, identifies the information that should be documented during a job hazard analysis, and provides example jobs and potential hazards that may be encountered.

The purpose of this training is to explain what a job hazard analysis is and offer guidelines to help you conduct a step-by-step analysis. This information should be used to analyze jobs and recognize workplace hazards. This course contains 3 lessons:

Lesson 1: Job Hazard Analysis Overview

Lesson 2: Job Hazard Analysis Forms

Lesson 3: Job Hazard Examples

Please refer to the Occupational Safety and Health Administration (OSHA) Website (Safety Manual) for a Job Hazard Analysis Form. It can also be found on the OSHA Website under Safety Manual and Forms. Even if your agency is not directly governed by OSHA, your state agency version of OSHA has most likely adopted the standards set forth by OSHA, which will be discussed in this training.

Outcomes

Upon completion of the course, participants will be able to:

- Explain what job hazard analysis is, and why it is important
- Describe the types of information that should be documented in a Job Hazard Analysis Form
- Given an example situation, list potential hazards

Target Audience

This training is designed for employers, foremen, supervisors, and employees. This training was developed by the Transportation Curriculum Coordination Council (TCCC) in partnership with AASHTO, NHI, and is recommended for TCCC levels II through IV.

Training Level: Basic

Fee: 2016: $0 Per Person; 2017: N/A

Length: 2 HOURS (CEU: 0 UNITS)

Class Size: MINIMUM: 1; MAXIMUM: 1

NHI Customer Service: (877) 558-6873 • nhicustomerservice@dot.gov

Course Number

FHWA-NHI-420018

Course Title

Instructor Development Course (3.5-Day)

The 3.5-day is geared to instructors who anticipate teaching from a complete set of training materials (instructor manuals, participant workbooks, and visual aids) developed by training professionals.

This Instructor Development Course (IDC) will provide new and experienced instructors the knowledge and skills to deliver more effective training. NHI defines training as a "demonstration of acquired skills and knowledge of adult learning principles which necessitates that learning outcomes be developed and their attainment be measured."

A skilled trainer, therefore, will emphasize the use of experiential learning techniques, such as problem solving analysis, discussion, question and answer sessions, group activities, demonstrations, role-plays, etc. In essence, these learning activities tap into the knowledge and skills that an adult learner brings to the classroom and have the goal of meeting both the learning outcomes and the participants' expectations.

Pre-Class Assignment:

Training Sessions: You must come prepared to present a 15-minute training session at the beginning of the workshop. The topic for your session should be job related; it can either come from a course you have taught, will be teaching, or are developing. The 15 minutes typically translate to about 5 to 7 minutes of content with time for exercises, activities and/or questions, etc. Visual aids, such as overhead transparencies or handouts should be brought with you. Please bring your own laptop computer if you are planning to do a PowerPoint presentation.

A word of caution, not all training facilities are equipped with the appropriate technical support for a PowerPoint presentation (i.e., in-focus projector or support software) or have the equipment to reproduce overhead transparencies. For this reason, we encourage you to make use of other types of visual aids, such as flip charts, write-on transparencies, and handouts. These nontechnical methods will NOT diminish, but enhance the value of your presentation. Use a holistic approach in your training.

Readings: Read the Instructional Systems Design (ISD) material posted on the NHI Web site. To access the material go to http://www.nhi.fhwa.dot.gov, under "About NHI" at the bottom of the page, select "NHI Philosophy on Learning", select "Adult Learning" (Print and Read), select "Instructional Systems Design (ISD)" (Print and Read). You will find printable downloadable files (PDFs) of all required readings and any other materials related to this course.

This course is part of the NHI Instructor Certification program. To learn more about NHI's Instructor Certification visit the NHI Web site at http://www.nhi.fhwa.dot.gov/resources/resources.aspx.

Outcomes

Upon completion of the course, participants will be able to:

- Explain the five steps in the ISD system
- Write a behavioral learning outcome
- Present, measure, and review a learning outcome
- Demonstrate at least two forms of interactivity and positive interpersonal skills
- List five training techniques (e.g., Do not talk to the flip chart; do not stand in front of the projector; and do not stand in one place)
- Demonstrate how to reach the three styles of learning
- Deliver a 15-minute training session that demonstrates adult learning principles

Target Audience

This course is intended for instructors who will be delivering interactive training to adult learners.

TRAINING LEVEL: Basic

FEE: 2016: $1130 Per Person; 2017: N/A

LENGTH: 3.5 DAYS (CEU: 2.1 UNITS)

CLASS SIZE: MINIMUM: 7; MAXIMUM: 12

NHI Customer Service: (877) 558-6873 • nhicustomerservice@dot.gov

Course Number

FHWA-NHI-420018A

Course Title

Instructor Development Course (4.5-Day)

This 4.5-day course prepares instructors who teach from a complete set of training materials (instructor manuals, participant workbooks, and visual aids) developed by training professionals. In addition, this course will teach instructors who need to create their own courses or modify existing courses, how to develop instructionally sound learning outcomes, instructor manuals, visual aids, exercises, workshops, and assessments. The course also uses practical techniques to reinforce the various skills need to develop sound course material.

The Instructor Development Course (IDC) will provide new and experienced instructors the knowledge and skills to deliver more effective training. NHI defines training as a "demonstration of acquired skills and knowledge of adult learning principles which necessitates that learning outcomes be developed and their attainment be measured."

A skilled trainer, therefore, will emphasize the use of experiential learning techniques, such as problem solving analysis, discussion, question and answer sessions, group activities, demonstrations, role-plays, etc. In essence, these learning activities tap into the knowledge and skills that an adult learner brings to the classroom and have the goal of meeting both the learning outcomes and the participants' expectations.

Pre-Class Assignment:

Training Sessions: You must come prepared to present a 15-minute training session at the beginning of the workshop. The topic for your session should be job related; it can either come from a course you have taught, will be teaching, or are developing. The 15 minutes typically translate to about 5 to 7 minutes of content with time for exercises, activities and/or questions, etc. Visual aids, such as overhead transparencies or handouts should be brought with you. Please bring your own laptop computer if you are planning to do a PowerPoint presentation.

A word of caution, not all training facilities are equipped with the appropriate technical support for a PowerPoint presentation (i.e., in-focus projector or support software) or have the equipment to reproduce overhead transparencies. For this reason, we encourage you to make use various types of visual aids, such as flip charts, write-on transparencies, and handouts to enhance your training session. These nontechnical methods will NOT diminish, but enhance the value of your presentation. Use a holistic approach in your training.

The Golden Rule for a Trainer/Instructor is: "Always be prepared to instruct."

Readings: Read the Instructional Systems Design (ISD) material posted on the NHI web site. To access the material go to http://www.nhi.fhwa.dot.gov, under "About NHI" at the bottom of the page, select "NHI Philosophy on Learning", select "Adult Learning" (Print and Read), and select "Instructional Systems Design (ISD)" (Print and Read). You will find printable downloadable files (PDFs) of all required readings and any other materials related to this course.

This course is part of the NHI Instructor Certification program. To learn more about NHI's Instructor Certification visit the NHI Web site at http://www.nhi.fhwa.dot.gov/resources/resources.aspx.

Outcomes

Upon completion of the course, participants will be able to:

- Explain the five steps in the ISD system
- Write a behavioral learning outcome
- Develop various types of visual aids
- Present, measure, and review a learning outcome
- Demonstrate at least two forms of interactivity and positive interpersonal skills
- List five training techniques (e.g., Do not talk to the flip chart; do not stand in front of the projector; and do not stand in one place)
- Demonstrate how to reach the three styles of learning
- Develop an appropriate assessment tool to measure learning
- Deliver a 15-minute training session that demonstrates adult learning principles

Target Audience

This course is intended for instructors who will develop and deliver interactive training to adult learners.

TRAINING LEVEL: Basic

FEE: 2016: $1310 Per Person; 2017: N/A

LENGTH: 4.5 DAYS (CEU: 2.7 UNITS)

CLASS SIZE: MINIMUM: 7; MAXIMUM: 12

NHI Customer Service: (877) 558-6873 • nhicustomerservice@dot.gov

Course Number

FHWA-NHI-420051

Course Title

Instructor Introduction to Video Conference Training (VCT) - Web-Based

This training is provided to you at no cost by the National Highway Institute.

This training provides basic information on video conference training, or VCT, and tips on how to present information effectively during a VCT session. This training will also discuss some of the specific challenges in communicating at a distance using video conferencing and how to meet those challenges.

Outcomes

Upon completion of the course, participants will be able to:

- Identify video conference equipment
- Describe how to prepare for a VCT
- List characteristics of effective instructors in a VCT environment
- Explain how to interact from a distance

Target Audience

This course is intended for instructors who will be delivering interactive video conference training to adult learners.

Training Level: Basic

Fee: 2016: $0 Per Person; 2017: N/A

Length: 1 HOURS (CEU: 0 UNITS)

Class Size: MINIMUM: 1; MAXIMUM: 1

NHI Customer Service: (877) 558-6873 • nhicustomerservice@dot.gov

NHI STORE PROVIDES RESOURCES AND REFERENCE MATERIALS

Created based on customer feedback, the NHI Store is an online resource that enables users to order course materials through the NHI Web site. These materials can be used to plan a workshop, support train-the-trainer programs, or gather highway-related reference materials. The NHI Store offers both electronic downloads and hard copy versions.

To search for and purchase NHI course training materials, please visit www.nhi.fhwa.dot.gov. Easy directions are provided for ordering and payment; special instructions are provided for FHWA employees.

If you are unable to find the training materials you need, please contact us at nhitraining@dot.gov.

The following pages list all materials available for purchase at the time this catalog was published. For the most up-to-date listing, visit the NHI Store at www.nhi.fhwa.dot.gov. Credit card payment is accepted.

LEGEND

PW - Participant Workbook
RM - Reference Manual
PP - PowerPoint Presentation
OM - Other Materials
EF - Electronic File

Course Number	Material Name	Format	Type	Price
130053	Bridge Inspector's Reference Manual-Compact Disc (November 2015)	Hard Copy	RM	$20.00
130053A	Bridge Inspection Refresher Training (August 2014)	Hard Copy	PW	$70.00
130053A	Bridge Inspector's Reference Manual-Compact Disc (November 2015)	Hard Copy	RM	$20.00
130054	Bridge Inspector's Reference Manual-Compact Disc (November 2015)	Hard Copy	RM	$20.00
130054	Engineering Concepts For Bridge Inspectors (September 2011)	Hard Copy	PW	$40.00
130055	Bridge Inspector's Reference Manual-Compact Disc (November 2015)	Hard Copy	RM	$20.00
130078	Fracture Critical Inspection Techniques for Steel Bridges (05/2016)	Hard Copy	PW	$50.00
130087	Guidelines For The Installation, Inspection, Maintenance And Repair Of Structural Supports For Highw	Hard Copy	OM	$50.00
130087	Inspection And Maintenance Of Ancillary Highway Structures-(March 2005)	Hard Copy	PW	$50.00
130088	Bridge Construction Inspection - Participant Workbook Volume 1 (March 2015)	Hard Copy	PW	$50.00
130088	Bridge Construction Inspection - Participant Workbook Volume 2 (March 2015)	Hard Copy	PW	$40.00
130091	Bridge Inspector's Reference Manual-Compact Disc (November 2015)	Hard Copy	RM	$20.00
130091	Underwater Bridge Inspection (January 2010)	Hard Copy	PW	$40.00
130091	Underwater Inspection of Bridges (June 2010)	Hard Copy	RM	$40.00
130091B	Underwater Bridge Repair (December 2009)	Hard Copy	RM	$40.00
130091B	Underwater Bridge Repair, Rehabilitation, and Countermeasures (December 2009)	Hard Copy	PW	$30.00

Course Number	Material Name	Format	Type	Price
130092	Fundamentals of LRFR and Applications of LRFR for Bridge Superstructures (September 2013)	Hard Copy	PW	**$40.00**
130092A	Load and Resistance Factor Rating for Highway Bridges (September 2013)	Hard Copy	PW	**$40.00**
130092B	Fundamentals of LRFR and Applications of LRFR for Bridge Superstructures (September 2013)	Hard Copy	PW	**$40.00**
130093	LRFD Seismic Analysis and Design of Bridges (July 2013)	Hard Copy	PW	**$50.00**
130093	LRFD Seismic Analysis and Design of Bridges (October 2014)	Hard Copy	RM	**$50.00**
130093	LRFD Seismic Analysis and Design of Bridges-Design Examples (July 2014)	Hard Copy	OM	**$50.00**
130093A	LRFD Seismic Analysis and Design of Bridges-Design Examples (July 2014)	Hard Copy	OM	**$50.00**
130095	LRFD and Analysis of Curved Steel Highway Bridges (February 2011)	Hard Copy	PW	**$70.00**
130095	LRFD and Analysis of Curved Steel Highway Bridges (February 2011)-Compact Disc	Hard Copy	RM	**$20.00**
130095A	LRFD and Analysis of Curved Steel Highway Bridges (February 2011)-Compact Disc	Hard Copy	RM	**$20.00**
130095B	LRFD and Analysis of Curved Steel Highway Bridges (February 2011)-Compact Disc	Hard Copy	RM	**$20.00**
130096	Design Criteria for Arch and Cable Stayed Signature Bridges (February 2012)	Hard Copy	RM	**$70.00**
130096	Design Criteria for Arch and Cable Stayed Signature Bridges (February 2012)	Electronic Copy	RM	**$20.00**
130096	Design Criteria for Arch and Cable Stayed Signature Bridges (March 2013)	Hard Copy	PW	**$40.00**
130102	Engineering for Structural Stability in Bridge Construction (04/2015)	Hard Copy	RM	**$50.00**
130102A	Engineering for Structural Stability in Bridge Construction (04/2015)	Hard Copy	RM	**$50.00**
131050	Asphalt Pavement In-Place Recycling Techniques (March 2013)	Hard Copy	PW	**$50.00**
132012	Soils And Foundations Workshop - Reference Manual Volume 1 (December 2006)	Hard Copy	RM	**$40.00**
132012	Soils And Foundations Workshop - Reference Manual Volume 2 (December 2006)	Hard Copy	RM	**$40.00**
132013	Geosynthetics Engineering Workshop (RM)	Hard Copy	RM	**$40.00**
132013A	Geosynthetics Engineering Workshop	Hard Copy	RM	**$40.00**
132013B	Geosynthetics Engineering Workshop	Hard Copy	RM	**$40.00**
132013C	Geosynthetics Engineering Workshop	Hard Copy	RM	**$40.00**
132013D	Geosynthetics Engineering Workshop	Hard Copy	RM	**$40.00**
132014	Drilled Shafts: Construction Procedures and LRFD Design Methods (May 2010)	Hard Copy	RM	**$50.00**
132033	Soil Slope and Embankment Design (September 2005)	Hard Copy	RM	**$40.00**
132035	Rock Slopes - Module 5 - Reference Manual	Hard Copy	RM	**$50.00**
132035	Rock Slopes - Module 5 - Student Exercises (August 1999)	Hard Copy	OM	**$50.00**

Course Number	Material Name	Format	Type	Price
132036	Earth Retaining Structures (RM)	Hard Copy	RM	**$50.00**
132036	Soil Nail Walls Reference Manual-GEC 7 (February 2015)	Hard Copy	RM	**$40.00**
132037	Shallow Foundations - Module 7 - Reference Manual	Hard Copy	RM	**$50.00**
132037	Shallow Foundations (April 2012)	Hard Copy	PW	**$40.00**
132040	Geotechnial Aspects of Pavements (June 2010)	Hard Copy	RM	**$40.00**
132041	Geotechnical Instrumentation - Module 11 - Reference Manual	Hard Copy	RM	**$50.00**
132042	Corrosion/Degradation of Soil Reinforcements for MSE/RSS (November 2009)	Hard Copy	RM	**$40.00**
132042	Design of Mechanically Stabilized Earth Walls and Reinforced Soil Slopes-Vol 1 (March 2012)	Hard Copy	RM	**$40.00**
132042	Design of Mechanically Stabilized Earth Walls and Reinforced Soil Slopes-Vol 2 (March 2012)	Hard Copy	RM	**$40.00**
132043	Design of Mechanically Stabilized Earth Walls and Reinforced Soil Slopes-Vol 1 (March 2012)	Hard Copy	RM	**$40.00**
132043	Design of Mechanically Stabilized Earth Walls and Reinforced Soil Slopes-Vol 2 (March 2012)	Hard Copy	RM	**$40.00**
132069	Driven Pile Foundation Inspection - Participant Workbook (July 2006)	Hard Copy	PW	**$50.00**
132069	Plan Set Handout Driven Pile Foundation Inspection Course (October 2002)	Hard Copy	OM	**$60.00**
132070	Drilled Shaft Foundation Inspection - Participant Workbook (December 2002)	Hard Copy	PW	**$50.00**
132070	Drilled Shaft Inspector's Course - Plan Set Handout	Hard Copy	OM	**$50.00**
132070B	Drilled Shaft Foundation Inspection - Participant Workbook (December 2002)	Hard Copy	PW	**$50.00**
132070B	Drilled Shaft Inspector's Course - Plan Set Handout	Hard Copy	OM	**$50.00**
132078	Micropile Design and Construction Reference Manual (December 2005)	Hard Copy	RM	**$30.00**
132081	Highway Slope Maintenance and Slide Restoration -- Participant Workbook	Hard Copy	PW	**$50.00**
132081	Highway Slope Maintenance and Slide Restoration -- Reference Manual	Hard Copy	RM	**$50.00**
132082	LFRD for Highway Bridge Substructures and Earth Retaining Structures (Feb 2012)	Hard Copy	RM	**$50.00**
132082	LFRD for Highway Bridge Substructures and Earth Retaining Structures (Feb 2014)	Hard Copy	PW	**$50.00**
132083	Implementation of LRFD Geotechnical Design for Bridge Foundations (February 2011)	Hard Copy	RM	**$20.00**
132094	LRFD Seismic Analysis and Design of Transportation Structures, Features and Foundations (Feb 2012)	Hard Copy	PW	**$50.00**
132094	LRFD Seismic Analysis and Design of Transportation Structures, Features and Foundations (Feb 2012)	Hard Copy	RM	**$75.00**
132094	LRFD Seismic Analysis and Design of Transportation Structures,...Design Examples (April 2012)	Hard Copy	OM	**$75.00**
132094A	LRFD Seismic Analysis and Design of Transportation Structures, Features and Foundations (Feb 2012)	Hard Copy	RM	**$75.00**

Course Number	Material Name	Format	Type	Price
132094A	LRFD Seismic Analysis and Design of Transportation Structures,...Design Examples (April 2012)	Hard Copy	OM	**$75.00**
132094B	LRFD Seismic Analysis and Design of Transportation Structures, Features and Foundations(August 2011)	Hard Copy	PW	**$50.00**
132094B	LRFD Seismic Analysis and Design of Transportation Structures, Features and Foundations(August 2011)	Hard Copy	RM	**$75.00**
132094B	LRFD Seismic Analysis and Design of Transportation Structures,...Design Examples (April 2012)	Hard Copy	OM	**$75.00**
133075	Freeway Management And Operations - Participant Workbook (August 2005)	Hard Copy	PW	**$50.00**
133075A	Freeway Management And Operations - Participant Workbook (August 2005)	Hard Copy	PW	**$50.00**
133078	Access Management Location and Design (February 2007)	Hard Copy	PW	**$50.00**
133099	Managing Travel For Planned Events - CD (September 2005)	Hard Copy	OM	**$20.00**
133099	Managing Travel For Planned Events - Participant Workbook (September 2005)	Hard Copy	PW	**$50.00**
133099A	Managing Travel for Planned Special Events	Hard Copy	PW	**$50.00**
133115	Advanced Work Zone Management and Design (August 2007)	Hard Copy	PW	**$20.00**
133115	Advanced Work Zone Management and Design (August 2007)	Hard Copy	RM	**$40.00**
133120	WZ Traffic Analysis Applications and Decision Framework-PW	Hard Copy	PW	**$50.00**
133121	Traffic Signal Design and Operations (Dec 2011)	Hard Copy	PW	**$50.00**
133122	Traffic Signal Timing Concepts (May 2014)	Hard Copy	PW	**$50.00**
133123	Implementing Successful Advanced Traffic Signal System Projects Including Adaptive Control	Hard Copy	PW	**$50.00**
133125	Successful Traffic Signal Management: The Basic Service Approach (May 2014)	Hard Copy	PW	**$50.00**
134005	VALUE ENGINEERING (February 2013)	Hard Copy	PW	**$30.00**
134005A	VALUE ENGINEERING (AUGUST 2010)	Hard Copy	PW	**$30.00**
134005B	VALUE ENGINEERING (February 2013)	Hard Copy	PW	**$30.00**
134005C	VALUE ENGINEERING (February 2013)	Hard Copy	PW	**$30.00**
134037A	Managing Highway Contract Claims: Analysis And Avoidance - Participant Notes (September 2004)	Hard Copy	PW	**$50.00**
134062A	Participant Workbook Volume I (November 2007)	Hard Copy	PW	**$40.00**
134062A	Participant Workbook Volume II (November 2007)	Hard Copy	PW	**$40.00**
134064	Transportation Construction Quality Assurance (June 2011)-1.5 Day Version	Hard Copy	PW	**$50.00**
134064	Transportation Construction Quality Assurance Reference Manual	Hard Copy	RM	**$50.00**
134064	Transportation Construction Quality Assurance Reference Manual	Electronic Copy	RM	**Free**
134064A	Transportation Construction Quality Assurance	Electronic Copy	RM	**Free**

Course Number	Material Name	Format	Type	Price
135010	River Engineering For Highway Encroachments: Highways In The River Environment (December 2001)	Hard Copy	OM	**$50.00**
135027	Errata for HEC-22 dtd September 2009 (Included in September 2013 Revision)	Electronic Copy	OM	**Free**
135027	Urban Drainage Design Manual, HEC-22 (Revised September 2013)	Hard Copy	RM	**$50.00**
135027A	Highway Stormwater Pump Station Design (HEC-24)	Hard Copy	OM	**$40.00**
135028	Highway Stormwater Pump Station Design HEC-24	Hard Copy	OM	**$50.00**
135041	One-Dimensional Modeling of River Encroachments with HEC-RAS (Jan 2016)	Hard Copy	PW	**$30.00**
135046	Evaluating Scour At Bridges, 5th Edition (HEC-18) (April 2013)	Hard Copy	OM	**$50.00**
135046	Stream Instability, Bridge Scour, and Countermeasures: A Field Guide for Bridge Inspectors (Feb2009)	Hard Copy	RM	**$20.00**
135046	Stream Stability at Highway Structures, 4th Edition (HEC-20)	Hard Copy	OM	**$50.00**
135047	Stream Instability, Bridge Scour, and Countermeasures: A Field Guide for Bridge Inspectors (Feb2009)	Hard Copy	RM	**$20.00**
135048	Countermeasure Design for Bridge Scour and Stream Instability	Hard Copy	OM	**$30.00**
135048	HEC-23 Bridge Scour And Stream Instability Countermeasures-Vol I	Hard Copy	RM	**$20.00**
135048	HEC-23 Bridge Scour And Stream Instability Countermeasures-Vol II	Hard Copy	RM	**$30.00**
135048	Stream Instability, Bridge Scour, and Countermeasures: A Field Guide for Bridge Inspectors (Feb2009)	Hard Copy	RM	**$20.00**
135056	Culvert Design for Aquatic Organism Passage: HEC-26, First Ed. (October 2010)	Hard Copy	OM	**$50.00**
135056	Hydraulic Design of Highway Culverts-HDS 5 (April 2012)	Hard Copy	RM	**$50.00**
135065	Introduction to Highway Hydraulics-(June 2008)	Hard Copy	OM	**$50.00**
135065	Introduction to Highway Hydraulics-HDS No. 4 (June 2008)	Hard Copy	OM	**$50.00**
135067	Highway Hydrology, Hydraulic Design Series No. 2, Second Edition - (October 2002)	Hard Copy	OM	**$50.00**
135082	HEC-25 (Volume 2)-Highways in the Coastal Environment: Assessing Exposure to Extreme Events	Hard Copy	RM	**$30.00**
135082	Highways in the Coastal Environment (HEC-25)	Hard Copy	RM	**$40.00**
135085	PLAN OF ACTION (POA) FOR SCOUR CRITICAL BRIDGES - CD (MAY 2007)	Hard Copy	PP	**Free**
135090	Hydraulic Design of Safe Bridges-HDS-7 (April 2012)	Hard Copy	RM	**$50.00**
136065	RISK MANAGEMENT (JULY 2013)	Hard Copy	PW	**$30.00**
136106	TRANSPORTATION ASSET MANAGEMENT (June 2013)	Hard Copy	PW	**$50.00**
136106A	Introduction to Transportation Asset Management (June 2013)	Hard Copy	PW	**$50.00**
137046	NHI Using IDAS Data	Electronic Copy	EF	**Free**
139004	Principles of Effective Commerical Motor Vehicle (CMV) Size and Weight Enforcement (Dec 2013)	Hard Copy	PW	**$50.00**
141043	Appraisal for Federal-Aid Highway Programs (May 2013)	Hard Copy	PW	**$30.00**

Course Number	Material Name	Format	Type	Price
141050	Introduction to Federal-Aid Right-of-Way Requirements for Local Public Agencies (August 2010)	Hard Copy	PW	**$50.00**
142005	NEPA And The Transportation Decision Making Process (11/2015)	Hard Copy	PW	**$50.00**
142042	Fundamentals Of Title VI / Environmental Justice PW (February 2007)	Hard Copy	PW	**$50.00**
142046	Bicycle Facility Design (July 2013)	Hard Copy	PW	**$50.00**
142046	Bicycle Facility Design (July 2013)	Electronic Copy	PP	**$50.00**
142047	Water Quality Management of Highway Runoff PW/RM	Hard Copy	PW	**$50.00**
142049	Beyond Compliance: Historic Preservation In Transporation Project Development - Exercise 4 (July 07)	Hard Copy	OM	**$20.00**
142049	Beyond Compliance: Historic Preservation In Transporation Project Development (July 2012)	Hard Copy	PW	**$50.00**
142049	Beyond Compliance: Historic Preservation In Transportation Project Development - Exercise 3(July 07)	Hard Copy	OM	**$20.00**
142049	Beyond Compliance: Historic Preservation In Transportation Project Development -Exercise 2 (July 07)	Hard Copy	OM	**$20.00**
142054	Design And Implementation Of Erosion And Sediment Control - Participant Workbook (December 2006)	Hard Copy	PW	**$30.00**
142054	Design And Implementation Of Erosion And Sediment Control - Reference Manual (December 2006)	Hard Copy	RM	**$30.00**
142055	Advanced Seminar on Transportation Project Development: Navigating the NEPA Maze (December 2008)	Hard Copy	PW	**$40.00**
152054	INTRODUCTION TO URBAN TRAVEL DEMAND FORECASTING (February 2012)	Hard Copy	PW	**$50.00**
231028	Using the AASHTO Audit Guide for the Procurement and Administration of A/E Contracts (Feb 2012)	Hard Copy	PW	**$35.00**
231029	Using AASHTO Audit Guide for Development of A/E Consultant Indirect Cost Rates (Feb 2012)	Hard Copy	PW	**$50.00**
231030	Using AASHTO Audit Guide for Auditing and Oversight of A/E Consultant Indirect Cost Rate (Feb2012)	Hard Copy	PW	**$50.00**
310110	Federal-Aid Highways-101 (April 2014)	Hard Copy	PW	**$50.00**
380005	Railroad-Highway Grade Crossing Improvement Program - Participant Workbook (July 2011)	Hard Copy	PW	**$50.00**
380034	Design Construction And Maintenance Of Highway Safety Features And Appurtenances - Participant Workb	Hard Copy	PW	**$60.00**
380034A	Design Construction And Maintenance Of Highway Safety Features And Appurtenances - Participant Workb	Hard Copy	PW	**$60.00**
380034B	Design Construction And Maintenance Of Highway Safety Features And Appurtenances - Participant Workb	Hard Copy	PW	**$60.00**
380069	Desktop Reference for Crash Reduction Factors (September 2007)	Electronic Copy	OM	**Free**
380069	FHWA Road Safety Audit Guidelines (June 2006)	Electronic Copy	OM	**Free**
380069	Road Safety Audits/Assessements Participant Workbook (August 2008)	Hard Copy	PW	**$50.00**

Course Number	Material Name	Format	Type	Price
380069	Road Safety Audits: Case Studies (December 2006)	Electronic Copy	OM	**Free**
380069	Toolbox of Countermeasures & Their Potential Effectiveness for Intersection Crashes (September 2007)	Electronic Copy	OM	**Free**
380069	Toolbox of Countermeasures & Their Potential Effectiveness for Pedestrian Crashes (September 2007)	Electronic Copy	OM	**Free**
380069	Toolbox of Countermeasures & Their Potential Effectiveness for Roadway Departure Crashes (Sept 2007)	Electronic Copy	OM	**Free**
380069	Traffic Signals (September 2007)	Electronic Copy	OM	**Free**
380071	Interactive Highway Safety Design Model (December 2013)	Hard Copy	PW	**$50.00**
380075	Desktop Reference for Crash Reduction Factors (September 2007)	Electronic Copy	OM	**Free**
380075	New Approaches To Highway Safety Analysis - Reference Manual (February 2006)	Hard Copy	RM	**$50.00**
380075	New Approaches to Highway Safety Analysis Participant Workbook (April 2011)	Hard Copy	PW	**$50.00**
380075	Toolbox of Countermeasures & Their Potential Effectives for Intersection Crashes (September 2007)	Electronic Copy	OM	**Free**
380075	Toolbox of Countermeasures & Their Potential Effectives for Pedestrian Crashes (September 2007)	Electronic Copy	OM	**Free**
380075	Toolbox of Countermeasures & Their Potential Effectives for Roadway Departure Crashes (Sept 2007)	Electronic Copy	OM	**Free**
380075	Traffic Signals (September 2007)	Electronic Copy	OM	**Free**
380076	Desktop Reference for Crash Reduction Factors (September 2007)	Electronic Copy	OM	**Free**
380076	Low Cost Safety Improvements Workshop - Participant Workbook (February 2010)	Hard Copy	PW	**$50.00**
380076	Toolbox of Countermeasures & Their Potential Effectiveness for Intersection Crashes (September 2007)	Electronic Copy	OM	**Free**
380076	Toolbox of Countermeasures & Their Potential Effectiveness for Pedestrian Crashes (September 2007)	Electronic Copy	OM	**Free**
380076	Toolbox of Countermeasures & Their Potential Effectiveness for Roadway Departure Crashes (Sept 2007)	Electronic Copy	OM	**Free**
380076	Traffic Signals (September 2007)	Electronic Copy	OM	**Free**
380077	Desktop Reference for Crash Reduction Factors (September 2007)	Electronic Copy	OM	**Free**
380077	Toolbox of Countermeasures & Their Potential Effectiveness for Intersection Crashes (September 2007)	Electronic Copy	OM	**Free**
380077	Toolbox of Countermeasures & Their Potential Effectiveness for Pedestrian Crashes (September 2007)	Electronic Copy	OM	**Free**
380077	Toolbox of Countermeasures & Their Potential Effectiveness for Roadway Departure Crashes (Sept 2007)	Electronic Copy	OM	**Free**

Course Number	Material Name	Format	Type	Price
380077	Traffic Signals (September 2007)	Electronic Copy	OM	**Free**
380090	Developing a Pedestrian Safety Action Plan Participant Workbook (January 2009)	Hard Copy	PW	**$50.00**
380095	Highway Design: Applying Flexibility & Risk Management (February 2016)	Hard Copy	PW	**$50.00**
380100	Interactive Highway Safety Design Model - Web-based course-Participant Workbook	Electronic Copy	PW	**$50.00**
N/A	FHWA-NHI-132037 Shallow Foundations	Hard Copy	RM	**$50.00**

NATIONAL HIGHWAY INSTITUTE (NHI)

Division of FHWA Office of Technical Services

1310 N Courthouse Road, Suite 300
Arlington, VA 22201
Phone: 703-235-0500 or Toll Free 877-558-6873
Fax: 703-235-0593

MAIN CONTACTS

Questions About?	E-mail	Telephone
NHI Training	nhitraining@dot.gov	703-235-0534
NHI Web site	nhiwebmaster@dot.gov	703-235-0556
Instructors	nhiinstructorliaison@dot.gov	703-235-0952
Materials	nhimaterials@dot.gov	703-235-0552

SUBJECT AREA CONTACTS

Questions About?	Contact
Asset Management	Tom Elliott
Business, Public Administration & Quality	Tom Elliott
Communications	Carol Keenan
Construction and Maintenance	Tom Elliott
Design and Traffic Operations	Carol Keenan
Environment	Heather Shelsta
Freight and Transportation Logistics	Carol Keenan
Geotechnical	Heather Shelsta
Highway Safety	Tom Elliott
Hydraulics	Carol Keenan
Intelligent Transport Systems (ITS)	Carol Keenan
Pavement and Materials	Tom Elliott
Real Estate	Heather Shelsta
Structures	Louisa Ward
Transportation Planning	Heather Shelsta

Name	Title	Email
Briggs, Valerie	Director	valerie.briggs@dot.gov
Elliott, Tom	Training Program Manager	thomas.elliott@dot.gov
Keenan, Carol	Training Program Manager	carol.keenan@dot.gov
Shelsta, Heather	Training Program Manager	heather.shelsta@dot.gov
Ward, Louisa	Training Program Manager	louisa.ward@dot.gov

www.ingramcontent.com/pod-product-compliance
Lightning Source LLC
Chambersburg PA
CBHW080648190526
45169CB00006B/2027

* 9 7 8 1 5 3 4 7 4 9 6 6 5 *